They Believed That?

They Believed That?

A CULTURAL ENCYCLOPEDIA OF SUPERSTITIONS AND THE SUPERNATURAL AROUND THE WORLD

William E. Burns

An Imprint of ABC-CLIO, LLC
Santa Barbara, California • Denver, Colorado

Library of Congress Cataloging-in-Publication Data

Names: Burns, William E., 1959- author.
Title: They believed that? : a cultural encyclopedia of superstitions and the supernatural around the world / William E. Burns.
Other titles: Cultural encyclopedia of superstitions and the supernatural around the world
Description: Santa Barbara, California : ABC-CLIO, [2023] | Includes bibliographical references and index.
Identifiers: LCCN 2022026261 | ISBN 9781440878473 (hardcover) | ISBN 9781440878480 (ebook)
Subjects: LCSH: Superstition—Encyclopedias. | BISAC: SCIENCE / History | RELIGION / Religion & Science
Classification: LCC AZ999 .B898 2023 | DDC 001.9—dc23/eng/20220603
LC record available at https://lccn.loc.gov/2022026261

ISBN: 978-1-4408-7847-3 (print)
978-1-4408-7848-0 (ebook)

27 26 25 24 23 1 2 3 4 5

This book is also available as an eBook.

ABC-CLIO
An Imprint of ABC-CLIO, LLC

ABC-CLIO, LLC
147 Castilian Drive
Santa Barbara, California 93117
www.abc-clio.com

This book is printed on acid-free paper ∞
Manufactured in the United States of America

Contents

Alphabetical List of Entries

Thematic List of Entries

Preface

This encyclopedia is dedicated to superstitious, supernatural, and discredited beliefs held in a variety of civilizations and societies from the ancient world to the nineteenth century, although there is also some discussion of the twentieth and twenty-first centuries. It does not, however, include the large body of supernatural beliefs central to organized religions, although religious topics do come up in passing. Obviously, no book could cover the vast range of beliefs held by human societies across the globe throughout thousands of years of recorded history. A whole library would be full to bursting. What this book contains is a selection of some of the most important, interesting, influential, or simply bizarre beliefs held by people from pole to pole. The emphasis is on the period before 1800, as in the nineteenth century the Industrial Revolution, the spread of mass literacy in many societies, and imperialism wrought vast changes on traditional cultures and beliefs. However, some beliefs documented in the period after 1800 are included when appropriate.

The encyclopedia was written in the age of COVID. Writing it would have been impossible without the internet, which gave me access to hundreds of documents and sources at a time when I could not visit libraries. I express appreciation to my family for providing support, as well as to the Gelman Library of George Washington University, my editor Kevin Downing, and Michael R. Lynn.

Introduction

This book is devoted to those human beliefs that fall in the "gray zone" between science, religion, and everyday life—call them superstitious, supernatural, magical, or just wrong. In an often incomprehensible world where lightning or plague could end life quickly or drought condemn a poor family to agonizing death, superstitious beliefs gave people a feeling of understanding or even control. However, superstition could also make the world an even more terrifying place, haunted by vampires and other monsters as well as ominous signs of a variety of shadowy future dangers.

"Superstition" does not have fixed meaning, but it has had different meanings in different cultures and epochs. One thing connecting different uses of superstition is that they are usually negative—superstition has had few defenders or advocates and is a concept defined principally by its enemies. Much of our knowledge of superstition in other times and places, particularly before the nineteenth century, comes from its opponents. Calling someone "superstitious" is not usually meant as a compliment, and to define a belief as "superstitious" is an exercise of cultural power. It is no coincidence that subordinated populations—women (particularly older women), the uneducated, the poor and working class, colonized populations, and members of marginalized ethnic groups—are particularly likely to be viewed by dominant groups as foolishly superstitious. A second is that superstition is defined as the opposite of something praiseworthy. In the West, superstition began as the opposite of true religion and became the opposite of true science. Despite the attempts of many to define it, however, "superstition" remains a fuzzy category, and what distinguishes a superstitious belief from a religious one or a belief that may be held for perfectly rational reasons but is just wrong, such as geocentrism, remains a subject to debate. The related category of "magic" is also difficult to define, with religion on one side and technology on another.

Superstition in Greece and Rome

Although what we would now call superstitious beliefs and practices are found in all human societies on record, the first people to discuss concepts such as that of superstition were the Greeks. The ancient Greeks and Romans defined "superstition" (originally a Latin word—the Greek concept was "deisdaimonia" or fear of the gods) as bad religion. The philosopher Theophrastus (c. 371–c. 287 BCE) defined it as "cowardice in relation to the supernatural" and painted the superstitious man as living a life dominated by fear. In Latin, "superstition" is literally a "standing over." Scholars are divided about the origin of the Latin term, but it may refer to the idea of superstition as an "old" thing remaining from earlier and presumably less enlightened times. Contempt for superstition did not prevent the Greeks and Romans from many practices we would define as "superstitious." Rome, in particular, was by our standards a superstitious culture, with numerous divination practices drawing on indigenous, Etruscan, and Greek traditions. Roman diviners examined the entrails of sacrifices, observed the flight of birds, and interpreted the meaning of lightning bolts. Not only individual lives but decisions made in the interests of the Roman people also were governed by superstitions as arbitrary as the value given to whether the sacred chickens ate on the eve of a battle. However, Rome did not lack skeptics including the orator and philosopher Marcus Tullius Cicero (106–43 BCE), author of an influential attack on divination, *On Divination.*

Plutarch of Chaeronea (c. 46 CE–c. 119 CE), a Greek philosopher living under the Roman Empire, was one of the most influential ancient writers on superstition. He defined it as an extreme in the field of belief. Many Greeks, including the philosopher Aristotle (384–322 BCE), defined virtue as lying between extremes, just as courage lay between cowardice and foolhardiness. Plutarch defined superstition as one bad extreme in religion, atheism being the other, and decorous worship of the gods, true "religion," lying in the middle. His essay on superstition is devoted to demonstrating that it is even worse than atheism, a system of false beliefs about the universe that does not produce the fear and terror felt by superstitious people. To believe that the gods cause evil and pain is to insult them far more than the atheist's claim that they do not exist. Since the gods and their power were everywhere in the universe, to fear them was far worse than to fear any particular thing. A person who feared the ocean could live far inland, but a person who feared the gods had literally nowhere to run or to hide. Even sleep, which offered peace from other fears, only tormented the superstitious person with evil dreams. The slave could even hope to escape a cruel master, but where could the gods be escaped? Even death offered no hope—Plutarch identified the fear of suffering after death as superstitious. Disasters caused by superstition were public as well as individual. The Athenian general Nicias (c. 470–413 BCE), by his superstitious inaction during a lunar eclipse, brought disaster not only to himself but also to the army of forty thousand men he led.

Like many Greeks and Romans, Plutarch disliked Judaism and identified Jewish observance of the Sabbath as superstitious. It is not known what he thought of Christianity, which goes entirely unmentioned in his voluminous surviving writings, but the later pagan opponents of the new cult claimed that it too was superstitious. Christians turned this around and claimed that the cults of the pagan gods, "idol worship," were superstitious. The Latin philosopher and bishop Augustine of Hippo (354–430), the most influential theologian of the Latin West, defined both "idolatry" and divination as superstitious, effectively relegating all pagan religion to the category of superstition. This was reflected in the law of the Christian Roman Empire, which increasingly defined, and condemned, non-Christian religion as superstition. The Christian church, in its never-ceasing struggles against those beliefs it defined as superstition, would frequently ascribe them to the corrupting influence of paganism, whether classically

Theophrastus on the Superstitious Man

"Superstition would seem to be simply cowardice in regard to the supernatural.

The Superstitious man is one who will wash his hands at a fountain, sprinkle himself from a temple-font, put a bit of laurel-leaf into his mouth, and so go about the day. If a weasel run across his path, he will not pursue his walk until someone else has traversed the road, or until he has thrown three stones across it. When he sees a serpent in his house, if it be the red snake, he will invoke Sabazius—if the sacred snake, he will straightway place a shrine on the spot. He will pour oil from his flask on the smooth stones at the cross-roads, as he goes by, and will fall on his knees and worship them before he departs. If a mouse gnaws through a meal-bag, he will go to the expounder of sacred law and ask what is to be done; and, if the answer is, 'give it to a cobbler to stitch up,' he will disregard the counsel, and go his way, and expiate the omen by sacrifice. He is apt, also, to purify his house frequently, alleging that Hecate has been brought into it by spells; and, if an owl is startled by him in his walk, he will exclaim 'Glory be to Athene!' before he proceeds. He will not tread upon a tombstone, or come near a dead body or a woman defiled by childbirth, saying that it is expedient for him not to be polluted. Also on the fourth and seventh days of each month he will order his servants to mull wine, and go out and buy myrtle-wreaths, frankincense, and smilax; and, on coming in, will spend the day in crowning the Hermaphrodites. When he has seen a vision, he will go to the interpreters of dreams, the seers, the augurs, to ask them to what god or goddess he ought to pray. Every month he will repair to the priests of the Orphic Mysteries, to partake in their rites, accompanied by his wife, or (if she is too busy) by his children and their nurse. He would seem, too, to be of those who are scrupulous in sprinkling themselves with sea-water; and, if ever he observes anyone feasting on the garlic at the cross-roads, he will go away, pour water over his head, and, summoning the priestesses, bid them carry a squill or a puppy around him for purification. And, if he sees a maniac or an epileptic man, he will shudder and spit into his bosom."

Source: Theophrastus. 1870. *The Characters of Theophrastus: An English Translation from a Revised Text*. London: Macmillan. Pp. 163–165.

Greco-Roman or one of the other pagan cultures of pre-Christian Europe, such as Celtic or Germanic paganism. The idea that Jewish observance of the law, necessary before the coming of Christ but not necessary after (according to Christians), was "superstitious" was also an important aspect of both the Christian concept of superstition and mainstream Christianity's growing hostility to Judaism.

Superstition in Ancient China

Greece and Rome's contemporary at the other end of Eurasia also had a concept of superstition.

Magical, divinatory, and superstitious practices have a long history in China, stretching to the "oracle bones" of the near-legendary Shang Dynasty, some of the earliest evidence for Chinese writing. Although classical Chinese thinkers did not have the association of superstition with the demonic characteristic of the Abrahamic faiths, they did distinguish between "superstitious" and acceptable beliefs. The Han Dynasty philosopher Wang Chong (25 CE–100 CE) devoted a considerable proportion of his essay collection Lunheng "Balanced Discourses" to denouncing beliefs he considered superstitious, such as the common belief in ghosts. Wang Chong's objections were broadly commonsensical, pointing out that the ghosts of murdered people should be repairing to magistrates to accuse their murderers but that there is no record of their having done so. Like his near-contemporary Cicero, Wang Chong also attacked the very common belief that celestial events were warnings of the anger of heaven, a belief shared by many cultures outside China. The cynical Wang Chong also claimed the belief that talent was rewarded with career success was another vulgar error.

Although a few philosophers admired his arguments, Wang Chong was largely forgotten after his death, and superstition in China flourished both among the common people and at the highest level of the government, where the Bureau of Astronomy, part of the Board of Rites, had the responsibility of determining lucky and unlucky days. One of the Five Classics of China, the *Yijing* or *I Ching*, was a text used for divination, and many other techniques for foretelling or influencing the future flourished among the great and common. The religious pluralism of China meant that there was no institution claiming the religious monopoly that Christianity did in the West and therefore little reason for a protracted campaign against superstition.

Superstition and Islam

Like Christians, Muslims identified superstition, identified as *shirk* or idolatry, with the time before their divine revelation. Superstition was a remnant of the polytheistic paganism practiced before Muhammad's conversion of Arabia to Islam, and thus was condemned by all true Muslims. The Islamic community had a rich culture of superstitions, some such as belief in the existence of jinn, backed by the authority of the Quran, and others such as the widespread use of dream interpretation, geomancy, astrology, and other divination methods less scripturally backed and more likely to be condemned by religious authorities. The use of talismans and amulets was also quite common. The Hanbali school of Sunni legal interpretation, one of the four major schools, is known for its strictness and severity. Hanbali scholars were particularly strongly opposed to practices deemed superstitious. This is particularly true of the Hanbali scholar Ibn Taymiyya (1263–1328), who was known to denounce the veneration of saints as shirk and also denounced more secular "superstitious" practices such as astrology, virtually ubiquitous in the Muslim world. In the early modern period when the Ottoman Empire ruled the Muslim Middle East, modern Turkey, and much of Europe, however, the official school of interpretation was the more tolerant Hanafi school, and superstitious beliefs and practices were widespread.

The Wahhabist movement of Sunni reform that emerged in eighteenth-century Arabia and became influential throughout much of the Sunni world drew upon Ibn Taymiyya and also vehemently opposed "superstitious" practice such as visitations to the shrines of Muslim saints. Indian Islamic reformers linked these and other "superstitious" practices to the corrupting influence of Hinduism, which does not make a strong distinction between superstitious and religious practices. However, many Sunni Islamic authorities endorse or tolerate practices that

Wahhabist and other "strict" interpreters of Islam would denounce as shirk.

Superstition in the Christian Middle Ages

The Christian Middle Ages continued the fight of late antique Christianity against pagan "superstition." Bishops issued decrees against belief in women who flew in the night, describing such activity as impossible. Thomas Aquinas, the greatest philosopher and theologian of the Latin West, adapted Plutarch's model to define superstition not as the opposite of atheism but of impiety. Aquinas was also a leader in the movement to define divination and other "magical" practices as based on an implicit or explicit pact with Satan or a lesser demon. This led to a greater hostility to magical practices on the part of Church authorities, although the foremost practitioners of many magical traditions in the Middle Ages remained the clergy, the only group literate and educated enough to carry on traditions such as alchemy, astrology, and necromancy.

The medieval church's campaign against superstition and magical beliefs was not restricted to learned magic, however. Preachers inveighed against superstitious practices persisting under the guise of religion, such as the veneration of the "dog-saint" Guinefort. Catholic writers in the Middle Ages and early modern period developed the idea that there were two legitimate categories of cause, the direct action of God or good and evil angels, or the actions of nature in its ordinary course. Attributing anything to any other kind of cause was superstitious and quite possibly a mask for demonic magic.

The Early Modern Period

The Church's war on superstition continued into the sixteenth century, but the situation was considerably complicated by the arrival of the Protestant Reformation. Protestant reformers defined many Catholic beliefs and practices, such as the consecration of holy water, as superstitious. Catholics were less likely to accuse their opponents of superstition, preferring to charge them with heresy. However, the Catholic Reformation intensified the Catholic campaign against superstition, further advanced by the immense number of new Catholics, with their own unique customs and beliefs, in Asia, Africa, and the Americas.

The English Protestant writer Francis Bacon (1561–1626), one of the founders of the modern scientific tradition, reframed superstition in a way that would be very influential. His essay on superstition laid the traditional charges against Catholic superstition, but added that there was a "superstition in avoiding superstition," a veiled thrust at the extreme Protestants of the day. Bacon was one of many writers on the subject in the sixteenth and seventeenth centuries, ranging from the Spanish Catholic priest Pedro Ciruelo (c. 1465–1548) to the French surgeon Ambroise Pare (c. 1510–1590) to the English natural philosopher Sir Thomas Browne (1605–1682). Browne's *Pseudoxia Epidemica* (1646) framed superstitions as "vulgar errors," to be refuted by close observations and scientific experiment. These practices had a long history before Browne, but Browne's synthesis was particularly influential. Browne also showed more interest in cataloging specific superstitions, including those that had no obvious religious application, than in denouncing the concept of superstition in the abstract.

Among both Protestants and Catholics, the campaign against superstition also took a violent form in the witch hunt that drew upon the earlier theories of the demonic origin of magic developed by Aquinas and others. Many of the religious and legal authorities who endorsed witch hunting saw it as part of a broader campaign against popular superstitious beliefs of dubious religious orthodoxy. (Browne believed in witches and on at least one occasion aided the prosecution in a witch trial.) Defensive magic was under particular suspicion, as both Catholic and Protestant religious authorities distrusted the idea that people could protect themselves against magic without seeking the aid of God. Protestants recommended prayer and repentance; Catholics added to this the use of "sacramentals" such as holy water and consecrated oil, but both were suspicious of such "superstitious" folk defenses as nailing a horseshoe over the threshold.

The scientific revolution of the seventeenth century gave science and reason more intellectual prestige. One product of this transformation is that superstition came to be increasingly defined in opposition to true science rather than true religion.

Superstition in the Enlightenment and Romantic Era

In the eighteenth-century Enlightenment, superstition was increasingly defined not as the opposite of atheism but of "enthusiasm" or "fanaticism." While the superstitious individual sought to reach God with various rituals or practices, the enthusiast sought a direct, unmediated connection with God. Both were to be condemned—the enthusiast as one who threatened the order of civil society and the superstitious person as one who threatened intellectual progress. The Scottish Enlightenment philosopher David Hume (1711–1776) viewed enthusiasm, although deplorable, as superior to superstition on the grounds that superstition encouraged timidity, while enthusiasm encouraged fearlessness. Superstitious people handed over power to priests; enthusiasts seized power for themselves.

Enlightenment radicals defined a broader range of beliefs, including mainstream religious beliefs, as superstitious. The French philosopher Voltaire (1694–1778) pointed out that many religions had denounced others as superstitious as a way of discrediting religion generally. For many, the solution was not a more vigorous enforcement of religious orthodoxy but broader education based on true science, philosophy, and reason.

Enlightenment opposition to superstition and magic did not mean that they disappeared in the eighteenth century; in fact, they flourished. In addition to the quasi-magical mythology and ritualism of the rapidly expanding movement of Freemasonry, the period also saw the spread of the Eastern European and New England vampire myths and the invention of animal magnetism and the tarot divination system. Even astrology, largely discredited in the early eighteenth century, was making a comeback by the end of the century.

Romantic nationalists emerging around the end of the eighteenth century were some of the first students of superstition to view it positively. They valued superstition as an expression of a nation's culture. The English Romantic poet John Clare (1793–1864) viewed superstition as part of a national heritage dating to the early Middle Ages: "Superstition lives longer than books; it is engrafted on the human mind till it becomes a part of its existence; and is carried from generation to generation on the stream of eternity, with the proudest of fames, untroubled with the insect encroachments of oblivion which books are infested with" (Clare, "Popularity in Authorship," p. 301). Folklore, including superstition, emerged as a field of scholarship in the nineteenth century, and many voluminous—and nonjudgmental—collections of superstition were printed. But, however nonjudgmental such collections were, they almost always continued to portray the superstitious as an "other" as the collector stood above superstition in his or her quest for knowledge.

Superstition and Its Enemies in Modern Times

Modern campaigns against superstition have been less likely to be motivated by the desire to purify religion (although the Wahhabist struggle continues) than by "modernizing" political ideologies such as liberal democracy, Marxism, or technocracy. The "Meiji reformers" of late nineteenth-century Japan sought to rid the country, and Japanese Buddhism, of superstition as part of their efforts to strengthen the nation. Chinese modernizers developed a concept of *mixin*, usually translated as "superstition," in order to attack popular religion and divination, as in the 1928 government decree, "Standards for Preserving and Abandoning Gods and Shrines," which defined many traditional Chinese religious practices as superstitious. It was followed by decrees against divination. This campaign intensified after the Communist takeover. There has even been a revival of interest in Wang Chong, although Chinese people, both in China and elsewhere, retain many traditional superstitious beliefs.

Colonial administrators in the extensive European empires of the nineteenth and twentieth centuries often saw attacks on "native" superstition as part of the "white man's burden" or "civilizing mission," the ideological constructs that justified imperial exploitation to members of the dominant culture. Colonizers also produced elaborate tomes cataloging and analyzing the superstitions of the people they ruled over, partially motivated by the belief that such knowledge would make colonial administration more effective.

In much of the world, the emphasis remains on understanding superstition rather than combating it. Psychologists, anthropologists, and historians study superstition in its local contexts and as an expression of the human propensity for making connections. The America behaviorist psychologist B. F. Skinner (1904–1990) even broadened the concept of "superstition" by extending it to animals in his 1947 paper "Superstition in the Pigeon"! Many books appear studying the superstitions of particular regions, occupations, and subcultures. Despite the best efforts of our modern-day Ciceros, Wang Chongs, Ibn Taymiyyas, and Sir Thomas Brownes, superstition will be with us for a long time to come.

Abracadabra. See Amulets and Talismans

Acupuncture

Acupuncture is the piercing of selected spots on the human body with sharp needles for therapeutic purposes. It is part of traditional Chinese medicine, and an early form appears in *The Yellow Emperor's Inner Classic*, generally considered the foundational text of the tradition. The *Classic* connected acupuncture to the flow of energy, or qi, through the body. The acupuncture needle cleared passages where qi was blocked or redirected qi flows. Gold and silver needles have been found in a tomb dating from the Han period. Over the centuries, a canonical set of 365 entry points on the body for acupuncture needles was established. These are represented in statues used for instructional and reference purposes. Mature acupuncture was described in the Ming Dynasty work, *The Great Compendium of Acupuncture and Moxibustion.* (Moxibustion is the application of fire or heat to acupuncture points.) Acupuncturists also believed that the effectiveness of the procedure was influenced by the time of day and the lunar cycle. During the Qing Dynasty (1644–1911) interest in acupuncture waned, and it was banned by imperial decree in 1822. Following the ban, official physicians stopped using acupuncture, but it continued among practitioners serving the poor and uneducated outside the framework of official medicine.

Acupuncture spread with other aspects of Chinese culture to Korea and Japan in the sixth century CE and to Vietnam by the tenth century. (Japan officially abolished the practice in 1876, as part of the modernization reforms of the Meiji period.) Knowledge of it first arrived in the West with the writings of Jesuit missionaries in the sixteenth century. The first description of acupuncture by a Western physician was the work of Willem Ten Rhijne (1647–1700), a Dutchman, who coined the Latin word "acupunctura" from which the English "acupuncture" is derived. Ten Rhijne, whose knowledge of acupuncture (and moxibustion, which he also discussed) was mostly derived from Japanese physicians, interpreted it in Western terms as allowing for the venting of noxious "winds." The circulation of Ten Rhijne's and other works on East Asian medicine raised awareness of acupuncture, and vogues for acupuncture appeared in many Western medical environments including London and Philadelphia in the early nineteenth century and France in the mid-twentieth century. The Western interpretation of "acupuncture" as it developed did not employ the system of acupuncture points in East Asian medicine. Physicians simply inserted a needle near the condition to be treated or wherever they thought best.

See also: Yellow Emperor

Further Reading

Cook, Harold J., ed. 2020. *Translation at Work: Chinese Medicine in the First Global Age.* Leiden: Brill.

Affair of the Poisons. See Black Mass

Alchemy

Alchemy, which at its most basic was the art of refining and mixing various substances, had a long and complex history, involving Chinese, Greek, Indian, Jewish, Arab, and European influences. Like modern chemistry, alchemy involved furnaces, beakers, and tubes, and the earliest laboratories were for alchemical work. Alchemists pioneered procedures such as distillation. Unlike modern chemistry, alchemy could also involve spells, prayers, incantations, and horoscopes taken at specific moments in the process. Alchemy also differed from modern chemistry in viewing certain states of matter as more perfect than others, as gold represented the perfect state of metal. Alchemists could help this perfection emerge by purging impurities and thus transforming "base metals" such as lead into gold. (The making of gold is also known as "chrysopoeia.") Some alchemists associated this process with the spiritual purification of the alchemist himself or herself. The macrocosm/microcosm analogy provided for a correspondence between the process of purification taking place in the outside universe and the purification of the alchemist.

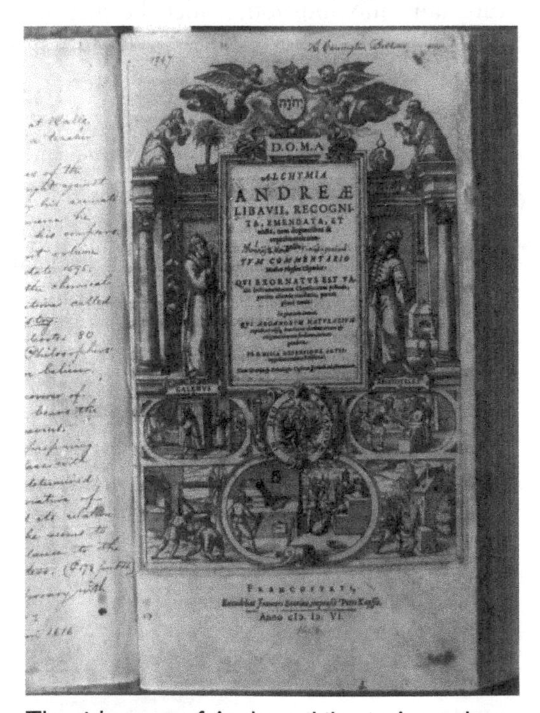

The title page of Andreas Libavius's work on Alchemy, 1606, places alchemy in a tradition of ancient wisdom. (Library of Congress, Prints and Photographs Collection, LC-USZ62-95269)

The origins of alchemy are obscure, but the earliest evidence comes from around the same time in Roman Egypt and Han Dynasty China, probably independently. Chinese alchemy was initially more oriented to health, and specifically the discovery of the "elixir of immortality." (Inferior but still powerful elixirs would grant long life rather than immortality.) Alchemy in China was associated with the religious philosophy known as Daoism, and much of the early Chinese alchemical literature was written by or ascribed to Daoist monks. (It was also ascribed, like many aspects of Chinese knowledge, to the mythical "Yellow Emperor.") Like Western spiritual alchemists, Daoist adepts thought the ideal alchemist would be spiritually pure. It was the goal of the Daoist alchemical adept to return substances to their original and most powerful form. Alchemy was viewed in the context of Chinese cosmology as a whole, including Yin/Yang theory and the five phases or elements. Cinnabar, an ore of mercury, was a common ingredient in Chinese alchemical preparations and could, according to some alchemists, be transmuted into a form of gold with the power to extend life. (The use of mercury, a poison, in elixirs meant that they may have shortened life rather than prolonging it.) The preparation of the elixir had to begin on a day deemed auspicious and be accompanied by rituals and invocations of supernatural beings. In addition to immortality, the elixir granted supernatural abilities like protection from demons.

One development in Chinese alchemy with few parallels in the West was the use of the alchemist's own body as a site for the production of alchemical elixirs. The goal remained to attain immortality through the creation of an alchemical embryo within the male adept's body as the nucleus for a new immortal body. Techniques for manufacturing the inner elixir included directed meditation and breath control. Some branches of inner alchemy included sexual practices. This tradition of inner alchemy was known as Neidan or "the way of the Golden Elixir" and was contrasted with Weidan or "external alchemy," the preparation of elixirs in the laboratory. Neidan originated around the eighth century CE and by the fifteenth century had supplanted Weidan as the most active Chinese alchemical tradition. A form of inner alchemy directed at practitioners with a female body emerged during the Qing Dynasty (1644–1912), although the texts were mostly written by men. This tradition, known as "Nudan," envisioned altering the female body in a male direction, with the cessation of menstruation (referred to as "beheading the Red Dragon") and the shrinkage of the breasts.

Indian alchemy, like Chinese, was oriented around the production of elixirs for consumption rather than gold-making. Fundamental to Indian alchemy is the dyad of male mercury and female sulfur. Both substances and human beings can be made perfect through the application of alchemical processes. The goal of the alchemist is to become like the "mercurial" god Siva, to possess the same powers of transmutation and perfecting. This led to some difficulties as mercury is rare in India and had to be imported; some alchemical texts refer to mercury compounds as "Chinese powder." Alchemy emerged in India somewhat later than in China or the West, and the first descriptions of alchemical processes date from around 900 CE. Even more than in China and the West, Indian alchemy was placed in a religious tradition and associated with the "tantric" worship of Siva and Sakti, the female divine principle.

Western alchemy initially drew on Egyptian metallurgy and Greek philosophy. It was associated with the quest for the "philosopher's stone" that could purify base metals into gold than with medicine. The earliest, partially legendary, Western alchemists date from the first few centuries of the Christian era—Pseudo-Democritus, Mary the Jewess, Cleopatra the Alchemist, and Zosimus of Panopolis. (The bain marie or "bath of Mary," a common piece of chemical equipment, is named after Mary the Jewess, also known as Mary of Egypt, and was, according to legend, invented by her, although there is evidence of similar devices centuries earlier. Mary the Jewess was also later identified with Miriam, the sister of Moses.) The writings of Hermes Trismegistus, which emerged out of the same Graeco-Egyptian milieu, were also influential in the history of Western alchemy. Alchemy was related to other magical and scientific systems; the seven metals in alchemy were systematically associated with the seven planets known to astrologers and the precious metals gold and silver with the luminaries, the Sun and Moon. Mercury was associated with the planet Mercury, hence its name. Copper was associated with Venus, iron with Mars, tin with Jupiter, and lead with Saturn. This meant that alchemical concepts could be expressed in astrological language. Thus, "the Sun exalted in Aries" could refer to gold warmed in the gentle heat of spring, the season of Aries.

After the fall of the Roman Empire, alchemy was carried on in the Islamic world. Some Muslim mystics found the spiritual aspects of alchemy as a process of purification congenial. A mysterious and secretive sect known as the Ikhwan al-Safa or "Brethren of Purity," which apparently began in tenth-century Basra in Iraq, created a massive and influential compendium of knowledge, the *Epistles of the Brethren of Purity*. The Ikhwan endorsed the transmutation of lead to gold and identified mercury and sulfur as the active principles in matter, both central ideas in Western alchemy. ("Mercury" and "Sulfur" are not simply the modern elements known by these names, but the passive and active principles in the material world.) The most legendary of the Islamic alchemists, the eighth-century Jeber ibn Hayyan, had been, according to tradition. a Sufi mystic. However, his approach to alchemy emphasized laboratory work as well as mysticism. He analyzed metals in terms of the Aristotelian categories of hot/cold and dry/wet. Work ascribed to him under the name Geber passed into the body of Western alchemical texts. As was the case with the Brethren and Jeber, many works in Islamic alchemy were anonymous or ascribed to legendary or fictitious characters. The Arabic Hermetic text the "Emerald Tablet" or *Tabula Smaragdina* was an account of the creation of the world often read as encoded directions for the creation of the philosopher's stone and influential among both Muslims and Christians. A more pragmatic version of alchemy, oriented more toward the production of useful substances than making gold or advancing spiritually, was practiced by the eminent physician Muhammad al-Razi. Some Arab chemists argued that the transmutation of metals was impossible and the alchemical enterprise a folly.

Alchemy reentered Europe along with other Graeco-Arabic science in the eleventh and twelfth centuries. Alchemy's most notable Christian practitioner in the Middle Ages was the English Franciscan friar Roger Bacon (c. 1219–c. 1292), whose optical and alchemical experiments won him a bad reputation as a magician. (According to legend, he built a brass talking head that would prophesy the future.) The reputation of alchemists as frauds, their boasts of perfecting the works of nature or God by human art, and the fact that their discipline was passed on outside the academy, all contributed to suspicion from Church and state, although the making of gold also attracted support from rulers. The Arabic practices of anonymity and pseudonymity continued among Medieval Christian alchemists. Numerous alchemical tracts were attributed to the eminent philosopher Albertus Magnus (1200–1280), which were not actually his work, for example, and there were Latin works falsely credited to Jeber Ibn Hayyan, referred to as "Pseudo-Geber." Alchemy was often associated with a closed system of communication, in which alchemical adepts did not publicize the processes they had discovered but passed them down to their students, often using a cryptic notation.

The advent of printing began to change alchemical communication by making alchemical texts more widely circulated, and many alchemical texts were published, most notably the vast compilations of *Theatrum Chemicum* (1602–1661) and *Theatrum Chemicum Britannicum* (1652). The new availability of the Hermetic tracts and the rise in popularity of Kabbalah among Christian philosophers helped move alchemy more in the direction of magic. Alchemists promoted the teaching of alchemical knowledge in schools and universities. Alchemical practice was very diverse in the sixteenth century, from the philosophical alchemists who interpreted alchemical substances allegorically and the alchemical process spiritually to more practical workers still seeking to transmute base metals into gold by preparing the "philosopher's stone." This was a goal for which early modern rulers, like medieval ones, were often willing to provide financial backing, and this form of alchemy was easily exploited by charlatans of the type satirized in Ben Jonson's (1572–1637) play *The Alchemist* (1610). Others applied alchemical principles to medicine. The leader in this endeavor, and the most influential alchemist in early modern Europe, was the German Theophrastus Bombastus von Hohenheim, better known by his nom de plume of Paracelsus (1493–1541). Paracelsus defined matter in terms of three principles or elements, the traditional sulfur and mercury along with salt—the "Paracelsian Triad." He was followed by the Paracelsians, or as they were also known, "chemical physicians." These alchemists were following in the footsteps of Chinese alchemists, although the degree of influence from China is not clear. In the sixteenth and seventeenth centuries, occult philosophers including Heinrich Cornelius Agrippa (1486–1535), John Dee (1527–1609), and Robert Fludd (1574–1637) integrated alchemy into their magical systems. Renaissance "spiritual alchemists" such as the English physician and antiquarian Elias Ashmole (1617–1692), compiler of the *Theatrum Chemicum Britannicum*, claimed that the highest form of the philosopher's stone, the "Angelical Stone," had the spiritual power to command angels and other spirits.

Both the alchemical process of creating gold and that of promoting long life attracted interest from rulers. Several Chinese Emperors were known to be seekers of the Elixir of Immortality, while Western rulers patronized alchemists with the hope of establishing a source of gold or chemical physicians with the hope of maintaining health. This could be very dangerous for the alchemist when the ruler decided that he had been defrauded.

Alchemists, unlike many natural philosophers, promoted an activist stance toward the natural world, striving not merely to understand it but also to control natural processes. Alchemical practice was compatible with different overall stances to the natural world, including scholastic Aristotelianism as well as more radical approaches such as atomism. Paracelsus and many others promoted a "Christian" alchemical philosophy as an alternative to "pagan" Aristotelianism.

Alchemists often employed a microcosm-macrocosm analogy, and many but not all alchemists saw natural substances as active, as opposed to the dead nature approach of the French philosopher Rene Descartes (1596–1650) and his "Cartesian" followers or the atomistic "mechanical philosophers" who reduced the universe to matter and motion. Alchemical processes were not generally referred to by mechanical terms. Instead, alchemists used language to envision chemical substances and processes as organic, such as "birth," "death," "resurrection," and "marriage." These terms were not always simply metaphors, but were considered accurate descriptions of processes. The close connection between alchemy and religion extended as far as identifying the Paracelsian triad with the Holy Trinity or reading the story of the death and resurrection of

Jesus Christ as a series of alchemical metaphors providing directions for the creation of the philosopher's stone. Mystical and millenarian alchemists such as the German Protestant Jakob Boehme (1575–1624) also identified alchemical processes with the Second Coming of Jesus Christ.

Alchemists fought bitter polemical wars among themselves as well as against the opponents of alchemy in general. Late seventeenth-century scientists such as Robert Boyle (1627–1691), who believed that he had transmuted elements, and Isaac Newton (1642–1727), who left voluminous manuscript records of alchemical experiments, were alchemical practitioners, combining chemical and mechanical philosophy. Newton's willingness to consider action at a distance in the form of universal gravitation, forbidden by the strict mechanical philosophy, can be traced in part to his alchemical studies. The general tendency of this period was for practical and spiritual alchemy to drift farther apart, the former being absorbed into chemistry, the latter into the magical tradition.

In the eighteenth century, "alchemy," identified primarily with gold-making, was distinguished from "chemistry." Alchemical gold-makers were considered con men, and the alchemical interests of scientific "heroes" such as Boyle and Newton were discreetly scrubbed from their official histories. Central Europe, where alchemy remained vital throughout the eighteenth century, was a partial exception where the spiritual and practical remained joined. Alchemical symbolism also played a major role in eighteenth-century Freemasonry. The nineteenth century saw a revival of alchemy as an occult discipline. Nineteenth- and twentieth-century occultists saw alchemy as entirely a spiritual discipline, separate from both practical chemistry and gold-making. This attitude was taken up in the influential work of psychologist Carl Jung (1875–1961) who drew on Chinese as well as Western alchemy.

See also: Astrology; Elemental Systems; Gold; Hermes Trismegistus; Iron; Kabbalah; Macrocosm/Microcosm; Natural Magic; Rosicrucianism; Silver; Weapon-Salve; Yellow Emperor; Yoga

Further Reading

Debus, Allen G. 1977. *The Chemical Philosophy: Paracelsian Science and Medicine in the Sixteenth and Seventeenth Centuries.* New York: Science History Publications; Janacek, Bruce. 2011. *Alchemical Belief: Occultism in the Religious Culture of Early Modern England.* University Park: Pennsylvania State University Press; Moran, Bruce T. 2005. *Distilling Knowledge: Alchemy, Chemistry and the Scientific Revolution.* Cambridge, MA: Harvard University Press; Nummedal, Tara. 2019. *Anna Zieglerin and the Lion's Blood: Alchemy and End Times in Reformation Germany.* Philadelphia: University of Pennsylvania Press; Pregadio, Fabrizio. 2019. *The Way of the Golden Elixir: An Introduction to Taoist Alchemy.* Third Edition. Mountain View, CA: Golden Elixir Press; Valussi, Elena. 2008. "Female Alchemy and Paratext: How to Read Nudan in a Historical Context." *Asia Major Third Series* 21: 153–193.

Alectorius

The alectorius or "rooster's stone" is one of the many stones believed to have magical or healing powers. The Roman natural historian Pliny the Elder (d. 79 CE), the earliest surviving writer on the stone, describes it as a crystalline object, about the size of a bean, found in the gizzard of a cock. It was associated with victory in all kinds of contests. According to Pliny and Damigeron, another ancient writer on stones, the legendary Olympic wrestler Milo of Crotona was never beaten due to his use of the alectorius. Damigeron suggested that in addition to giving victory to wrestlers, gladiators, and soldiers, it could restore deposed kings to their kingdoms and make both men and women more sexually attractive. To use the stone, it could be held in the mouth or worn in clothing.

The alectorius reappears in the writing of the Bishop of Rennes, Marbodus (1035–1123) who, in addition to repeating much of the ancient material on Milo of Crotona and the virtues of the stone, added the idea that when held under the tongue it could quench thirst. Marbodus also innovated by saying that it was found in the stomach of a capon, a castrated rooster. The rooster had to be castrated when three years old and the stone harvested four years later. The stone reappears in many other medieval and Renaissance writers, including the philosopher and natural magician Albertus Magnus (c. 1200–1280), who at one point describes it as coming from the craw and at another the liver. He suggested that the older the better the rooster from which the stone is to be taken. Albertus also mentions a stone found in the head of a rooster, although it is not clear if this stone, which Albertus calls "Radaim," is a different stone or another name for the alectorius. The Dutch physician and natural magician Levinus Lemnius (1505–1568) suggested that the source of the stone is the semen of the capon, which is denied a natural egress and thus hardens into a stone. Although the gizzard remained the most common source, it was also described as coming from the intestines.

Medieval writers also had suggestions on how to use the alectorius, which could bring victory to feudal knights as well as ancient gladiators. One fifteenth-century English work on stones, the North Midland Lapidary, recommended wearing the stone as an amulet in a gold ring with nine religious inscriptions. Another procedure was to break the stone into small pieces and then drink them in a glass of wine. In addition to bringing victory in battle, medieval writers also suggested the alectorius was an aphrodisiac. Its medical virtues included helping with bladder stones, a common ailment of the time, and promoting conception and lactation. The celebrated Danish astronomer Tycho Brahe (1546–1601) carried one to which he credited his good luck when gambling. Belief in the aphrodisiac quality of the alectorius is found as late as the eighteenth century.

See also: Amulets and Talismans; Aphrodisiacs; Bezoars; Luck; Precious Stones; Swallow Stones

Further Reading

Duffin, Christopher John. 2007. "Alectorius: The Cock's Stone." *Folklore* 118: 325–341.

Alicorn. See Unicorn

Amber

Amber, the solidified and fossilized resin of pine trees, can persist for millions of years. Its striking qualities, including its ability to be electrically charged when rubbed, have helped make it the subject of many legends and superstitious beliefs. Belief in its medical and magical qualities goes back to the ancient Mediterranean. The Roman natural historian Pliny the Elder (d. 79 CE) described northern Italian countrywomen as wearing amber necklaces to protect themselves from diseases of the throat including tonsillitis and goiter. Pliny also claimed that amber was a powerful form of defensive magic, protecting from charm and sorcery. Amber was also believed to protect against nightmares.

Theories of amber's origins varied, from mythological explanations that it was originally the tears of the gods or the rays of the setting Sun to the more prosaic theory recounted by Pliny that it originated in the dung of birds. Amber was frequently identified as the "Lynx stone," supposedly the petrified urine of the lynx. Pliny ascribed this theory (which he did not himself endorse) to Demostratus, who distinguished between red amber formed from the urine of the male lynx and white amber formed from the urine of the female.

Although some early Christian writers condemned belief in amber's protective power as superstitious, it remained popular throughout the Middle Ages and Renaissance. The natural philosopher Albertus Magnus (1200–1280) recommended wearing it to protect against immoderate sexual desire and to alleviate the pain of childbirth. He also claimed that burning amber drove away snakes. Wearing an amber amulet on the wrist or just inhaling the scent of amber protected against plague. Amber also had the power to reveal unchastity—the Renaissance Italian physician Camillus Leonardus, author of *Speculum Lapidum* (1502; English translation *The Mirror of Stones*, 1750), recommended placing amber on the left breast of a sleeping wife to make her confess her evil acts. An alternative suggested by Leonardus was to soak a piece of amber in water for three days and then show it to a woman, who if adulterous would be immediately compelled to urinate. In Scotland, amber beads called "lammers" were used to treat disorders of the eye in both people and livestock.

Amber was also taken internally in the form of "oil of amber" to treat disorders of the brain. William Salmon (1644–1713), an English quack, recommended drinking powdered amber in a small amount of wine every day for a week to treat epilepsy. The sixteenth-century English physician Thomas Raynalde claimed that amber had the power to "restrain blood," and thus, inserted into a woman's vagina along with other substances with the same power, it would correct excessive menstrual bleeding. In Renaissance Italy, drinking the water in which an amber rosary had been dipped strengthened the heart. (Amber was a common material in rosaries.) Many early modern medical writers claimed that mixed with other ingredients to form a poultice directly applied to skin, amber could treat many conditions, including gout, syphilis, and palsy of the tongue. The smoke from burnt amber was also medically beneficial or could even drive away evil spirits. Belief in the medical power of amber, frequently associated with the sale of amber necklaces, persists to the present day.

See also: Ambergris; Amulets and Talismans; Blood; Defensive Magic; Menstruation; Snakes

Further Reading

Duffin, Christopher John. 2013. "History of the External Pharmaceutical Use of Amber." *Pharmaceutical Historian* 43: 46–53; King, Rachel. 2012. "'The Beads with Which We Pray are Made from It': Devotional Ambers in Early Modern Italy." In Wietse de Boer and Christine Gottler, eds. *Religion and the Senses in Early Modern Europe.* Leiden, Brill. Pp. 153–175.

Ambergris

Ambergris or "gray amber" (not to be confused with true amber) is a substance produced in the digestive tracts of sperm whales and valued by perfumiers as a fixative that allowed scents to last longer. It was also consumed in candies and sherbets. Defecated or vomited, it can float in the ocean for years; also, it can be extracted from dead whales by whalers or scavengers. Its mysterious origin led to a host of legends. Medieval French people believed in a variety of amber, from which it has received its name. Medieval and early modern Arabs, including the eminent physician Ibn Sina (980–1037), believed it to be produced by fountains under the sea. One Byzantine writer, the eleventh-century botanist Simeon Sethus, agreed, as did Garcia d'Orta (1501–1568), a Portuguese physician resident in India. Persians thought it to be a kind of solidified-sea foam, or perhaps produced by a large fish. Indians thought it came from a gum tree that grew in Arabia, and was thence carried into the sea by the rains, or that it was originally a mixture of honey and beeswax produced by bees in Africa. The Franciscan friar and missionary Gregorio de Bolivar in the early seventeenth century believed it was a form of bitumen spontaneously produced by sponges on the sea floor and that the ambergris that floated to the sea surface untouched was of superior quality than that that had passed through the guts of whales. Seventeenth-century French physicians thought that ambergris was originally honeycomb. The monk and encyclopedist Bartholomaeus Anglicus (c. 1203–1272) believed it to be literally the excess sperm produced by copulating whales, which then drifted in the ocean until it solidified. The Chinese referred to it as the spittle of dragons, while the natives of the Maldive Islands in the sixteenth century told the Portuguese it was solidified bird dung. By the sixteenth century, some writers were on the right track, suggesting that the substance was actually formed in the bowels of whales, perhaps after eating a specific herb. In the early eighteenth century, New England whalers, who had just begun to hunt sperm whales, insisted that the guts of the whale produced ambergris. However, some European savants still insisted that it was originally produced in underground fountains and only showed up in the guts of whales because the whales had swallowed it. The matter was not finally settled until the nineteenth century.

As early as the ancient Egyptians, ambergris was believed to have medical uses. Medieval Arab physicians recommended it for sore throats, heart disease, and mental troubles. Medieval Europeans believed it to be effective for epilepsy, colds, and heart troubles. Its sweet smell led some Europeans to believe that it protected against the Black Death. Ambergris was also used as an aphrodisiac and a cure for impotence. It was a key ingredient in "Lapis de Goa," a kind of artificial bezoar manufactured in the Portuguese Indian colony of Goa from the mid-seventeenth to the mid-nineteenth century. The Chinese, who valued ambergris highly, believed it to be good for the heart and brain.

See also: Amber; Aphrodisiacs; Bezoars

Further Reading

Azzolini, Monica. 2017. "Talking of Animals: Whales, Ambergris and the Circulation of Knowledge in Seventeenth-Century Rome." *Renaissance Studies* 31: 297–318; Chardin, John. 1988. *Travels in Persia 1673–1677.* New York: Dover; Dannenfeldt, Karl H. 1982. "Ambergris: The Search for Its Origins." *Isis* 73: 382–397.

Amulets and Talismans

Amulets and talismans are small charms usually carried on the body for various magical purposes, including strength, luck, the prevention or cure of disease, and protection against the evil eye and other forms of harmful magic, or demons. Amulets were also used to encourage fertility or were placed on or held by pregnant women for ease and safety in delivery. There are many ways to distinguish between an amulet and a talisman. Some claim that an amulet is defensive magic, while a talisman is carried to enhance the wearer's power. Another distinction is between talismans that are made out of one piece of material and amulets that are composite. However, the terms are also used interchangeably. Amulets have a long history, with small, amulet-like ornaments found in Paleolithic sites and evidence of their magical uses going back to ancient Egypt and Mesopotamia. They are found today in a broad range of cultures and traditions.

Amulets are frequently made out of rare, beautiful, or exotic substances such as gems, jewels, or organic "stones" such as the alectorius. Amber was another precious material used for amulets. They could be carried separately on the body or worn in rings or other pieces of jewelry. Although many amulets are originally made for that purpose, coins and other objects could also be made into amulets. One example is the "touch pieces" given by the king of England after applying the royal touch for scrofula. These coins would be drilled with a hole in the middle and worn around the neck and were believed to have their own protective power. Parts of plants or animals could be used as amulets; a famous example is the rabbit's foot whose association with luck in the West has persisted for many centuries. Dried cauls were considered amulets against drowning. The Roman natural historian Pliny the Elder (d. 79 CE) describes many amulets made of animal parts used to cure or prevent fevers including a viper's head wrapped in a linen cloth and the muzzle and ear tips of a mouse wrapped in red cloth, the mouse to be set free after its parts were removed. Since these organic amulets, such as the ones of paper or parchment, would decay over the centuries, fewer survived than amulets made of stone or metal. African American root workers identified roots that could be carried for positive effect, as John the Conqueror root was carried for virility and good fortune. Human body parts could also become amulets, such as the Hand of Glory.

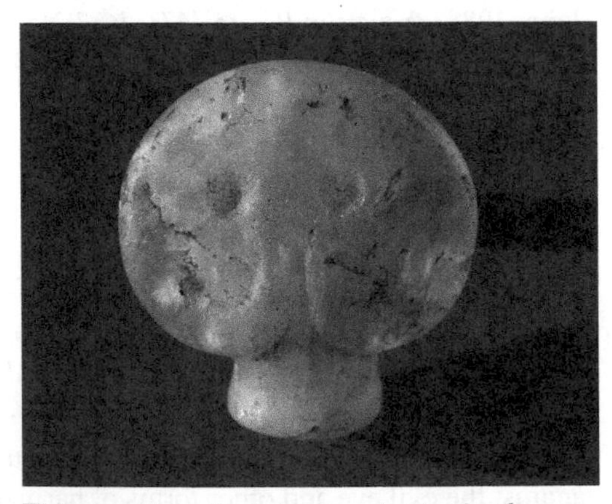

This amethyst Egyptian amulet in the form of a bull's head is over 5,000 years old. (The Cleveland Museum of Art, John L. Severance Fund)

Amulets could be not only individually made but also mass-produced; the eighteenth-century Benedictine abbey of Scheyern in Bavaria did a brisk business in little lead crosses believed to protect against witchcraft and harmful magic. Woodblock printing could be used to reproduce amulets, as could more efficient techniques such as metal stencils. Amulets were brought together into collections, which amplified the power of the individual amulets. These include the medicine bags of Native Americans or the mandinga pouches of Brazil.

Some persons were viewed as particularly skilled in the making of amulets. These could be particularly holy people such as Christian or Buddhist monks or Sufi sages or the marabouts of West Africa. They could also be magical specialists such as sorcerers or astrologers. Amulet-making could also be a hereditary trade with the ability to do

The Pedlar of Spells by Lu Yu (1125–1209 CE)

"An old man selling charms in a cranny of the town wall. He writes out spells to bless the silkworms and spells to protect the corn. With the money he gets each day he only buys wine. But he does not worry when his legs get wobbly, For he has a boy to lean on."

Source: Waley, Arthur. *A Hundred and Seventy Chinese Poems.* New York: Knopf, 1922. P. 152.

so considered to run in some families, such as the "ritual masters" of Vietnam.

In ancient Egypt, amulets in the form of scarabs, a beetle considered sacred, were particularly popular beginning around 2000 BCE. Although the vast majority of scarabs were worn by the living, to protect against misfortune, there was also a type of scarab amulet known as the heart scarab that was placed over the heart of a mummy to protect it in the afterlife. The scarab, carved with a passage from the Book of the Dead, would protect the deceased against the possibility that their heart would witness against them in the afterworld. Less common were amulets made in the shape of the ankh, the symbol of life.

Writing was frequently incorporated into amulets, particularly the sacred writing of Abrahamic religions. The Lord's Prayer appears in early Christian amulets and Quranic verses, the names of God, or traditional sayings of Muhammad in Islamic ones. Many Jewish and Islamic amulets were entirely representations of the written word, without any symbols. The prohibition on idolatry in Judaism and Islam made textual amulets less suspect than representational ones. The principles of gematria for Jews and lettrism for Muslims, the idea that God had created the world out of letters, were easily adaptable to the needs of amulet-makers and users. The numerical value of a piece of writing in an amulet could structure its creation. Writing did not always have to be sacred or even have meaning; the nonsensical word "abracadabra" appears in directions for making an amulet against malaria, by the pagan physician Quintus Sammonicus (d. 212 CE). It would have a long history in the making of amulets, reappearing in the seventeenth-century Plague of London. Amulets could also skirt the prohibition on representation by featuring abstract geometric designs, such as the various versions of the seal of Solomon.

Religiously based amulets also appeared in Buddhism, where small representations of the Buddha or of revered Buddhist figures often appeared in amulet form. Hindu amulets included representations of Hindu deities. In addition to religious representations, amulets could also be endowed with sacred power by rituals and prayers carried out in their manufacture or by the holiness, real or claimed, of their maker. The distinction between magical amulets and sacred objects could be a subtle one.

Astrological symbols and representations also frequently appeared on amulets. Astrologers frequently manufactured amulets as well as drawing horoscopes. The theory of the efficacity of such talismans is that they draw down powers associated with the planets and bind them in the amulet. Timing was crucial in the making of an astrological amulet—the planets, particularly the one represented on the amulet, had to be in the proper positions for the amulet to be effective.

The eye frequently appears on amulets. Blue eyes usually made of glass, referred to as *nazars*, are very common in the Middle East, Greece, Turkey, and elsewhere and protect against the evil eye, an example of like curing like. If a representation of an eye is put on the back of a stylized hand, the resulting object is referred to as a hamsa and is believed to have power against a wide range of supernatural evils, from the evil eye to the demoness Lilith. The popularity of the nazar and hamsa across religious communities in the Middle East is an example of how amulets could cross cultural and religious barriers. (These crossings were sometimes contested. Sixteenth-century Spanish Christian authorities, believing that the hamsa was inextricably bound to the Islamic and Jewish cultures, would forbid its display or use.) Another popular body part for representation in amulets in some cultures was the male genitals. Phallic amulets were particularly common in the Roman Empire and were believed to protect against bad luck and ill-health. They were particularly commonly carried by soldiers. Phallic amulets are also found in other places, such as Thailand.

In China, amulets in the form of flat discs or squares of bronze or brass resembling coins are common, believed to bring good fortune such as prosperity and success in the imperial examinations and to ward off evil. They carried zodiacal symbols, representations of legendary animals, and quotations from the classics. Some also carried Daoist or Buddhist elements. Chinese influence has resulted in the creation of similar types of small metal amulets in Korea, Japan, and Vietnam. Another type of Japanese amulet is the omamori, a short prayer written on paper and then sealed in a small drawstring bag. Shinto and Buddhist versions are commonly sold at temples. Rather than general good fortune, they usually have specific purposes such as prosperity, protection from evil, educational success, or love. Korean written amulets for similar purposes use larger pieces of paper and display rather than hide the writing. Vietnamese also employ paper in addition to metal amulets for such purposes as defending against malevolent ghosts who bring diseases of humans or livestock or for success in examinations. Vietnamese paper amulets can be carried on the person or displayed in the home. The communist government of Vietnam tried and failed to ban amulets.

Abrahamic religious tradition has been suspicious of amulets, particularly ones that do not incorporate texts or other specific religious references. The Talmud endorsed amulets while distinguishing between "proven" amulets, which could be worn on the Sabbath, and unproven amulets that could not. Proven amulets were those that had healed a person three times. Given the Talmud's endorsement, Jewish authorities could not forbid Jews to make or carry amulets, but they sought to regulate the trade. Medieval rabbis put forth rules by which amulet-making would be conducted in a religiously appropriate way with rules for writing amulets analogous to those applying to writing sacred books. Medieval Jewish authorities also warned against the use of specifically Christian amulets, indicating that some Jews were at least willing to use them.

In the Christian tradition, Roman Catholicism and Orthodoxy have been friendlier toward the treating of sacred items such as rosaries as amulets than Protestantism, although the position of the Church is that such devices do not accomplish their ends of their own accord, but as a result of the user's faith and the power of God. One of the most common specifically Catholic amulets is the St. Christopher medal, believed to protect travelers from harm. Use of the St. Christopher medal has spread outside the Church. Orthodoxy has a tradition of using clerically manufactured cloth amulets to protect from evil, pouches containing dirt from the grave of a saint or the remnants of items used during a service. Small ikons can also be used as amulets. Protestant theologians and church authorities have generally condemned any use of amulets, although that has not stopped Protestant believers from using them.

Islamic religious authorities have taken different positions on amulets. Hadiths, traditions attributed to Muhammad, have condemned amulets, but some authorities say that these refer only to amulets dating from the pagan era in Arabia, and strictly Islamic amulets remain acceptable, as long as the believer continues to trust in God rather than thinking the amulet has autonomous power. Some venerated early Islamic leaders are portrayed as wearing such amulets. Other authorities condemn all amulets. In particular, the Salafist tradition in Islam, often labeled as "Islamic fundamentalism," has condemned the use of amulets.

See also: Alectorius; Amber; Astrology; Defensive Magic; Dogs; Evil Eye; Execution Magic; Foxes; Hand of Glory; Hares and Rabbits; Hyenas; Incantation Bowls; Kabbalah; Lettrism; Lilith; Love Magic; Luck; Mandinga Pouches; Natural Magic; Otoliths, Fish; Precious Stones; Rats and Mice; Relics; Royal Touch; Sabianism; Snakes; Solomon; Spiders; Swallow Stones; Toads

Further Reading

Baldwin, Martha R. 1993. "Toads and Plague: Amulet Therapy in Seventeenth-Century Medicine." *Bulletin of the History of Medicine* 67: 227–247; Francis, Doris, ed. 2007. *Faith and Transformation: Votive Offerings and Amulets from the Alexander Girard Collection.* Santa Fe, NM: Museum of New Mexico Press; Mommersteeg, Geert. 1988. "'He Has Smitten Her to the Heart with Love': The Fabrication of an Islamic Love-Amulet in West Africa." *Anthropos* 83: 501–510; Thuat, Vũ Hong. 2008. "Amulets and the Marketplace." *Asian Ethnology* 67: 237–255.

Aphrodisiacs

Aphrodisiacs are those substances believed to naturally excite love and desire, particularly when consumed as food and drink. The distinction between those substances that are naturally aphrodisiac and the ingredients in love-philters that employ magic for sexual or romantic purposes is an unstable one and does not work the same way in all cultures. The term "aphrodisiac" is derived from Aphrodite, the Greek goddess of love and sex, although it was not coined until the eleventh century CE. An enormous range of foods have been described as aphrodisiac across cultures, leading some scholars to claim that every food has been claimed at some time or another to be aphrodisiac. Aphrodisiacs were viewed frequently not just in the context of provoking desire but also of increasing fertility.

A common belief is that plants or animals are aphrodisiac due to a shape resembling male or female genitals, an example of the "doctrine of signatures" that was widely applied to medical uses for plants and animals generally. The banana is one example of a food widely believed to be aphrodisiac based on its phallic appearance. The oyster owes its legendary reputation as an aphrodisiac to its resemblance to an aroused women's genitals. The orchid is named after the ancient Greek word for testicle. Greeks, including botanists such as the first-century Greco-Roman physician Dioscorides, believed that the tuberous roots of the orchid, the part resembling testicles, was an aphrodisiac. Dioscorides named many other plants as provoking sexual desire, including the turnip and the leek. He also named one animal, a small lizard called the skigkos or skink. Animals thought to have particularly powerful sexual drives, such as goats or sparrows, were often considered to be sources of particularly aphrodisiac meat. Consuming animal testicles was also recommended to both men and women.

In Galenic medicine, substances that aroused or maintained sexual desire did so due to their composition, conceived of in terms of the four elements, rather than their shape. Plants with an "airy" quality were likely to be aphrodisiacs. Erections were believed to be, in part, the result of the penis inflating with "wind," so windy foods, which also provoked flatulence, were believed particularly important for men.

In terms of the four Aristotelian qualities (hot/cold, moist/dry), hot foods were believed to be the most powerful aphrodisiacs. Hot foods and to a lesser extent moist foods functioned as aphrodisiacs for both men and women. Information about aphrodisiac qualities of "hot" plants was common in early modern herbals. Because it was hot, some writers recommended meat in general, regardless of its source, as an aphrodisiac. Although there was little fear of men becoming overheated, some writers suggested that women should not consume too many hot foods, lest they become overheated and thus infertile. Heat in the male was particularly valuable not just for producing a child but also a male child, as males were believed to naturally have more heat.

Medieval Arab physicians and botanists developed the scattered references made by the Greeks to aphrodisiac substances into a herbal medicine oriented to sexual desire in the context of sexuality in general, including such topics as fertility and pleasure, as well as contraceptives and abortifacients. European conceptions of the aphrodisiac were more limited to sexual desire and fertility. (Sexual pleasure was important for both male and female in conception; as in the Galenic tradition that dominated medicine in both the Christian and Islamic worlds, sexual climax in both partners was believed necessary for conception.)

The onslaught of new commodities arriving in Europe from the Americas and the Indian Ocean included many substances that from their exoticism were identified as aphrodisiacs. Many of the newly more common spices, such as pepper, ginger, and cinnamon, were believed to be "hot" and therefore aphrodisiac. Chocolate, already considered an aphrodisiac in Mexico where it originated, was also considered an aphrodisiac in Europe. When it initially arrived in Europe in the sixteenth century, the potato was originally thought to be an aphrodisiac, due to its phallic resemblance as well as its exoticism. This explains the line William Shakespeare gives to Sir John Falstaff in *The Merry Wives of Windsor*, "Let the sky rain potatoes" (V.v 18).

The European Enlightenment resulted in the rejection of the alleged aphrodisiac qualities of many plants by the medical establishment, although the truffle was an exception. However, aphrodisiacs continued to play an important role in culture. In the nineteenth century, there was an increased emphasis on exoticism in the marketing of aphrodisiacs, as sexual exuberance was associated with non-European cultures. Although many "simple" substances were believed to be aphrodisiacs, there were also "compound" recipes combining many substances purported to have an aphrodisiac effect. (Compounds had the advantage that they could be secret and proprietary. Proprietary medicines for sexuality and fertility appeared in eighteenth-century England and would later be a staple of the patent-medicine industry.)

In the Indian tradition, milk and sugar are considered to be aphrodisiacs, recommended in the *Kama Sutra* as the best of aphrodisiacs. Herbs such as garlic and wild asparagus are also recommended. Early Chinese medical writings contain little on aphrodisiacs, but in the Middle Ages, possibly influenced by contact with Indian Ayurveda through Buddhism, interest increased. In traditional Chinese medicine, spices are often combined and applied to the male and female genitals directly rather than consumed. Ginseng is considered a particularly powerful aphrodisiac. Cinnamon and pepper are also effective. Although rhinoceros horn is used in traditional Chinese medicine, the idea that its primary use is as a male aphrodisiac is a myth. It is mostly used to treat fevers.

See also: Alectorius; Ambergris; Bananas; Chocolate; Elemental Systems; Goats; Herbs; Love Magic; Mandrakes; Saffron

Further Reading

Evans, Jennifer. 2014. *Aphrodisiacs, Fertility and Medicine in Early Modern England.* Woodbridge, Suffolk: Boydell & Brewer; Moore, Alison M. Downham and Rashmi Pithavadian. 2021. "Aphrodisiacs in the Global History of Medical Thought." *Journal of Global History* 16: 24–43; Steavu, Dominic. 2017. "Buddhism, Medicine, and the Affairs of the Heart: Āyurvedic Potency Therapy (Vājīkarana) and the Reappraisal of Aphrodisiacs and Love Philters in Medieval Chinese Sources." *East Asian Science, Technology and Medicine* 45: 9–48.

Apparitions

An apparition is one of the most mysterious events that humans experience—the appearance of a person, animal, plant, supernatural being, or object without any apparent material component. An apparition can be viewed or described as a waking dream. Apparitions can be perceived not only as indistinct or transparent but also as solid, three-dimensional objects indistinguishable from actually existing phenomena. An apparition can appear to a group or to an individual, although apparitions that appear to a single individual with no other witnesses are more likely to be written off as hallucinations or lies. There are also occasions when apparitions are manifest to more than one but not all members of a group.

Apparitions can be of a wide range of beings. Ghosts are frequently perceived as apparitions. However, apparitions of sacred personages are also common, such as the apparitions of the Virgin Mary in Catholicism. Evaluating the validity of apparitions, which are often associated with prophecies or warnings to the faithful, is a function of the Church hierarchy. Celestial phenomena such as rainbows and parhelia can also be perceived as apparitions, although the presence of natural-philosophical explanations usually takes them out of the category.

An apparition can be viewed as a prodigy or a presage of future events. One example is the "battle in the sky," a vision of a large-scale armed conflict. These frequently occurred in wartime and were viewed as omens of a battle in the location where they were witnessed. In some cases, the aerial battles received very detailed descriptions in accordance with the military technology and tactics of their day. Another common apparition was that of a person that was known to be a long distance away, usually a spouse or relative. These apparitions were also particularly common in wartime and were strongly associated with the death of the individual in the apparition. Although battles in the sky and apparitions of the dead usually took place far away from the actual fighting, apparitions also appeared on the battlefield itself. A famous but probably spurious example is the "Angels of Mons" in World War I. Rumors of supernatural bowmen appearing on the British lines very early in the war circulated in 1915, probably based on a published story by the Welsh author Arthur Machen (1863–1947), "The Bowmen," about spectral bowmen appearing to aid beleaguered British troops.

The investigation of apparitions was an important part of nineteenth- and twentieth-century psychical research. The Society for Psychical Research launched an investigation that established that there were no eyewitnesses to the "Angels of Mons," and they are generally now considered a hoax. The UFO visions of the twentieth century have much in common with the tradition of apparitions.

See also: Dreams; Ghosts; Prodigies; Rainbows

Further Reading

Christian, William A., Jr. and Gabor Klaniczay, eds. 2009. *The "Vision Thing": Studying Divine Intervention.* Budapest: Collegium Budapest; Clarke, David. 2002. "Rumours of Angels: A Legend of the First World War." *Folklore* 113: 151–173; Hardison, S. Alexander. 2015. "On the 'Types' and Dynamics of Apparitional Hallucinations." *Paranthropology: The Journal of Anthropological Approaches to the Paranormal* 6: 65–74.

Ars Notoria

The Ars Notoria or Notary Art is a form of medieval European ritual magic aimed at gaining knowledge and intellectual skills such as memory or rhetorical persuasiveness through prayer, fasting, and the contemplation of figures included in manuscripts. It was a learned art associated with the biblical King Solomon and ultimately with God, from whom Solomon's wisdom was derived. It was also ascribed to the semilegendary first-century CE pagan sage Apollonius of Tyana.

Whole subjects ranging from arithmetic and astrology to law, medicine, and philosophy could be learned not through years of study but in as little time as a month through the Ars Notoria. The prayers were supposed to be performed at particular times of day and phases of the Moon. (Despite this interest in lunar phases, astrology played only a minor role in the Notary Art.) The figures included prayers written out in complex designs such as spirals and circles and enclosed or adorned with images of animals, people, or angels. (These were the *notae* that were the namesake of the Notary Art.) The prayers included the names of angels and names of God, often in foreign languages. They were also sometimes addressed to the Virgin Mary, and works of the Ars Notoria were sometimes bound together with works of Marian devotion. Despite the Church's opposition to the Ars Notoria, it was certainly possible that its practitioners, predominantly monks and clerics, saw that what they were doing was a form of Christian piety.

The Ars Notoria was widely condemned by medieval theologians and clerics, including the great Scholastic philosopher Thomas Aquinas who discussed it in his condemnation of superstition. Practitioners were seeking knowledge, a laudable goal for Thomas, by the inappropriate means of fasting and the contemplation of figures, rather than the appropriate means of diligent study. There was also a strong possibility that the foreign names in the prayers were those of demons rather than angels. Thomas's discussion shows a familiarity with the Ars Notoria gained either from examinations of manuscripts himself or from discussions with knowledgeable persons. Because the practitioner was striving for good things in themselves and risked no soul other than his own, it was not considered to be as much of a menace as necromancy or other branches of ritual magic.

Although the Middle Ages were the heyday of the Ars Notoria, manuscripts continued to be circulated into the early modern period. It was listed on the sixteenth-century Catholic Index of Forbidden Books, and there was a Latin edition with English translation published in the seventeenth century.

See also: Solomon

Further Reading

Fanger, Claire, ed. 1998. *Conjuring Spirits: Texts and Traditions of Medieval Ritual Magic.* Phoenix Mill: Sutton; Skinner, Stephen and Daniel Clark, eds. 2019. *Ars Notoria: The Grimoire of Rapid Learning by Magic with the Golden Flowers of Apollonius of Tyana.* Translated by Robert Turner. Singapore: Golden Hoard Press.

Asses. See Horses and Donkeys

Astrology

Among the most complex, sophisticated, and versatile of divinatory sciences was astrology, the "science of the stars." The idea that celestial bodies held meanings for humans was rooted in realities such as the weather, the tides, and the rough correspondence of the lunar and menstrual cycles. The earliest known attempt to go beyond these natural relations was in ancient Babylon in the early second millennium BCE. Babylonian astrologers sought to identify and interpret celestial omens for their impact on the king and society as a whole. Eclipses, comets, and a variety of other celestial phenomena were viewed as carrying meanings, sometimes very precise ones. Astrology was a specialized practice, the domain of a particular group of scribes. The Babylonians and their Middle Eastern successors viewed astrology as divine wisdom that the gods had given to humanity. Omens were messages from the gods, and bad omens did not lead to inexorable doom if the proper expiatory sacrifices were made. The Babylonians were responsible for the invention of basic astrological concepts such as the Zodiac. By the fifth century BCE, there is some evidence that astrology was spreading beyond the court.

The conquest of the Middle East by Alexander the Great in the late fourth century brought Greek astronomy into contact with Babylonian astrology. The blending of these two along with elements from Egypt created "Hellenistic" astrology. Hellenistic astrology emphasized the "aspect"—the angular separation between planets. (The astrological planets included the Sun and Moon as well as Mercury, Venus, Mars, Jupiter, and Saturn.) The significant aspects were conjunction, where two planets appeared in the same region of the sky, opposition, where they appeared in opposite regions, and separations by a sixth, quarter, or third of the sky. Hellenistic astrologers developed the concept of the house—twelve unmoving regions of the sky reckoned from the horizon. Each house ruled a particular aspect of life, such as travel, business, children, or health. The first house or "ascendant" was particularly important, because the first house ruled the person for whom a horoscope was being drawn. For millennia, people were astrologically identified with the zodiac sign that occupied the ascendant rather than with their sun sign as they do now. Numerous systems for dividing the sky into houses have been put forth over the history of astrology.

Cicero on the Origins of Astrology

On Divination by the Roman philosopher and orator Marcus Tullius Cicero is one of our main sources on the divination techniques of all kinds used by the ancients. Like modern scholars, he ascribes the origin of astrology to ancient Mesopotamia, crediting this to the wide view of the heavens on the Mesopotamian plain.

"Now I am aware of no people, however refined and learned or however savage and ignorant, which does not think that signs are given of future events, and that certain persons can recognize those signs and foretell events before they occur. First of all—to seek authority from the most distant sources—the Assyrians, on account of the vast plains inhabited by them, and because of the open and unobstructed view of the heavens presented to them on every side, took observations of the paths and movements of the stars, and, having made note of them, transmitted to posterity what significance they had for each person. And in that same nation the Chaldeans—a name which they derived not from their art but their race—have, it is thought, by means of long-continued observation of the constellations, perfected a science which enables them to foretell what any man's lot will be and for what fate he was born."

Source: Cicero. 1923. *On Divination* I.2. Translated by William Armistead Falconer. Cambridge, MA: Loeb Classical Library.

Hellenistic astrologers also developed the three main branches of astrology—the natal, deducing the fortunes of an individual from the position of the stars at his or her birth, the mundane, relating the history of the world to the courses of the celestial bodies, and the katarchic or electional, the finding of the right astrological moment for a given action such as the marriage or the laying of a building's cornerstone or the founding of a city. Although Hellenistic astrology was widely used by kings and political leaders, it was also available to private individuals, who consulted astrologers for the drawing of horoscopes and the giving of advice on such matters as business and marriage.

The influence of Hellenistic astrology was not limited to the Middle East. It was adopted, along with other elements of Greek culture, by the Roman Empire, where it largely replaced earlier systems of divination such as hepatoscopy. (It was forbidden to calculate the horoscope of the emperor, as being able to predict the emperor's day of death could encourage assassinations. This evolved into the doctrine that the emperor was not subject to the stars.) Roman writers wrote astrological works in Latin, drawing on Greek sources. The most influential astrological writer under the Roman Empire was a Greek, however—Claudius Ptolemy (c. 100 CE–c. 170 CE), a great astronomer whose astrological work, the *Tetrabiblos* or "Four Books," was considered the authority in the field for centuries. Ptolemy saw astrology as the working out of physical forces associated with the planets rather than as the work of planetary gods. Hellenistic astrology also influenced Iranian and Indian astrology. (Despite the opposition to astrology found in some early Zoroastrian texts, astrology flourished in the Iranian world.) Astrological works based on the writing of the Hellenistic astrologer Dorotheus of Sidon (first century CE), whose approach to astrology was much more "magical" than Ptolemy's, were found in the libraries of medieval China and Japan.

Astrology flourished in India while developing in a different way than it did in the Middle East or Europe. Even before contact with the Hellenistic tradition, from which it derived much of its astrological vocabulary, Indians had developed a system based on twenty-eight "lunar mansions"—the areas of the sky occupied by the Moon throughout its cycle. This idea was incorporated with Hellenistic ideas. Indian astrologers also treated "Rahu" and "Ketu"—the points where the orbit of the Moon intersects the ecliptic—as planets in their own right. They also reckoned the zodiac differently than Westerners. Indian astrologers viewed the zodiac in terms of the actual constellations, the "sidereal zodiac" rather than the Western "tropical zodiac" defining the Sun's position at the spring equinox as the first degree of Ares regardless of its relation to the constellations. Indian astrologers treated the stars as revealing an individual's karma, the consequences of their actions in this and previous lives, rather than exercising an independent power. Astrology became a recognized profession in India, and its practitioners eventually formed a hereditary *jati* or "caste."

Astrology in China was principally carried on by a government bureaucracy, the Bureau of Astronomy, a part of the Board of Rites. The principal astrological function of the Bureau was to determine auspicious and inauspicious days. Astrology was treated as a state secret in China. However, as elsewhere in East and Southeast Asia, there was also a type of astrology that was practiced much more openly. This type of astrology was introduced and

A fifteenth-century Italian representation of astrology as part of a series on the liberal arts. (The Cleveland Museum of Art, Dudley P. Allen Fund)

practiced by Buddhist monks, who introduced astrological ideas from the Hellenistic tradition, as well as Iranian and Indian concepts. Tibetan Buddhists had a mostly Indian-derived astrological tradition with some Chinese and indigenous elements, which they claimed had been originally taught by the Bodhisattva Manjushri. In early Buddhism, astrology was considered too worldly to be practiced by monks, but this idea was marginalized in a few centuries. Increasingly, Buddhists came to believe that astrology was necessary in order to determine the proper time to perform ceremonies when they would be most effective.

The adoption of Christianity in the Roman Empire had surprisingly little impact on astrology, despite several imperial laws against it and other forms of divination. The Christian attitude toward astrology was complex, as the legend of the Star of Bethlehem that announced the birth of Jesus could be seen as astrological and the Three Magi from the East who came to the Christ Child were traditionally viewed as astrologers. The apocalyptic element of Christianity also caused Christians to look to the heavens for the signs of the coming end times. However, the association of the planets with pagan gods caused some Christians to be suspicious of astrology and to believe it to be demonic in origin. The collapse of Roman power in Western Europe in the fifth and sixth centuries meant that astrology disappeared along with astronomy, as there were simply not enough educated people to carry on the tradition. In Byzantium, the surviving Eastern Roman Empire, astrology continued to be practiced, but the astrological tradition was carried on most vigorously in the Islamic world.

Building on Hellenistic and Iranian traditions, the astrologers of the Islamic world (not all of whom were Muslims themselves) wove astrology into theories of the cosmos and human history. A particularly important idea, originating in Iran, was that of the importance of the "great conjunctions" of Jupiter and Saturn, the two slowest moving of the known planets. The conjunctions took place roughly every twenty years, and their passage through the zodiac was held to determine much of human history. Conjunctions were associated with plagues, wars, new religions, and the rise and decline of empires and cities. Islamic rulers born near the time of a conjunction would adopt the title "Lord of the Conjunction." Numerous Greek, Persian, and Indian astrological works were translated into Arabic. The date for the founding of Baghdad, the capital of the Abbasid Caliphate, was determined by a team of astrologers including the Persian Jew Mashallah ibn Athari (740–815). Astrologers in the Islamic world also attempted to provide a physical basis for astrology by speculating that the stars were the causes of generation and corruption.

Astrologers were ubiquitous in medieval Islamic society. They were frequently portrayed in the art of the period, accompanied by the instrument that became the tool and symbol of their profession, the astrolabe, invented in the Hellenistic world but improved on in the Islamic world. They produced several introductions and handbooks to astrology in Arabic. Although some Muslim religious scholars opposed astrology on the grounds that allowing power to the stars impinged on the omnipotence of God, their opposition was seldom effective. The great observatories of the Islamic world such as Maragha in Persia and Samarkand in Central Asia were designed to produce accurate star charts for the benefit of astrologers. Like Christians, Muslims included astrology in their apocalyptic literature.

It was from the Islamic world, particularly Islamic Spain, that astrology was reintroduced into the Latin West. In the great translation movement beginning in the eleventh century, both originally Arabic and originally Greek works were translated into Latin. In medieval Europe, astrology was studied in universities and practiced in courts. Rulers married, consummated their marriages, baptized their children, and declared their wars at the times their astrologers thought best, although, like the Romans, they penalized those who would attempt to calculate the time of a ruler's death. "Horary" astrologers answered questions by taking a horoscope of the moment that the question was asked. Some religious authorities argued that astrology denied humans free will, but astrologers defended astrology by saying the stars impelled people in certain directions but did not compel them—"agunt non cogunt." The friar and philosopher Roger Bacon (c. 1219–c. 1292) defended astrological magic in that it relied purely on the influences of the stars and did not invoke or deal with demons in any way, as did forbidden forms of magic.

Astrology reached new heights of popularity and influence in the Renaissance and Reformation period. The revival of classical knowledge led to a revived interest in astrology as well as many other subjects, as well as movements to "purify" astrology of additions made in the Islamic period by returning to the Greek sources, particularly Ptolemy. Various kinds of philosophy that challenged the university establishment were associated with astrology. The Protestant Reformation, which aroused interest in the apocalypse, also raised interest in astrology. The printing press allowed for the wide circulation of astrological information, particularly in the form of the annual almanac. The chaos and uncertainty of the great religious conflicts of the era, such as the French Wars of Religion, the Thirty Years' War, and the British Civil Wars, saw outpourings of astrological predictions and analyses.

By the end of the seventeenth century, the public role of astrology was beginning to diminish in the West. Some regimes opposed it as politically destabilizing. Others viewed it as unscientific. The rise of heliocentric astronomy was not fatal to astrology, as the positions of the planets relative to the Earth did not change. The Scientific Revolution saw several attempts to reform astrology in accordance with the "new philosophy." None were particularly successful, and the idea that astrology was "superstitious" and archaic spread among the educated population of Europe. Astrology disappeared from almanacs.

Astrology was closely allied with medicine, and many physicians were also astrologers. It was believed that astrology was necessary for treatment, as without astrology the physician could not know the right time to administer a medicine. Astrology was also useful in diagnosis. Taking a horoscope of when the patient took to his or her bed—"decumbiture"—and of the moment when the patient's urine was presented to the physician—"uroscopy"—were both established techniques of medical astrology. In the Galenic tradition that dominated both Islamic and Christian medicine, there was the concept of "critical days" that decided the course of an illness. The concept of critical days had originated in Greek medicine before astrology played a role in Greek thought, but Galen, followed by medieval and Renaissance physicians, gave it an astrological interpretation. Astrologically minded physicians produced case histories that related every twist or turn of an illness to the changes in the celestial bodies. The signs of the zodiac governed the parts of the human body, and the "Zodiac Man"—a drawing of a man with each body part identified with a sign—was a staple of popular astrological books in the early modern West. Medical astrology dealt with mass death as well as the medical history of individuals. When Philip VI of France asked the medical faculty of the University of Paris to explain the Black Death that struck Europe in 1347, they claimed the ultimate cause was a conjunction of Saturn, Mars, and Jupiter in Aquarius in 1345.

Another area closely associated with astrology was agriculture, the primary economic activity of most settled societies throughout history. Celestial phenomena, particularly the phases of the Moon, were held to indicate the right time for planting, harvesting, and other agricultural operations. Astrology also affected agriculture through the influence of the planets on the weather. Some people who opposed astrology in general were willing to accept agricultural astrology as a natural phenomenon.

Astrology was also closely identified with other magical systems, including alchemy, geomancy, lettrism, necromancy, Hermeticism, and Kabbalah. Ceremonial magicians had to pick the right astrological moment to carry out their rituals. The astrological planets and signs provided ways to classify sets of seven and twelve. Metals were identified with planets, as the Sun was with gold or Mars with iron. Herbs were also identified with planets, which provided the skillful practitioner with information as to which conditions they were useful in treating.

See also: Alchemy; Amulets and Talismans; Comets; Divination; Eclipses; Geomancy; Herbs; Hermes Trismegistus; Kabbalah; Lettrism; Luck; Macrocosm/Microcosm; Moon; Natural Magic; New World Inferiority; Prodigies; Rosicrucianism; Sabianism; Solstices and Equinoxes; Sun

Further Reading

Burns, William E. ed. 2018. *Astrology Through History: Interpreting the Stars from Mesopotamia to the Present Day.* Santa Barbara, CA: ABC-CLIO; Kotyk, Jeffrey. 2018. "Japanese Buddhist Astrology and Astral Magic." *Japanese Journal of Religious Studies* 45: 37–86; Maxwell-Stuart, P.G. 2010. *Astrology from Ancient Babylon to the Present.* Chalford, Gloucestershire: Amberley; Saif, Liana. 2015. *The Arabic Influences on Early Modern Occult Philosophy.* New York: Palgrave Macmillan.

Atropaic Magic. See Defensive Magic

B

Bananas

The banana, one of the Earth's most commonly consumed fruits, has become the object of superstitious beliefs in many cultures. The banana has been identified, in Jewish, Christian and Islamic traditions, as the forbidden fruit of the Garden of Eden. The Quran gives a lengthy description of the tree of the forbidden fruit that strongly resembles a banana tree. The banana appears as the "fruit of Eden" in several medieval Jewish texts. The Swedish scientist Carl Linnaeus (1707–1778), the inventor of the modern system of biological classification, referred to this belief in giving the banana the scientific name *Musa Sapientium*, "Musa" being Arabic for banana and "Sapientium" Latin for of the wise or knowledgeable. The banana's cousin, the plantain, was *Musa Paradisiaca.* The leaves Adam and Eve covered themselves with after discovering their nakedness may well have been meant to be banana leaves, which are far larger and more suited to the purpose than fig leaves. The word "fig" referred to bananas as well as figs in many ancient texts.

Due to their shape, bananas are frequently associated with male virility or fertility. In East Africa, where multiple varieties of banana form a staple of the diet, the Mpologoma banana is associated with lions and consuming it is believed to assist male potency. East Africans also believe that a woman burying her placenta under a banana tree will bring prosperity to the family. In the Ayurvedic medicine of India, bananas are associated with water and Earth and are used to treat skin disorders. Ayurvedic medicine also considers using banana flowers and stem as treatment of diabetes. Bananas are also associated with fertility in Indian culture and incorporated into wedding ceremonies. *The Golden Wheel Dream-book and Fortune-teller*, a nineteenth-century American magical guide, suggests dreaming of bananas points toward marriage and fertility, as well as prosperity.

Bananas are associated with both good and bad luck. Among Indigenous Hawaiians, it was considered bad luck to dream of bananas. In Islamic dream interpretation, by contrast, eating a banana meant profits from business. Bananas on saltwater fishing boats are considered bad luck, a superstition that dates back to the eighteenth-century Caribbean but has spread to many other areas, including Alaska, Hawaii, and New Zealand. In some places, this prohibition has spread to freshwater fishing or small craft in general or extended to banning bananas for onshore breakfast the day of a fishing trip or even to wearing clothing from the chain Banana Republic. In the Caribbean, though, it was also considered good luck to slice into a banana near the stem and see a y-shaped mark.

See also: Aphrodisiacs

Further Reading

Koeppel, Dan. 2009. *Banana: The Fate of the Fruit that Changed the World.* New York: Plume.

Banshee. See Death; Fairies

Barghest. See Dogs

Barnacle Goose

In medieval Europe, the barnacle goose was a goose believed to grow from a barnacle, usually one growing on a tree. The barnacle itself was believed to be a fruit of the tree. (In one version of the story, those geese that fell in the water lived and those that fell on the ground died or never became animate.) The theory may have originated as an explanation for geese that never appeared to lay eggs or nurture the young. These geese were actually migratory and nested in the arctic, but this explanation was not known until the early modern period. The barnacle itself had a strong resemblance to the neck of a goose.

The barnacle goose raised several questions of meaning and classification. The twelfth-century bishop Giraldus Cambresis viewed the ability of the goose to emerge from a barnacle as evidence of God's ability to create Adam from the dirt. Some argued that since the goose was a fruit rather than a bird, its consumption was permissible during Lent, although this never became official doctrine. The natural philosopher Albertus Magnus (1200–1280) was skeptical of the barnacle goose, but belief remained popular in the Middle Ages.

In the late Middle Ages, barnacle geese were particularly strongly associated with Scotland. In 1527 the Scottish humanist and principal of Aberdeen University Hector Boece (1465–1536) set forth a new theory of the barnacle goose, which was that the geese were engendered by worms within seaweed and brought to life by the power of the ocean. Another Scotsman, Sir Robert Moray (1609–1673), the first president of the Royal Society, endorsed the barnacle theory in a paper published in 1677, claiming to have seen embryonic geese in barnacles he had found on a dead fir tree. Moray's account, which did not extend to claiming that he had seen a goose actually emerge from the barnacles, was challenged at the time. In the seventeenth and eighteenth centuries, however, the story was less commonly believed, partly because Dutch sailors had identified nests of the geese in the far north. It remained a live topic, however, until 1751 when an English physician named John Hill (1716–1775) published a denunciation of the Royal Society, *A Review of the Works of the Royal Society of London*, in which ridicule of Moray's paper on barnacle geese figures heavily.

Further Reading

Goodare, Julian and Martha McGill, eds. 2020. *The Supernatural in Early Modern Scotland.* Manchester: Manchester University Press; Pastore, Christopher L. 2021. "The Science of Shallow Waters: Connecting and Classifying the Early Modern Atlantic." *Isis* 112: 122–129.

Bathing

Bathing, the immersion of the usually unclothed body in water or, rarely, another liquid, has a variety of cultural meanings other than cleanliness and has attracted many superstitious or supernatural beliefs. In Judaism, Islam, Hinduism, and other religious traditions, ritual bathing and washing is an important part of purification. Traditional Chinese religion also enjoined bathing as part of purification before performing a sacrifice or carrying out other ritual activity. The connection between bathing and purity was such that in many cultures "impure" menstruating women were forbidden to bathe.

Christianity, which originated in a Greco-Roman pagan world where the communal bathhouse was an important center of leisure activities, inverted the relation between bathing and purity. It reacted to the sensual culture of the Roman bathhouse with a violent suspicion of bathing and an exaltation of the unwashed body as pleasing to God. (This was compatible with the belief in the cleansing waters of baptism.) In the Middle Ages, this was secularized into the idea that washing was unhealthy, particularly after the coming of the Black Death in 1347. Physicians stated that baths, particularly hot baths, opened the pores and made the body more vulnerable to disease. Water was generally considered unhealthy, and body odors were covered up with perfumes and scents among the classes that could afford it. Medieval and early modern Christians believed bathing to be particularly suspect in that it was associated with Islamic culture, as Muslims had continued and expanded the bathing culture of the ancient Mediterranean. (For Muslims, dreaming of taking a bath was associated with positive things such as recovery from illness and release from prison.) The idea that water and bathing were usually dangerous and unhealthy was beginning to be challenged in the seventeenth century, but elements of it persisted into the nineteenth century in Europe and its colonies.

Many cultures have ascribed medical virtue to bathing in specific springs or bodies of water. Hot springs have been particularly valued as have "sulfur springs." These are springs with water having a high sulfur content, distinguishable by the smell. Chinese sources dating as early as the Han Dynasty recommend bathing in sulfur springs to treat hemorrhoids, baldness, ulcers, and skin diseases as well as many other medical conditions. From the time of the fifth-century BCE Greek physician Hippocrates, the "Father of Medicine," Greco-Roman physicians saw bathing in and drinking specific types of spring water as vital to maintaining the balance of the body's humours. Hippocrates distinguished between springs based on whether they emerged from earthy hills, a positive quality, or from stone, a negative one. The best springs faced east toward the rising Sun, the worst toward the south. The idea of health-giving springs also has a biblical root in the pools of Siloam and Bethesda where Jesus is described in the Gospel of John as having performed healing miracles. The belief in healthful spring waters persisted in the West even in times when bathing and even drinking water generally was viewed with suspicion. In seventeenth-century England, bathing in warm springs was generally believed to promote fertility in women (though not in men) by "opening up" the body.

The ultimate extension of the idea of life-giving spring waters was the legend of the Fountain of Youth. The legend of a fountain that preserves youth and health and provides those who bathe in it with exceptionally long lives is found in the work of Greek historian Herodotus and continues to appear in classic and medieval literature. The location of this fountain is usually in the East. During the early sixteenth century, possibly influenced by similar Native American legends, some Spaniards came to believe there was a fountain of youth in Florida. The Spanish explorer of Florida Juan Ponce de Leon (1474–1521) was described by later writers as searching for the Fountain of Youth, but there is no direct evidence of this.

Bathing in substances other than or additional to water was sometimes believed to have particular effects. According to legend, Queen Cleopatra of Egypt (69–30 BCE) bathed in ass's milk to cultivate beauty, and the notorious "blood countess" of Hungary, Elizabeth Bathory (1560–1614), was believed to bathe in the blood of virgins to maintain her youth. In both cases, however, the story did not emerge until long after the deaths of the women. The "seven-flower bath," a bath in water in which seven flowers of different plants and different colors float, is believed to bring good fortune in East Asian cultures.

See also: Blood; Humours, Theory of; Menstruation; Saffron

Further Reading

Ashenburg, Katherine. 2008. *The Dirt on Clean: An Unsanitized History.* New York: North Point; Schafer, Edward H. 1956. "The Development of Bathing Customs in Ancient and Medieval China and the History of the Floriate Clear Palace." *Journal of the American Oriental Society* 72(2): 57–82.

Benandanti

In the sixteenth century, the small Venetian territory of Friuli on the borders of the German, Italian, and Slavic worlds had a reputation for magic. Friulians who emigrated to Venice frequently practiced as cunning folk and magical healers. The Friulian *benandanti* or "good walkers" were men and women, mostly of peasant background, born with cauls, traditionally a sign of supernatural power, who battled against witches. Details of their activities differ somewhat among the different benandanti who described them to the Roman Inquisition, but the common elements were that benandanti armed with fennel stalks fought in defense of the harvest against male and female witches armed with sorghum stalks. Those called to become benandanti began around the age of twenty and could retire around the age of forty. Both the benandanti and the witches were organized in military units with banners and captains, and both were usually described as going to the battle in spirit while their bodies were asleep. Some benandanti claimed to have been originally recruited to the company by angels.

Also in spirit, the benandanti were sometimes claimed by themselves or others to have witnessed processions of the dead and to know the final fate of departed souls. Some benandanti informed people of the fate of their departed relatives for a fee or practiced as cunning folk or witchfinders, and many peasants and workers who were not benandanti themselves credited them with the supernatural power to heal victims of witchcraft and to recognize witches. Indeed, some witches spoke of fearing the power of the benandanti.

The benandanti defined themselves as good Catholics and opponents of evil witches, while the Roman Inquisition, along with the foot soldiers of Catholicism, the confessors, preachers, and parish priests, defined them as witches. The first encounter between Inquisition and benandanti dates to 1575, but the first indication that the Inquisitors were winning the battle occurred in 1618. A thief and accused witch named Maria Panzona, identifying herself as a benandanti, admitted spontaneously, without leading questions from the Inquisitors, that she was a witch and derived her powers from the devil. Panzona later retracted this confession and affirmed herself to be a traditional benandanti, a good Catholic, and opponent of witches, but she was eventually sentenced to three years' imprisonment. Subsequent accused benandanti gave descriptions of the sabbat more or less coinciding with the picture held by the Inquisitors, although they sometimes emphasized their own separateness from the witches who also attended. By mid-century, the Inquisition's theory of witch benandanti had definitely vanquished the traditional picture, but the Inquisition's loss of interest in witch persecution meant that the remaining benandanti themselves suffered relatively little.

See also: Caul; Defensive Magic; Witchcraft

Further Reading

Ginzburg, Carlo. 1985. *The Night Battles: Witchcraft & Agrarian Cults in the Sixteenth & Seventeenth Centuries.* Translated by John Tedeschi and Anne Tedeschi. New York: Penguin.

Bezoars

Bezoars—large and indigestible obstructions found in the stomach and intestines of people and animals, particularly ruminants such as goats—were thought by many to have healing qualities.

The medical use of bezoar has a long history in the Middle East and Persia; the word "bezoar" is derived from the Persian "padzahr," meaning antidote. Early modern Persians used bezoars to treat poisons and fevers and to stimulate the body. They believed bezoars were more powerful if they came from animals that lived in hot and dry countries and fed on herbs. The best bezoars were to be found in the bodies of goats from the province of Khorassan. The bezoar was administered by two or three grains being scraped off with a knife and given with a spoonful of rose water. So popular was the bezoar as a medicine that bezoars were counterfeited with rosin and wax. In Tibet, when properly prepared, the bezoar was believed to be powerful not just against disease and poison but even against demonic or other supernatural attacks.

Belief in the medical use of the bezoar was introduced in medieval Europe through the writings of Arab physicians such as the Spaniard Ibn Zuhr (1094–1162), who had a high reputation in the field of poisons and their antidotes and claimed that bezoar was a powerful antidote. The bezoar as an antidote was also endorsed by the eminent Jewish physician and philosopher Moses Maimonides (1138–1204), who claimed to have verified its effectiveness on his patients who had been bitten by poisonous animals. Bezoar was recommended as a treatment for the Black Death after it arrived in Europe in 1347. The opening of shipping routes to the East at the end of the fifteenth century and the growing import trade in Indian bezoars that followed increased interest, soon to be further enhanced by the Spanish discovering bezoars in their newly conquered American provinces. Renaissance Europeans mostly thought of the bezoar as a universal antidote for poisons, although it was also believed to have power against fevers and other diseases. Bezoars, rather ordinary-looking objects, were even adorned with gold and gems befitting their value. Drinking vessels were made out of bezoars to protect the drinker against poison. Like other antidotes such as unicorn's horns, bezoars were tested by being given to condemned criminals who drank poison in exchange for a pardon, if they survived. In one of the best-known examples, the French surgeon Ambroise Pare (1510–1590) described an experiment to test the power of a bezoar in which a cook found guilty of stealing was poisoned and then treated with a bezoar stone with a promise of a reward if he survived. The experiment failed, and the man died of the poison. Such was the demand for bezoars in Europe that Jesuits in the Portuguese colony of Goa in India sold an artificial bezoar known as the "Lapis de Goa" or

A seventeenth-century Indian Bezoar stone with a finely wrought silver case and stand. (The Metropolitan Museum of Art, New York, Gift of Mr. and Mrs. Gordon S. Haight, 1980)

"Stone of Goa" that combined pieces of bezoar with other ingredients including ambergris and musk. The Lapis de Goa was manufactured from the mid-seventeenth to the mid-nineteenth century, by which point its market had shrunk to what remained of the Portuguese Empire. European belief in the power of the bezoar diminished in the seventeenth and eighteenth centuries.

Bezoar is known in China as Houzao and is used in traditional Chinese medicine to treat fevers and respiratory diseases, particularly in children. This idea seems to have entered Chinese medicine from India, and to this day, many of the bezoars used by traditional Chinese practitioners originate from India. The Indigenous people of the Brazilian interior used the bezoars of deer to treat indigestion.

See also: Ambergris; Goats; Unicorns

Further Reading

Chardin, John. 1988. *Travels in Persia 1673–1677.* New York: Dover; Duffin, Christopher John. 2010. "Lapis de Goa: The 'Cordial Stone.'" *Pharmaceutical Historian* 40: 22–30; Fabián, Omar. 2019. "The Allure of the Bezoar Endures." *Materials Research Society Bulletin* 44: 968; Rankin, Alisha. 2021. *The Poison Trials: Wonder Drugs, Experiment and the Battle for Authority in Renaissance Science.* Chicago: University of Chicago Press.

Bhoot. See Ghosts

Bibliomancy

Bibliomancy is a form of divination by selecting a passage from a book at random and applying it to the query. Simpler forms of bibliomancy involved simply reading the passage chosen at random and applying it; more complex forms involved using multiple passages or manipulating the letters of the text. A physical book was not always necessary; diviners could also ask a child to recite the last passage he or she had learned or draw a meaning from the first word they had heard on entering a church.

In the Mediterranean, bibliomancy has a history dating to the classical world. The Greeks and Romans used the works of their classical epic poets, Homer and Virgil, for bibliomancy. The use of Virgil, which is far better documented than the use of Homer, had a specific name, *Sortes Vigilianae* or Virgilian lots, and was ascribed to several Roman Emperors by their biographers. It drew power from the common belief that Virgil was a magician as well as a poet.

In Abrahamic traditions, sacred books—the Torah, the Bible, the Quran—were popular for bibliomancy. References to the use of the Torah for divination are found as early as the Maccabean period. Using sacred books for divination not only imbued the process with divine power, it also meant that bibliomancy, unlike other divination methods, could be presented as a pious attempt to ascertain the will of God rather than an impious attempt to gain forbidden knowledge. Keen opponents of other forms of divination, such as the Jewish physician and philosopher Moses Maimonides (1138–1204), accepted the legitimacy of bibliomancy using sacred texts. Among the Hasids, a pietistic Jewish movement that arose in Poland in the eighteenth century, the use of bibliomancy spread to other sacred texts, including the writings of Hasidic leaders and sages. Contemporary Hasids of the Lubavitch sect use the published letters of their former leader Menachem Mendel Schneerson (1902–1994) for bibliomancy.

The use of sacred texts did not end the use of nonsacred texts, however. The use of Virgil continued in the Middle Ages and actually boomed in the Renaissance. As late as the seventeenth century, an ominous passage from Virgil was believed to herald the violent death of King Charles I of Great Britain. In the early modern Persianate world, the Quran was supplemented as a bibliomantic text with the *Divan* of the widely revered poet Hafiz (1325–1390), a collection of short lyric poems known as ghazels. The *Divan* was used in a form of group bibliomancy, in which different ghazals were assigned to different members of the group. Another popular text for bibliomancy in the early modern Persian and Ottoman empires was the *Falnama* or *Book of Omens*, which combined text and elaborate images, often of prophets or sages as well as astrological images. Use of the *Falnama* extended to the social and political elite, and beautiful and expensive copies made for courtly users survived. More informal methods of bibliomancy can use any book, the picking of a random book from a shelf or library being incorporated into the divinatory process.

See also: Divination

Further Reading

Bar-On, Shraga. 2021. “If You Seek to Take Advice from the Torah, It Will Be Given: Jewish Bibliomancy through the Generations.” In Josefina Rodríguez-Arribas and Dorian Gieseler Greenbaum, eds. *Unveiling the Hidden—Anticipating the Future: Divinatory Practices between Qumran and the Modern Period.* Leiden: Brill. Pp. 161–191.

Black Mass

The Black Mass was an inverted parody of the Catholic Mass, usually carried out as a magical ritual by a corrupted priest. Primarily a French phenomenon, it originated in parody Masses of the Middle Ages and the use by witches and sorcerers of consecrated hosts. There is evidence of a parody Mass performed by a priest at a "sabbat" or gathering of witches in southwestern France in 1594, involving black communion wafers and consecrated wine replaced by what was allegedly the devil's urine. The most influential episode in the early history of the Black Mass (although the term itself was not used until the nineteenth century) was the "Affair of the Poisons." Beginning in 1677, investigators discovered that nobles at the court of the French King Louis XIV (r. 1643–1715), possibly including the king's mistress Madame de Montespan (1640–1707), were involved with street magicians and witches for both poisoning their rivals and working love charms. The Black Masses were celebrated by a priest named Etienne Guiborg. Guiborg used the bodies of women and girls, quite possibly including Montespan, as the altars on which he invoked demons. In another ceremony Guiborg claimed to have participated in, menstrual blood, semen, and dried powdered bat were mixed in a chalice. Guiborg then blessed the mixture, and it was used as a love potion. Another ceremony to create a love potion involved the blessing of a placenta. The most horrifying ceremony Guiborg was accused of participating in was the sacrifice of a baby. The baby's throat was cut, and the blood collected in a chalice. Continuing the parody of the Mass, Guiborg dipped a host in the blood. The purpose of these ceremonies seems to have been to win back the king's love for Montespan and to destroy her rival, Madame Fontanges, who did indeed sicken and die. There were also charges that Black Masses were said for the death of the king, but this is less likely as the king's death would bring no benefit to Montespan. The persons of lower social status involved with the Black Masses were tortured and executed. Guiborg was imprisoned for life and Montespan marginalized at court. Knowledge of the affair was suppressed. In 1682, the king issued a decree against fortune-tellers, witches, and poisoners. This date is usually taken for the end of the Affair of the Poisons.

Another version of the Black Mass is the Mass of Saint Secaire supposedly performed in Gascony. (The earliest evidence of the Mass of Saint Secaire comes from nineteenth-century folklorists, and some believe it to be fictional.) The inversion of the Catholic Mass was central to the Mass of Saint Secaire, which could only be said by an evil priest in a ruined church and involved triangular and black rather than round and white hosts. (Saint Secaire is not a recognized saint, and the origin of the term is unknown.) The usual purpose of the Mass of Saint Secaire was the killing of an enemy.

In the Enlightenment and the nineteenth century, the Black Mass was performed, or at least described, more for its blasphemy than for an attempt to win magical power. The Marquis de Sade (1740–1814) frequently employed elements of the Mass in his writings on sexual debauchery and cruelty. The publicizing of the extensive documentary record of the Affair of the Poisons in the second half of the nineteenth century led to a revived interest in the Black Mass among French occultists, decadents, and Satanists, although they may have talked and written about the Mass more than actually performed it.

See also: Blood; Death; Excrement; Left-Handedness; Love Magic; Menstruation; Witchcraft

Further Reading

Tucker, Holly. 2017. *City of Light, City of Poison: Murder, Magic and the First Police Chief of Paris.* New York and London: Norton.

Blood

No part of the body carried more meanings than blood. Blood was a life-giving fluid, but it also stood for violence and death. Blood was also sacred in many traditions, but at the same time, it was associated with evil monsters like the vampire.

In the Galenic medicine that emerged in the ancient Mediterranean to dominate the Christian and Islamic medical worlds, blood was one of the four humours. Those dominated by blood were "sanguine," identified with healthy fatness, good cheer, lustiness, and courage—and sometimes with red hair. The quality of the blood—whether it was thick and abundant or weak and watery—influenced health, as did the temperature of the blood, which should be neither too cool nor too hot; early modern Persians believed excessive consumption of dates among people not used to them would overheat the blood, leading to ulcers and loss of vision. But an overabundance even of good blood could be dangerous. This was the rationale of bloodletting, the deliberate removal of blood from the body in order to relieve the excess and bring the humours back into balance. Spring was the season of blood, as summer was of choler, fall of black bile, and winter of phlegm. Bloodletting was therefore most effective when performed in the spring. For physicians who followed the tradition of ancient medicine set forth by Galen, blood appeared in the body in the guise of many other fluids. For women, menstruation cleansed the body of impure blood. Semen was an altered, or "coagulated," form of blood, which made ejaculation dangerous for men who were losing the equivalent of blood. Milk was also considered a form of blood.

A Rain of Blood and the End of the World in Virginia, 1881

"Mr. James M. Quillen arrived here to-day from Nickelsville, and makes a statement, corroborated by the Mayor and other prominent citizens of that place, in which he affirms that yesterday, about midday, a strange cloud was seen hovering over a half acre field on the farm of Dr. Abram Saylor, in the lower end of Russell. A few minutes after a red shower began to fall, and covered the ground and clothes of those who stood beneath with a red substance, which could not be told from blood. Mr. Quillen's shirt front and hat were covered with what appeared to be blood-stains. The shower lasted about a minute, and the red mass came down in a slow and fine drizzle. The cloud then rolled off gradually. The singular part of the occurrence is that save in this one place the sky was clear. The phenomenon causes intense excitement among the colored and ignorant white people, many affirming the approach of the end of the world."

Source: St. Louis Globe-Democrat January 21, 1881. As quoted in Maxwell, Tom. 2012. "For the Scrutiny of Science and the Light of Revelation: American Blood Falls." *Southern Cultures* 18: 99.

In contrast to the Galenist belief in balancing blood with the other humors, Johannes Baptista van Helmont (1579–1644), a physician and follower of the German medical revolutionary Paracelsus (1493–1541), believed that blood was the carrier of life and bloodletting never a good idea—the less blood in the body, the less life. The English physician William Harvey (1578–1657), the discoverer of the circulation of the blood, although no Paracelsian, also believed blood carried a spirit endowing the body with life. Youthful blood was particularly healthy, and some argued that an infusion of the blood of a young person could reverse the ravages of old age. In Greek mythology, the sorceress Medea restored youth to Aeson, the father of her lover Jason, by drawing out his blood and replacing it. The notorious serial murderer Elizabeth Bathory (1560–1614), the "blood countess" of Hungary, was said to have preserved her youth by bathing in the blood of virgins, although the earliest evidence for this story came over a century after her death. The application of blood was also believed to lighten the skin. It was also believed possible to treat humans with animal blood. This belief led to the first blood transfusions in England and France in the late seventeenth century. Feverish, mentally disturbed, and angry people were transfused with the blood of lambs and calves, considered gentle and placid creatures (and leading one satirist to speculate on the possibility of patients growing coats of wool). The transfusions ended abruptly when one French patient died shortly after a transfusion of calf's blood.

Despite the belief in the life-giving properties of blood, the drinking of human blood, outside the consumption of tiny amounts for the purpose of blood brotherhood rituals, was regarded as a form of cannibalism and rejected by most cultures. There were a few exceptions, such as the Roman natural historian Pliny the Elder (d. 79 CE) recommending drinking the blood of gladiators as a treatment for epilepsy, but generally blood drinking was shunned. Various monsters, including the Western vampire, the Malay penanggalan, the Filipino manananggal, the Japanese nukekubi, and many others, were depicted as blood drinkers. European witches were depicted as nourishing their familiars with their own blood or sometimes as sucking blood from victims themselves. The Dominican friar and witch hunt preacher Bernardino of Siena (1380–1444) charged witches with drinking the blood of children. Drawing blood from a witch was regarded as effective defensive magic. This sometimes extended to drinking the witch's blood. In nineteenth-century northern Europe, drinking the blood of an executed criminal was believed to be a treatment for epilepsy and other medical conditions.

In addition to medical meanings, blood had religious meanings. The religious meaning of blood was particularly important in Christianity, as Christians were saved by "the blood of the Lamb." For Catholics, the blood of Christ is ritually consumed in the Mass, in which wine is transformed into the blood, even though it appears to the senses like wine. (The association between wine, particularly red wine, and blood predates Christianity, however.) This is the doctrine of "transubstantiation," literally a change in substance. Emerging in the Middle Ages, it came to be perceived as the key issue dividing Catholics from Protestants, for whom the wine remained wine while representing Christ's blood. Catholics also believed that the bread of the Host was transformed into Christ's flesh, and they told stories of how stabbed consecrated hosts bled. However, the idea of the redemptive blood was hardly limited to Catholics; it was memorably expressed by the English Protestant hymn-writer William Cowper (1731–1800): "There is a fountain filled with blood/Drawn from Immanuel's veins/And sinners, plunged beneath that flood/Lose all their guilty stains." Not only Christ's blood but also the blood of the saints and martyrs could work miracles.

Judaism and Islam prohibit the consumption of any type of blood. In kosher or halal slaughter, every drop of blood must be drained from the animal. Blood was not necessarily impure in the Jewish context, however, as it was also used in rites of purification. Sacrificial blood was used to purify altars and sacred spaces. Blood was smeared on doorposts and lintels as a way to ward off harm from those within. Ritual bleeding was practiced in Islam, as some pious Shia Muslims cut themselves on the sacred day of Ashura, commemorating the death in battle of Muhammad's grandson Husayn in a ritual known as tatbir. Tatbir remains controversial, however, and many Shia religious leaders have condemned or questioned it.

The Mexica, or Aztecs, of pre-Spanish Mexico also gave a religious meaning to blood. (Chocolate had the connection to blood for them as red wine had for Westerners.) They believed that Huiztipochtli, the god of warfare and the Sun, the analogy in the cosmos to the heart in the body, was nourished by human blood that contained the principle of life. This was the rationale for the sacrifice of human captives to the Sun, although humans were sacrificed to other gods as well. Living people could also voluntarily cut themselves from the ears, tongue, lips, and even the genitals as well as other parts of the body and offer the blood that sprang from the cuts to the gods. The shedding of blood was an important part of sacrifice in many cultures; the ancient Greeks spoke of the "bloody altars."

Blood was also intimately associated with war and violence. Warriors were "men of blood," and Charles I of England was described by his Parliamentarian opponents as "the man of blood Charles Stuart." This did not prevent his supporters from crowding the execution block after his death to dip their handkerchiefs in his blood (it was claimed that the king had bled twice as much as an ordinary man, presumably so that everyone could get a generous amount of royal blood). It was later claimed that the king's blood healed the sick and injured, proving God's support for the royal cause. Lakes, rivers, great floods, and oceans of blood were associated with the Christian apocalypse described in the biblical Book of Revelation. Even the Moon would appear as red as blood. This association applied to more mundane conflicts as well. Europeans and their American descendants believed that rains of blood from the heavens presaged great battles or other significant events. Rains of blood are surprisingly common in the annals of the medieval and early modern world and even into the nineteenth century. A rain of "divine blood" in Germany at the beginning of the sixteenth century left garments marked with red crosses.

There were several recorded rains of blood in the nineteenth-century United States, mostly in the South. In his confessions, the American slave revolt leader Nat Turner described seeing evidence of a "bloody dew," which he identified as the blood of Christ falling from heaven as an inspiration for his revolt in 1831. South Carolina and Massachusetts suffered heavy rains of blood in 1841, which even left traces of bloody flesh on the ground in South Carolina. Rains of blood fell on North Carolina in 1850, 1876, and 1884. The Southern domination of rains of blood led some northerners to see them as a sign of the depravity of southern slavery, although no rains of blood were reported anywhere in the country during the Civil War. A rain of blood in Jersey City in 1844 was identified as a sign of the end of the world by the apocalyptic movement led by William Miller, the "Millerites."

Blood was associated with personal as well as mass violence, in early modern Europe; the wounds of murder victims were believed to open and bleed anew in the presence of the killer. This idea persisted in some legal proceedings into the eighteenth century and in popular belief into the nineteenth century.

The sharing of blood creates a relationship between two people, as in the widely spread concept of "blood brotherhood" created by the mingling of blood through a cut or drinking wine containing a drop of blood from each "brother." The ancient Chinese tradition was that of two people drinking animal blood together to seal an oath. Bleeding for something was a sign of commitment, whether to another person or a larger cause. In China, Buddhist scriptures were sometimes literally "written in blood," or rather a mixture of the blood of the copier and ink. The writers seem to have believed that the merit to be gained by copying scriptures would be further enhanced by using their own blood, drawing on both the connection of blood with commitment and the ascetic qualities of drawing blood from one's own body. A legendary figure called Gu Xin copied scriptures in blood and accumulated merit that it restored eyesight to his blind mother. (In Chinese culture, which placed a heavy emphasis on devotion to one's parents, Gu Xin was proving himself as a good son as well as a good Buddhist.) In the early modern European witch hunt, the "Satanic Pact" by which a witch bound himself or herself to Satan's service was frequently represented as being written in blood or sealed by the shedding of blood.

The nineteenth-century American grimoire *The Golden Wheel Dream-book and Fortune-teller* states that dreaming of vomiting blood is a good omen, betokening wealth to the poor and children to the childless.

See also: Amber; Blood Libel; Chocolate; Defensive Magic; Evil Eye; Execution Magic; Hares and Rabbits; Humours, Theory of; Mandinga Pouches; Menstruation; Necromancy; Prodigies; Royal Touch; Satanic Pact; Tea; Vampires; Weapon-Salve; Witchcraft

Further Reading

Biale, David. 2007. *Blood and Belief: The Circulation of a Symbol between Jews and Christians.* Berkeley: University of California Press; Camporesi, Piero. 1995. *Juice of Life: The Symbolic and Magic Significance of Blood.* Translated by Robert R. Barr. New York: Continuum; Davies, Owen and Francesca Matteoni. 2017. *Executing Magic in the Modern Era: Criminal Bodies and the Gallows in Popular Medicine.* Cham: Palgrave Macmillan; Jortner, Adam. 2017. *Blood from the Sky: Miracles and Politics in the Early American Republic.* Charlottesville: University of Virginia Press; Kieschnick, John. 2000. "Blood Writing in Chinese Buddhism." *Journal of the International Association of Buddhist Studies* 23: 177–194; Maxwell, Tom. 2012. "'For the Scrutiny of Science and the Light of Revelation': American Blood Falls." *Southern Cultures* 18: 93–107; Nicolson, Marjorie Hope. 1965. *Pepys' Diary and the New Science.* Charlottesville: University Press of Virginia.

Blood Libel

The blood libel is the myth propagated by Christian anti-Semites that Jews ritually murder Christian children—in some versions, in order to use their blood to make matzos. (The consumption of blood, along with murder, is actually forbidden by Jewish law.) The idea of Jewish ritual murder has roots in pagan Greek anti-Semitism. There is evidence of Greeks in Alexandria in Egypt, a city known for the fierce hostility of its Jewish and Greek population, accusing Jews of kidnapping one Greek a year and sacrificing him while vowing eternal hatred toward the Greeks. However, the idea does not seem to have been widespread in the ancient world as few references to it survive.

The belief in Jewish human sacrifice reemerged in the Middle Ages in the context of the commonly held Christian belief in Jewish responsibility for the death of Jesus. (In all of the high-profile blood libel cases, the alleged victims were boys, like the Christ child.) The first documented appearance of the blood libel was in the English city of Norwich in 1144, following the mysterious disappearance of a Christian youth named William. The story originally charged the Jews of Norwich simply with murder, but as it developed, elements of ritual killing were added to the charges. Local authorities stepped in to protect the Jewish population. William became the object of a local cult as "St. William of Norwich" but was never recognized by the Church as a saint. His story was heavily promoted in a work by a Benedictine monk called Thomas of Monmouth, *The Life and Miracles of St. William of Norwich.* Thomas charged the Norwich Jews with fixing a crown of thorns to William's head and crucifying him in imitation of Christ's passion, an element that would reappear in blood libel accounts. He also claimed that Theobald, a Jewish convert to Christianity, had told him that the Jews sacrificed a Christian every year in hopes of being restored to the Holy Land.

Several similar cases followed, one of the most publicized being that of another youth, Hugh of Lincoln, in 1255. The entire Jewish community, not just in the city of Lincoln but throughout all of England, was blamed for Hugh's death. The growth of blood libel accusations went along with increased hostility toward the Jews by the English people, Church, and State, culminating in England being the first country in medieval Europe to expel its Jews in 1290.

Blood libel accusations quickly extended to the European continent, with the first recorded occurrence being that of a case in Blois, France, in 1171 that led to the execution of over thirty Jews. Numerous other beliefs were attached to the blood libel, among them was one that Jewish men "menstruated" through hemorrhoids and required the blood of Christians to replace that which they lost. Simon of Trent was a two-year-old Christian boy whose mysterious death in 1475 led to a highly publicized trial in the Italian city of Trent. Under torture, Jews confessed to drinking the child's blood and sprinkling dried blood on matzos and in wine while mocking Christ. Fifteen Jewish men were executed for Simon's murder. Simon was admitted to the Roman Martyrology, an official list of Christian martyrs used in Catholic liturgy. He was not removed until 1965. Although Simon never became a saint in the eyes of the church, he was the object of a local cult. Cults of children allegedly murdered by Jews persisted into the twentieth century. The cult of Andreas Oxner, a three-year-old Tyrolean child murdered in 1462 who was over a century later identified as a victim of the Jews, persisted until 1994, when the Catholic Bishop of Innsbruck forbade it.

From their antisemitic origin, accusations of blood drinking and the ritual murder of children extended to other marginalized groups in late medieval and early modern Christian society, including heretics and witches. Blood libel accusations against the Jews have continued to the present day and have extended from Catholic to Orthodox Christian and, beginning in the nineteenth century, Islamic communities.

See also: Black Mass; Blood; Menstruation

Further Reading

Hsia, R. Po-Chia. 1992. *Trent 1475: Anatomy of a Ritual Murder Trial.* New Haven, CT: Yale University Press; Rose, E. M. 2015. *The Murder of William of Norwich: The Origins of the Blood Libel in Medieval Europe.* Oxford: Oxford University Press.

Brownies. See Fairies

Brunonianism

Brunonianism was a medical theory named after the Latin version of the name of its originator, John Brown (1736?–1788). Brown, a medical student at the University of Edinburgh, one of Europe's leading medical faculties, presented his medicine as following the canons of Enlightenment science, striving toward general principles and laws rather than "multiplying entities" as did the elaborate disease classification schemes of his professors at Edinburgh. In his Latin treatise *Elementa Medicinae* (1780), Brown held that all diseases could be reduced to either an excess or a deficiency of excitability in the body. (An expanded English translation of this work, *Elements of Medicine*, was published by Brown in 1788, shortly before his death.) Excitability, the ability to respond to stimuli, diminished over the course of a human life but never changed its nature. Interaction between excitability and outside stimuli constituted life. Diseases caused by a deficiency in excitability—asthenia—were more common than diseases caused by overexcitability—sthenia. Brown was influenced by his own gout, for which the usual treatment, abstemiousness in food and drink, had failed. He identified gout as an asthenic disease for which the proper treatment was stimulants including rich food, alcohol, and opium. Switching to this regimen was followed by a period of remission in Brown's gout, leading to his belief in its effectiveness. Although traditional physicians sometimes prescribed these things, the Brunonians got a reputation for overprescribing, and Brown himself became an opium addict, in part to deal with the pain caused by the return of his gout.

Since all diseases were either sthenic or asthenic, disease symptoms or clinical histories were of little importance in diagnosis—in fact, misleading. The job of the Brunonian physician was to maintain the patient at a proper state of excitability by manipulating the stimulation to which the patient was subject. The fact that all diseases had a single, quantifiable cause offered physicians the possibility of certainty in diagnosis—a major element in Brunonianism's appeal.

Brunonian medicine was taken up by younger Edinburgh physicians, who were resentful of the university medical faculty and the town's medical elite. It spread to England where it became identified with radical materialism and republicanism. On the European continent, Brunonianism was most influential in Italy and Germany. Joseph Frank (1771–1842) introduced it at the Pavia Hospital around 1792 and Adalbert Marcus (1753–1816) at the showcase hospital in the German town of Bamberg. Brunonianism also influenced romantic *Naturphilosophie*. Some early enthusiasts, including Frank, eventually decided that the system was too dogmatic and reductionist.

See also: *Naturphilosophie*

Further Reading

Bynum, W. F. and Roy Porter, eds. 1988. *Brunonianism in Britain and Europe*. Medical History, Supplement 8. London: Wellcome Institute for the History of Medicine; Risse, Guenter B. 1970. "The Brownian System of Medicine: Its Theoretical and Practical Implications." *Clio Medica* 5: 45–51.

C

Cats

Cats have lived with humans for thousands of years, and in that time, they have accrued a great deal of mythology. This mythology is ambiguous, and cats are viewed in both a positive and a negative light. In ancient Egypt, cats were sacred to the goddess Bast, frequently depicted with the head of a cat, and it was forbidden to kill one. The cat was viewed as a protector of the household against rodents. Cats were even given funerals, mummified and buried in cat cemeteries, some of which contained many thousands of cats. In medieval and early modern Europe, by contrast cats had strong negative associations. Cats were thrown in the fire on St. John the Baptist's Day. Cats were frequently depicted as associating with witches, carrying them to their meetings, or sabbats, and having their form taken by witches, Satan, or demons. In early modern England, cats were among the small animals identified as demonic familiars, or "imps," in witch trials. The nineteenth-century American grimoire, *The Golden Wheel Dream-book and Fortune-teller*, states that cats in dreams signify misfortune, enemies, and false friends.

Cats are frequently associated with luck, good and bad. Black cats are associated with bad luck in North America and Europe, but this is reversed in Britain, where black cats are often, but not always, associated with good luck. The time and place of encountering a cat also mattered. To encounter one when setting out in the morning was frequently bad luck.

Cats were also believed to be weather predictors. In late medieval France, a cat that sat in a window and licked first its backside and then its foot was believed to be a predictor of rain. In early modern England, there are sources saying that both restful cats and active cats presage rain.

In medieval Europe, cats were employed both by learned magicians and by witches. The idea that witches worship Satan in the form of cat at the sabbat goes back to descriptions of medieval heretical meetings. It was embodied in the papal decree *Vox in Rama* issued around 1233, which identified "Luciferian" heretics in Germany as worshipping a demonic cat that they kissed under its tail. But cats also played a major role outside the sabbat, in less glamourous day-to-day witchcraft. Witches were often believed to transform themselves into the likeness of cats, and in that form, they killed babies and children by biting or scratching them. This idea can be found both in common folklore and the writings of demonologists, as can stories of how wounds made on cats later showed up on the witch who had taken the cat's shape. The cat attained its greatest connection with witchcraft in England, where its main role was as a familiar. Cats were also associated with the demonic in Ireland, where it was a custom when entering a house to call down blessing on all of its inhabitants, except the cat.

Learned magicians also made use of cats. In the early fourteenth century, a group of French necromancers were tried by the

Cats in South India

"The sight of a cat, on getting out of bed, is extremely unlucky, and he who sees one will fail in all his undertakings during the day. 'I faced the cat this morning' or 'Did you see a cat this morning?' are common sayings when one fails in anything. The Paraiyans are said to be very particular about omens, and, if, when a Paraiyan sets out to arrange a marriage with a certain girl, a cat or a valiyan (a bird) crosses his path, he will give up the girl. I have heard of a superstitious European police officer, who would not start in search of a criminal, because he came across a cat."

Source: Thurston, Edgar. 1912. *Omens and Superstitions of Southern India.* London: T. Fisher Unwin. P. 57.

church for planning to evoke a demon inside a circle made from strips of cat's skin. A medieval German magician's manual speaks of writing spells in the blood of a cat.

The parts of a cat's body or items associated with it also had magical or healing power. In medieval Europe, the feces of a cat were held to be a cure for baldness, while in eighteenth-century France, mixing a cat's feces with wine and drinking it was a cure for colic. In Ireland, the blood of a black cat cured erysipelas and other diseases, and black cats had their tails cut off bit by bit to provide the required blood. The bile of a male cat mixed with the fat of a white hen would produce an ointment, enabling the user to see what others could not. In eighteenth-century England, sore eyes could be cured by rubbing them with the tail of a black cat, while in France during the same period, sucking the blood from a tomcat's tail helped people recover from bad falls. Cat's ears were among the ingredients used by obeah practitioners in the British Caribbean for working magic. Although the famous "hell-broth" brewed by the three witches in William Shakespeare's *Macbeth* contains no body parts from a cat, the propitious time for its brewing is marked by a "brinded" (striped) cat mewing three times.

While Christian Europe was the center of cat-hating, cats had a generally positive image in other cultures, including East Asia and the Islamic world. The "beckoning" cat with one paw raised is believed to bring prosperity in Japan. The Maneki-neko, a statue of a beckoning cat placed in front of a business to attract customers, first appeared in the mid-nineteenth century and has since spread to other East Asian cultures. Muhammad, founder of Islam, was believed to have been fond of cats. A hadith, or tradition, shows Muhammad cutting off a sleeve rather than disturbing a sleeping cat. Muhammad is supposed to have forbidden the killing of cats. Cats in Islam are viewed as clean animals. A cat eating food or drinking water does not make it ritually impure. This contrasts with Muslim suspicion of dogs, often viewed as unclean.

See also: Dogs; Luck; Mummies; Obeah; Rats and Mice; Witchcraft

Further Reading

Darnton, Robert. 1984. *The Great Cat Massacre and other Episodes in French Cultural History.* New York: Basic Books; Engels, Donald W. 1999. *The Classical Cats: The*

Rise and Fall of the Sacred Cat. London and New York: Routledge; Kieckhefer, Richard. 1989. *Magic in the Middle Ages.* Cambridge, UK: Cambridge University Press.

Caul

The caul is a remnant of the amniotic sac. Some babies are born with a caul over their head or, very rarely (about one in every eighty thousand births), completely enclosed in the caul. Although not dangerous, the caul has provoked many superstitions, some about persons born with the caul and some about the magical properties of the caul itself. One belief, which dates at least as far back as ancient Rome, is that a lawyer who carries a caul into court will win his cases. The early Christian church frowned on caul beliefs as superstitious; St. John Chrysostom (347–407) denounced a clergyman named Praetus who bought a caul from a midwife (midwives were frequently identified as caul-sellers) to procure good fortune. Nonetheless, caul beliefs continued among Christians. In the Middle Ages and Renaissance, it was believed that cauls increased in power if they were baptized or had Masses said over them—a practice strongly condemned by the Church.

To be born with a caul was generally considered a positive sign. In Scandinavia, to be born with the caul gave great advantages in life, including overall success and protection against magic or weapons. In late medieval France, a man born with a caul who carried it with him could not be killed in battle. In early modern Ruthenia, a boy who kept his caul and wore it with his clothing would become a bishop. The Dutch physician Levinus Lemnius (1505–1568), author of *Occult Miracles of Nature* (1559), was skeptical of the powers of the caul, which he thought was a natural phenomenon, but he suggested that those who did believe in its powers distinguished between a reddish caul, which betokened good fortune for the infant, and a blackish one which betokened the opposite. Some traditions asserted a lifelong connection between the person born with the caul and the caul itself, in which the condition of one corresponded with that of the other. In eighteenth-century England, the caul of a healthy person was believed to be firm and that of an unhealthy person limp.

The caul could be associated with a special supernatural destiny or power, which could be positive or negative. In early modern Friuli, persons born with cauls became benandanti, protectors of the community against evil witches. Scottish Highlanders believed that being born with a caul was one way to gain the Second Sight. Among African Americans in the Southern United States, there was a common belief that those born with the caul could see, hear, or communicate with ghosts. In Poland, those born with the caul would become vampires unless the caul was dried, ground up, and fed to the child on his or her seventh birthday. Reddish cauls were particularly strongly associated with vampirism.

One common European belief concerning cauls was that a person carrying a caul could never drown. This led to a market in dried cauls in seafaring communities. This was sometimes extended to the belief that a ship with a caul on board could not be shipwrecked.

See also: Amulets and Talismans; Benandanti; Luck; Lycanthropy; Pregnancy and Childbirth; Second Sight; Vampires

Further Reading

Barber, Paul. 1988. *Vampires, Burial and Death: Folklore and Reality.* New Haven, CT: Yale University Press; Ginzburg, Carlo. 1985. *The Night Battles: Witchcraft &*

Agrarian Cults in the Sixteenth & Seventeenth Centuries. Translated by John Tedeschi and Anne Tedeschi. New York: Penguin.

Changeling. See Fairies

Chelidonius. See Swallow Stones

Chiromancy. See Palmistry

Chocolate

Chocolate, made from the cacao bean, was a drink developed in Central America, from which it spread north to the Valley of Mexico. (Solid chocolate was not developed until the eighteenth century.) Mexicans associated chocolate with blood as a carrier of a divine, life-giving essence. It was prescribed for hemorrhages and other losses of blood. *Achiote*, a herb sometimes added to chocolate, gave it a reddish appearance, reinforcing the similarity to blood. Chocolate played a role in religious rites. It was one of the substances sacrificed to the gods and accompanied the dead on their journeys to the afterlife. Chocolate was also used to cool fevers and as a male aphrodisiac. The Northern Lacandon people, a group formed from displaced Native Americans after the Spanish conquest, believed that dreaming of frothy chocolate presaged dying from foaming at the mouth.

Spaniards in the Americas acquired the habit of chocolate-drinking from the Native Americans. They treated it as a stimulant. It was eroticized, made part of courtship and considered to be an active ingredient in love potions. It could cheer on a miserable day or fire up a preacher before the delivery of a sermon. By the end of the sixteenth century, chocolate had made its way across the Atlantic to Spain. The philosopher and medical writer Olivia Sabuco de Nantes Barerra (1562–c. 1646), in her *Colloquies on the Nature of Man* (1588), the first mention of cacao in a printed European text, suggested that like almonds and hazelnuts cacao was good for the brain, and if eaten by nursing mothers, it would help an infant's brain develop. More influential, although his work remained in manuscript for many decades, was the physician Francisco Hernandez de Toledo (1514–1587) who had journeyed to Mexico to investigate the medical uses of its plants. He classified chocolate in Aristotelian terms as "cold" and "wet," suggesting that he had received the Native idea that the beverage was useful in cooling the "hot" disease of fever. He also endorsed chocolate as an aphrodisiac and warned of the dangers of overconsumption. Subsequent medical writers in the seventeenth century, when the first works devoted entirely to chocolate began to appear, suggested that chocolate aided the digestion and was both a general and an erotic stimulant. Some warned that overconsumption, particularly in the evening, could lead to insomnia.

Chocolate arrived in Europe outside Spain in the mid-seventeenth century and was initially viewed with suspicion. The French author Mme. de Sevigne (1626–1696) claimed that a pregnant noblewoman's excessive imbibing of chocolate had caused her to give birth to a black baby. The strong flavor of chocolate was also believed to make it an excellent vehicle for poison. However, chocolate also maintained its positive associations. In mid-seventeenth century Britain, the "Indian Nectar" was marketed as a provoker of male virility. Chocolate continued to be associated with sociability; the nineteenth-century American grimoire *The Golden Wheel Dreambook and Fortune-teller* claimed that dreaming of chocolate signified trouble caused by gossip.

See also: Aphrodisiacs; Blood; Herbs

Further Reading

Camporesi, Piero. 1994. *Exotic Brew: The Art of Living in the Age of the Enlightenment.* Translated by Christopher Woodall. Cambridge, UK: Polity Press; Norton, Marcy. 2008. *Sacred Gifts, Profane Pleasures: A History of Tobacco and Chocolate in the Atlantic World.* Ithaca, NY and London: Cornell University Press.

Christopher, St. See Dogs

Chudail. See Ghosts

Coca Leaf. See Herbs

Coffee

Coffee, a stimulant beverage originating in the highlands of Ethiopia, has attracted many superstitious and legendary beliefs. It was associated with religious figures—one sixteenth-century Ethiopian text claims that the Three Magi had drunk coffee to stay awake on their journey to visit Jesus. In the Islamic world, beginning in the early sixteenth century, shortly after coffee had been introduced by Sufis, there was some controversy over whether coffee should be prohibited as an intoxicant like wine. Muslims also legitimized it by association with religious figures. It was claimed that coffee had been given by the Angel Gabriel to Muhammad, founder of Islam, to energize him before a battle. Gabriel was also supposed to have given coffee to King Solomon, a figure of particular importance in both Ethiopia and the Islamic world. (One religion that did manage to successfully ban coffee was the Church of the Latter-Day Saints, which includes it along with tea in a blanket prohibition on "hot drinks.")

Arab physicians in the Galenic tradition identified coffee as "cold and dry," somewhat counterintuitive in classifying a beverage commonly consumed hot. Anti-coffee writers in the Middle East charged coffee with killing sexual desire. One writer, Da'ud al-Antaki, claimed that consuming coffee with milk could cause leprosy. However, al-Antaki also claimed coffee offered the health benefits of protecting from smallpox and measles.

One widespread superstition is that in a cup of coffee, bubbles moving toward the drinker means good luck or specifically coming into money and if moving away it means bad luck. In the Middle East, spilling coffee is considered good luck. Coffee grounds can also be used for divination in the manner of tea leaves, a practice known as tasseography that probably originated in Ethiopia. The method of brewing coffee used in the former Ottoman territories, such as Greece, the Balkans, and Turkey, leaves grounds in each cup. Amateur and professional fortune-tellers take advantage of this by inverting the cup and placing it over a saucer. The grounds are allowed a few minutes to dry, and the diviner interprets the future of the drinker in the size, shape, and position of the blotches left by the grounds in both cup and saucer. Coffee is widely consumed in militaries, and like many items in the military, it has attracted superstitions. In the U.S. Navy, it is considered bad luck for a person's coffee mug to be washed out, to the point where washing out another person's coffee mug can be considered a hostile act.

When coffee arrived in Europe in the seventeenth century, it was stigmatized as Oriental and faced competition from established businesses selling alcoholic drinks. There was also political opposition based on the idea that men gathering in coffeehouses criticized or even plotted against the government. The anti-coffee campaign in London, embodied in the anonymous pamphlet *The*

Women's Petition against Coffee, suggested that coffee caused impotence, possibly drawing on Islamic ideas about coffee's opposition to sexual desire. However, the degree to which this belief was actually held is unknown—certainly not enough to prevent coffee and coffeehouses from becoming popular. A similar but much better-organized campaign was launched by the American breakfast food tycoon C. W. Post (1854–1914), who promoted his coffee substitute made from grain, Postum, by claiming that real coffee was unhealthy and stunted growth while creating jittery nerves. Although it never rivaled coffee in the market, Postum became popular among Mormons.

See also: Divination; Herbs; Solomon; Tea

Further Reading

Hattox, Ralph S. 1985. *Coffee and Coffeehouses: The Origins of a Social Beverage in the Medieval Near East.* Seattle and London: University of Washington Press; Morris, Jonathan. 2019. *Coffee: A Global History.* London: Reaktion Books.

Comets

The dramatic appearance of comets and their relative rarity when viewed with the naked eye has invested them with social and cultural meaning in many cultures. Unlike planets, they were seldom deified, although there was a god of comets in Oceana, known as Wahieloa or Rongo-mai. Comets were viewed as heralds of disaster by ancient Mesopotamian, Indian, Greek, and Chinese civilizations among others. The Romans saw a comet as the sign of the death of an emperor or other great personage to the point that they even joked about it. Chinese interest in cometary omens led to the most precise and complete record of cometary appearances from the ancient and medieval world, dating to 613 BCE. Daoist historian and astronomer Li Chunfeng (602–670) pointed out that a comet resembled a broom, and like a broom, it came to sweep away old things. Halley's Comet's appearance in 1066 was associated with the Norman Conquest of England the same year. The Aztec ruler Moctezuma II feared that a spectacular comet that appeared over the Aztec capital city of Tenochtitlan in 1519 presaged the coming of a foreign invader, a fear vindicated by the arrival of Spanish conquistador Hernan Cortes shortly after. In the West, the idea of comets being a warning survived the coming of Christianity and was endorsed by many who denied the validity of traditional astrology. In Zoroastrianism, the dominant religion of pre-Islamic Iran, the end of the world would happen when a celestial body called Gochihr, usually identified as a comet, struck the Earth and melted it. Comets were not always negative; however, they could also be seen as an endorsement of a powerful ruler. Some Christians even claimed that the Star of Bethlehem that heralded the birth of Christ had been a comet.

Shakespeare on Comets as Signs

"When beggars die, there are no comets seen; The heavens themselves blaze forth the death of princes."

Source: William Shakespeare. *Julius Caesar* II.2.30–31.

Natural philosophers in the West debated what comets actually were, regardless of what they signified. The Greek philosopher Aristotle (384–322 BCE) claimed that comets were exhalations of flaming gas from the Earth, not heavenly objects. Comets, which appeared and disappeared, waxed and waned, were incompatible with Aristotle and his followers' view of the heavens as perfect and unchanging. The opposing view, held by the Roman philosopher Seneca (4 BCE–65 CE) among others, was that comets were heavenly objects such as stars and planets. Neither interpretation precluded the idea that comets were divine signs associated with great events. Aristotle's view dominated the European Middle Ages, but in the sixteenth century, the belief that comets were celestial objects began to overcome its rival as part of an overall realization that the heavens were indeed changeable.

Muslims referred to comets as "stars with tails" or "stars with locks of hair" although those terms could also refer to meteors or novas, and the distinction is not always clear in Islamic astronomical records. Muslims adopted the idea that the comet was a herald of the birth or death of the great. Comets were associated with the deaths of Muhammad and the fourth caliph, Ali. There was also a tradition in Islam of sometimes associating comets with positive developments such as a fruitful harvest. Muslim religious authorities discouraged believers from interpreting comets as signs of the apocalypse, but such interpretations did circulate.

Astrologically, comets could be interpreted like planets and other celestial phenomena, according to sign and house as well as the aspects they formed with other bodies. Other factors, such as the comet's shape, color, and direction of the tail, were more unique to comets. The Hindu astrological classic, the *Brithat Samhita* by Daivajna Varahamihira (505–587), classified comets and their implications for the world by shape and color. A comet with three tails and three colors would herald the end of the current cycle of the world. Astrologers held comets to affect (usually adversely) those countries and areas of life governed by the signs they traversed. They could also have bad effects on those areas to which their tails were pointed or those under the nucleus. The shape of comets also had meaning, although they were capable of various interpretations. Comets in the shape of swords, for example, presaged violence. The course of a comet, whether east to west or west to east, was also important, as was the area of the sky in which it was first visible. Diego Rodriguez (c. 1596–1668), holder of the chair of mathematics at the University of Mexico City, linked the path of the comet of 1652 through the Mexican sky with the Virgin Mary's triumph over the dragon of the Book of Revelation in the form of the constellation Medusa.

In the seventeenth century, the analysis of cometary paths in terms of Newtonian mechanics demonstrated that comets returned predictably. In 1705, the English scientist Edmond Halley (1656–1742) made the first Western prediction of a comet's return, claiming that the comet of 1682 would reappear in 1758. The vindication of his prediction after his death was a great triumph of Enlightenment science, and the comet has become known as "Halley's Comet." The fact that comets were no longer sudden and mysterious visitors, however, did not end their astrological and religious uses. Astrologers were used to dealing with periodic phenomena in their work, and if anything, the fact that comets were now regular and predictable like the movements of planets made them easier to deal with. However, since the discovery of comet regularity coincided with the decline of Western astrology, the new knowledge was not fully incorporated into astrological theory or practice. Even for religious interpreters in the eighteenth century, the comet was less likely to be viewed as a divine sign and more a physical object, even a missile, launched by God and aimed at the Earth or a celestial body as Gochihr was for Zoroastrians. English physicist Isaac Newton (1642–1727) believed that comets could play an important role in the apocalypse not by smashing into the Earth directly, but by plunging into the Sun, causing its fire to destroy the Earth. Newton's friend Halley suggested that Noah's Flood could have been caused by a comet crashing into the Earth. These were not merely scientific speculations, but widely supported by religious leaders and the general populace. The founder of Methodism John Wesley (1703–1791) warned in a sermon that God could punish human sin by destroying the Earth with a comet, and popular misunderstanding of a paper given at the French Royal Academy of Sciences in 1773 precipitated a panic as French people feared the Earth's imminent annihilation by comet.

See also: Astrology; Prodigies

Further Reading

Cook, David. 1999. "A Survey of Muslim Material on Comets and Meteors." *Journal for the History of Astronomy* 30: 131–160; Genuth, Sara Schechner. 1997. *Comets, Popular Culture, and the Birth of Modern Cosmology.* Princeton, NJ: Princeton University Press; Yeomans, Donald K. 1991. *Comets: A Chronological History of Observation, Myth and Folklore.* New York: Wiley.

Curses

The idea of using magical or supernatural power to cause harm to a specific individual or group, to "curse" them, is an old one. Curses were an alternative to physical violence in situations where that was impossible or disadvantageous. In addition to cursing their enemies, people also invoked conditional curses on other people, as a guarantee that they would fulfill obligations. Such curses only took effect if the cursed person violated an agreement or failed to fulfill their obligations. People even invoked curses on themselves as part of taking an oath. Curses could be pronounced by supernatural entities as well as by people.

Curses appeared on ancient Egyptian tombs, threatening the destruction of those who failed to properly maintain them or on those who entered the tomb without being purified. (The legend of a "mummy's curse" directed at any and all who disturb a tomb, pyramid, or mummy is a modern invention.) Ancient Mesopotamian literature, including the *Epic of Gilgamesh,* is full of curses. Many thin sheets of lead inscribed with curses, "curse tablets," survive from the time of the Roman Empire. Curses are directed against opponents in lawsuits, thieves, business rivals, and persons who had spurned the curser sexually or romantically. Curse tablets, like other ancient curses, invoked the power of gods and demons to carry out the will of the curser. The evil eye can be regarded as a specialized form of curse, although different from traditional curses in that the caster was not always aware of it.

Curses are ubiquitous in the Hebrew Bible, beginning with what became the paradigm curse for Jews and Christians, the curse God laid on Adam and Eve during their expulsion from Eden. Humanity's very existence is under a curse. The curse laid on Eve, and through her, on all women, is particularly strong. Menstruation is referred to idiomatically as the "curse of Eve." The serpent who tempted Eve and Adam and Eve's son Cain were also recipients of divine curses. Another biblical curse that would have a long history was the "curse of Ham" (actually laid by Noah on Ham's son Canaan). Ham was the one of Noah's three sons who revealed his father's nakedness. As Ham was considered the ancestor of African peoples, the curse of Ham would be viewed as a curse on all African and African-descended people, frequently invoked to justify enslavement of Africans by Jews, Christians, and Muslims.

Despite the apostle Paul's admonition to "bless and curse not," the coming of Christianity did not mean the end of cursing. Jesus himself uttered some curses, the most famous being that on the barren fig tree. Cursing was of particular value for church communities and individuals who could not defend themselves with force or call upon others to do so. Cursing in this sense was considered the inverse of blessing. Elaborate rituals for cursing developed in the medieval church, specifying the miserable fate of the one cursed against both before and after death and frequently extending to associates as well. Such curses could be directed against a specific individual, or they could be open-ended. An open-ended curse would be directed against any individual harming a monastery or its property.

Cursing continued to have a central cultural role in the early modern period. Early modern witch trials frequently occurred because a person thought a witch had cursed them, often after they had denied the alleged witch charity. One reason women were identified as witches so frequently is that verbal retribution such as cursing was believed characteristically female, as opposed to physical retribution that was male. Not all curses were issued by women, however, or were even religiously problematic—the inscription on the grave of William Shakespeare calls for a curse on anyone who moves his body.

As does Christianity, Islam discourages cursing. A hadith, or tradition, attributed to Muhammad, founder of Islam, forbids Muslims from cursing one another. Curses, another hadith asserts, ascend to heaven and return to Earth, but if the person or thing cursed does not deserve it, they rebound upon the curser. The prohibition has not prevented cursing from becoming a significant practice in the Islamic world. The division between Sunni and Shia Muslims is delineated and reinforced by the common custom of pronouncing curses against the other side's heroes. This began with the cursing of Ali, Muhammad's son-in-law regarded by Shia Muslims as his true heir. The cursing of Ali was introduced by his successful rival, the founder of the Umayyad Dynasty of caliphs, Muawiyah (602–680). Muawiyah ordered that Ali be cursed at Friday prayers. Shia Muslims responded by cursing the first three caliphs, revered by Sunnis, and Aisha, wife of Muhammad and enemy of Ali. This was made state policy by Ismail I (1487–1524), the founder of the Safavid Dynasty in Iran, with those refusing to curse the first three caliphs being sometimes punished with death. So seriously was this cursing taken that in the peace treaty between the Sunni Ottomans and the Safavids in 1555 the Ottomans insisted that the Safavids cease this practice, a clause the Safavid monarch agreed to but did not carry out. Muawiyah's son and successor Yazid (646–683), held responsible for the death of Ali's son Husayn, was widely cursed by both Sunnis and Shiahs, although there was some debate regarding the propriety of cursing a Muslim such as Yazid among Sunni jurists. Islam also developed a form of debate in which the debaters invoked God's curse on themselves if they were wrong.

The ability to curse effectively is associated with supernatural power in many societies. In addition to medieval monks and early modern witches, other supernatural specialists known for their ability to curse included shamans, West Indian "obeah-men," and the juju specialists of Africa. Indigenous Australian society has a type of shaman called a kurdaitcha who inflicted a death curse through a ritual known as "bone-pointing." After a ritual, a specially prepared bone was pointed at the victim, along with the pronunciation of a brief curse. Bone-pointing was believed to be highly effective.

See also: Death; Doll Magic; Evil Eye; Luck; Obeah; Menstruation; Race; Witchcraft

Further Reading

Goldenberg, David. M. 2003. *The Curse of Ham: Race and Slavery in Early Judaism, Christianity, and Islam.* Princeton, NJ: Princeton University Press; Little, Lester K. 1993. *Benedictine Maledictions: Liturgical Cursing in Renaissance France.* Ithaca, NY: Cornell University Press; Mikati, Rana. 2019. "Cross My Heart and Hope to Die: A Diachronic Examination of the Mutual Self-Cursing (*mubāhala*) in Islam." *Journal of the American Oriental Society* 139: 317–331.

Cynocephali. See Dogs; Monsters

D

Death

As a universal and terrifying event, death has attracted many superstitious and supernatural beliefs, in many ways serving as the foundation for religion. Death is one of the most common events foretold by omens and a common subject for queries in divination. There are also many beliefs surrounding a dying or recently dead person and the right way for them to be treated by the living. Murderers have also been able to avail themselves of the supernatural to cause the death of another.

Many animals have been credited with being heralds of death. In the West, deaths are particularly likely to be presaged by creatures that are black, from the cawing of crows, rooks, and ravens (the death bird *par excellence*) to the appearance of black dogs or cats. (This is reversed for horses, as white horses are associated with death in many parts of Europe and America. A hearse drawn by white horses, for example, might lead to another death.) The hooting of owls, nocturnal birds, was also claimed to be a herald of death. Europeans who settled in New England, possibly influenced by Native American beliefs, added the whippoorwill to the list of death-bringing ominous birds. Crowing or cackling hens, particularly in the morning, could be a sign of an approaching death. In Japan, having a weasel cross a person's path would bring death in three years unless they retraced their steps. Animal behavior could change after a death as well as before it; in South India, it was believed that tame pigeons or other birds would leave a house after the funeral of one who lived there.

Among the best-known predictors of death in England was the deathwatch beetle. This creature made tapping noises in the wood of old houses, which were believed to predict the death of a resident, although it is actually a mating call. One variation on this belief was that the rappings only presaged death if they occurred in sets of three. The English "debunker" Sir Thomas Browne describes in his *Pseudoxia Epidemica* (1646) how he identified the insect responsible for it and demonstrated the connection by keeping the creatures in small wooden boxes. Knocking their heads against the walls of the boxes, they made a recognizable "deathwatch" sound.

Supernatural creatures as well as animals could herald death; among the best known is the Irish banshee, literally a "fairy woman" whose keening heralded an imminent death. Legendary banshees were associated with particular families, and the death of a particularly powerful or important person could be heralded by the cries of many banshees. Similar legends exist in other Celtic societies, such as Wales and Highland Scotland. There were legends among some Irish American families of banshees crossing the Atlantic and plying their trade in the New World. The appearance of some ghosts was also believed to indicate that death was near. In many cultures, a phantasm or apparition of a person could appear to a close relative as a sign of the person's death. "Corpse-lights," blue flames visible over cemeteries and church yards, were also heralds of death—a small flame of a child's, a large one of an adult's. In Heian Japan, a "human fire" was believed to depart from the body of a person about to die.

Even an object as mundane as a loaf of bread could warn of an imminent death. A common European and Euro-American superstition was that a crack in the crust as bread was removed from the oven meant that someone was going to die. Many of these superstitions originated in a time when death, particularly the death of infants and small children, was a common household event and even relatively commonplace events could be considered signs of an approaching end.

Some beliefs encouraged people to avoid specific actions that would result in death. In the West, there is a very common belief that if thirteen sit at one table for a meal, one will die within the year. This belief is often associated with the Last Supper, when Christ sat down with His twelve apostles and was crucified shortly thereafter. However, its origin is unclear. There are numerous variations on which individual of the thirteen faces doom, from the last one to sit down, to the first one to arise, to the oldest, and to the youngest. In England, carrying a garden tool throughout the house would lead to the death of an inhabitant. In Japan, if a person saw a hearse, they had to hide their thumb (the "parent finger") or one of their parents would die.

The deaths of the mighty were particularly likely to be heralded by omens. Spectacular phenomena such as comets were not linked to the death of ordinary people, but of the great ones of the Earth. Eclipses were also sometimes linked with the deaths of the powerful. Under the Roman Empire, the inclusion of death omens became a standard part of imperial biography, and the association of spectacular events with imperial death was so accepted that at least one emperor joked about it. The idea that the deaths, like other life events, of the great were marked by prodigies persisted not only into the early modern period but also met with some resistance. Muhammad, founder of Islam, denied that a solar eclipse that appeared on the day of the death of his infant son Ibrahim was an omen, warning his followers against belief in death omens.

There were numerous beliefs about the timing of deaths. The idea that deaths in a family come in threes was common in the modern West and survives in the current belief that the deaths of celebrities come in threes. The idea that certain years were particularly dangerous is also widespread. In the West, the sixty-third (nine × seven) year was the "grand climacteric," when a person was particularly likely to die. Japan has a more complicated system called yakudoshi, with multiple fatal years that are different for men and women. A common belief among people who live near the sea, including the British, New Englanders, and Japanese, is that deaths occur at ebb tide.

In China, there is a group of superstitions associating death with the number four. This is because the pronunciation of "four" is similar to that of "death." In modern times, this is expressed by omitting a fourth floor in buildings and endeavoring to omit it from addresses and other identification numbers. Sets of four or designating items in a series as the fourth are also to be avoided. Prices set by Chinese businesses are less like to end in four than in other numbers. The Chinese Communist Party's campaign against the "Gang of Four" following the death of Mao Zedong in 1976 made use of this association. The connection between "four" and "death" also exists in Japan and Korea, due to the influence of Chinese vocabulary.

Death, whether the querent's own or another's, was one of the most common subjects of divination. To predict the death of princes, however, raised questions of encouraging rebellion and assassination and was often forbidden. This was particularly a problem for astrologers, who had several methods for predicting deaths. In the Roman Empire, it was worthy of the death penalty to attempt to forecast the death of the emperor. (There was an official doctrine that the emperor alone among humanity was not subject to the influence of the stars.) Prognosticating the death of the ruler continued to be an offense, sometimes punishable by death, into the Middle Ages. The astrologer William Lilly (1602–1681), an opponent of Charles I of Great Britain, laid out the stellar configurations pointing to the king's death but let astrologically knowledgeable readers work it out for themselves. Modern astrologers do not generally use astrology to predict deaths, as it is regarded as unethical. Palmistry or chiromancy, divination by the lines on the hand, is another form of divination associated with the prediction of death.

There are also superstitions about what behavior or conditions are appropriate around dying people, corpses, or funerals. The idea that mirrors must be covered or turned to the wall in the home of a dying or recently dead person is a common one, sometimes extending to pictures. (Mirrors are also connected with death through the belief that to break one leads to death.) The idea that doors and windows, and in some cases boxes and drawers, must be opened to ease the passage of the dying is also common. In England, there was a belief that placing a dying person under a beam would ease their death. Much of Northern Europe and North America had the belief that clocks in the house must be stopped at the time of death. Some superstitions are about preventing one death from being followed closely by others. One superstition is that if there is a rain into an open grave, a death will follow. (Another version of this legend is much more positive; rain into a grave means that the dead person has gone to heaven.) Beliefs that a grave must not be left open on Sunday on the pain of another death are also found in North America and Europe.

There are also beliefs about how a dead body should be treated to ensure a happy afterlife or to prevent it from being resurrected as a monstrous creature such as a vampire. In Eastern and Southeastern Europe, the vampire heartland, there were beliefs about preventing the creation of a vampire through the dismemberment or disembowelment of a corpse. Pieces of a dead body could also have magical power, as in the case of a hand of glory. Mummification was one way of securing a happy afterlife. The direction in which a corpse was facing when buried was also sometimes considered relevant to its future happiness.

Sometimes the death of an enemy or a person standing in one's way is desired, and there are many magical schemes in many cultures to procure that end. Witches were frequently charged with murder by magic. In aboriginal Australia, a type of magical specialist called a kurdaitcha was believed to be able to cause a person's death through a ritual called "bone-pointing." Many magical rituals could be adapted for working a person's death or, as in the case of the possibly legendary French variation on the Black Mass, the Mass of Saint Secaire, were actually devised for that purpose.

There are numerous signs of death in dreams, although to dream of death itself is not always one of them. In Western dream interpretation, to dream of death signifies a major change in one's life. The greater tarot card Death, represented as a skeleton, also signifies major change. In Islamic dream interpretation, to dream of death signifies religious failure. The range of dream images signifying death was vast. Pennsylvania Germans, for example, thought dreaming of muddy water or dreaming of a wedding could signify death. Dreaming of the painless extraction of a tooth could presage the death of a friend.

See also: Astrology; Black Mass; Blood; Cats; Comets; Curses; Defensive Magic; Divination; Dogs; Doll Magic; Dreams; Eclipses; Execution Magic; Fairies; Ghosts; Hand of Glory; Horses and Donkeys; Mirrors; Mummies; Necromancy; Palmistry; Prodigies; Ravens and Crows; Tarot; Vampires; Witchcraft

Further Reading

Bergen, Fanny D., W. M. Beauchamp and W. W. Newell. 1889. "Current Superstitions I: Omens of Death." *Journal of American Folklore* 2: 12–22; Fogel, Edwin Miller. 1915. *Beliefs and Superstitions of the Pennsylvania Germans.* Philadelphia: American Germanica Press; Geneva, Ann. 1995. *Astrology and the Seventeenth Century Mind: William Lilly and the Language of the Stars.* Manchester: Manchester University Press; McManners, John. 1981. *Death and the Enlightenment: Changing Attitudes to Death among Christians and Unbelievers in Eighteenth-Century France.* New York: Oxford University Press.

Decay of Nature

An idea found in several cultures is that the natural world was "decaying" with the passage of time, resulting in more disasters such as plagues and irregularities such as freaks and prodigies. For medieval and early modern Christians, this was linked with the idea that with the fall of man nature had fallen as well, losing the original perfection it had had in the Garden of Eden. The shrinking of the human life span from the centuries-long lives of the patriarchs of the Book of Genesis to contemporary times was one example of the decay of nature. The idea also drew on the ancient pagan descriptions of a golden age in the past. A degenerating nature could be expected to produce such phenomena as plagues and monstrous births, which in turn were used as evidence of nature's degeneration. The corruption of nature was not only great but also continuously increasing. Decay-of-nature theorists tended to speak of nature in harsh and critical tones, to find its aberrations not worthy of admiration, in fact disgusting.

The degeneration of nature was linked with the apocalypse in Christian thinking, as nature's errors and catastrophes were linked with the "signs" in the Book of Revelation and the other apocalyptic books of the Bible. Degeneration could also be linked with old age in a human, as the growing number of "errors" of nature were analogous to the growing number and prominence of infirmities in an aged person. The Church of England Bishop Godfrey Goodman (1583–1656), the author of *The Fall of Man, or the Corruption of Nature, proved by the Light of our Natural Reason* (1616), poured particular scorn on those he identified as "philosophers" who viewed errors of nature of increasing nature's beauty by contrast. He pointed out that the uniformity of the Sun was superior to the variation of plants. Goodman's most influential adversary, the Reverend George Hakewill (1578–1649), author of *An Apologie of the Power and Providence of Almightie God* (1627), argued that prodigies were signs directly from God, and the decay-of-nature thesis gave too much responsibility to nature, marginalizing the power and providence of God.

The traditional Chinese version of the decay-of-nature theory was cyclical rather than linear like the Christian theory. Both disasters such as plagues, earthquakes, famines, and floods and aberrant events such as monsters and comets were signs of the ruling dynasty having lost the mandate of heaven. By comparison, when a dynasty first came to power, the restoration of good weather and fertility was a sign of its possession of the mandate.

See also: Comets; Earthquakes; Macrocosm/Microcosm; Monsters; Prodigies

Further Reading

Harris, Victor. 1966. *All Coherence Gone: A Study of the Seventeenth-Century Controversy over Disorder and Decay in the Universe.* London: Cass.

Defensive Magic

Defensive, protective, or apotropaic magic is magic that protects the user against hostile magic or malevolent supernatural forces or even simple bad luck. Defensive magic can be used by an ordinary person or by a magical specialist, such as the "cunning folk" of early modern England or the obeah men and women of the Caribbean. It can take the form of an act or an object. Defensive magic can be religious in nature, involving the invocation of protective gods or saints, inscribing sacred texts on amulets, or employing sacred objects such as relics or holy water. It can also be secular, involving nonreligious acts or the use of everyday objects such as glass bottles. Many Christian and Muslim religious authorities have been suspicious of secular defensive magic, regarding it as indicating a failure to trust in God or, even worse, an implicit or explicit reliance on the power of Satan. Catholics suggested prayer and the use of relics or "sacramentals" such as consecrated wafers, holy water, and blessed oil, while Protestants believed that prayers and repentance were the only true protections. However, for ordinary people, use of defensive magic may have mitigated feelings of helplessness in dealing with the terrifying forces of the unknown supernatural.

The varieties of defensive magic are as multiplicitous as magic itself. Some substances, such as rowan-wood, iron, silver, amber, garlic, salt, or cowrie shells, were believed to be intrinsically protective or anti-magical and were frequently employed in defensive magic. Some protective objects were ordinary objects such as horseshoes repurposed for magical defense, while others, such as Brazilian Mandinga pouches or the incantation bowls of pre-Islamic Mesopotamia, were carefully made for that specific purpose. Sometimes defensive charms incorporated elements of what they were defending against—charms against the evil eye, for example, frequently incorporated images of eyes. A few forms of defensive magic were aggressive and even violent—in much of early modern Europe, for example, drawing blood from a witch was considered to be a form of protection. (Some clerical authorities thought that physically attacking a witch was the most licit form of defensive magic, as it did not involve relying on magical powers that were ultimately satanic.) Violent attacks on witches with an original purpose of neutralizing the witch's harmful magic could easily escalate into assault and murder. Other defensive techniques were much more passive—archeologists are discovering the frequency with which Europeans buried dead animal parts, items of clothing, and other objects within a house or its foundations to protect it and the dwellers therein from evil forces. The threshold of a house was a particularly good place to employ defensive magic, whether an animal buried under it or a horseshoe hung above it. Although the mezuzah, the case containing a scroll with Torah verses fixed to the doorpost of a Jewish home, was primarily placed as a matter of religious obligation, it was also believed to have protective power.

Bewitchment created magical links that could be exploited by victims and their families. In German villages during the European witch hunt burning, the excrement of a bewitched person was thought to burn the witch herself, revealing her identity. Excrement was frequently associated with counter-magic. A common practice in some parts of England was burying or burning "witch bottles" flasks containing some of the victim's urine along with hair or fingernail clippings and sometimes thorns, pins, and nails. It was thought that heating the witch bottle would cause the witch intense pain when she urinated and force her to free her victim. Alternatively or if heating the bottle failed, it could be buried as a protective charm. There is evidence of buried witch bottles in England as late as the nineteenth century, long after the end of the witch hunt. Alternatively, things associated with the witch herself rather than her victim could be attacked, for example, burning thatch from a witch's house. Brooms, often associated with witches, were sometimes credited with anti-witch powers. Sweeping salt with a broom was an effective counter to witch-sent disease in the French-speaking territory of Lorraine, and in England, laying a broom across the threshold was thought to prevent a witch from leaving the victim's house.

Although many forms of defensive magic were restricted to local areas or cultures, others were practiced over wide areas and crossed cultural and religious boundaries. The hamsa, a symbol combining a hand and an eye, was used to protect against the evil eye and other forms of harmful magic throughout the Mediterranean and the Middle East by Muslims, Christians, and Jews. Muslims call it the Hand of Fatima after the prophet Muhammad's daughter, Jews the Hand of Miriam after Moses's sister, and Christians the Hand of Mary after Jesus's mother. (The associations of the hamsa with Islam and Judaism were strong enough, though, that Christian Spain banned it in 1526.) The hamsa itself, however, had roots before any of these religions, in the ancient culture of polytheistic paganism. It is also used among Hindus in India. Other forms of defensive magic, such as the singa, a carved figure of a being part-human and part-reptile, are found in a specific culture. The singa is characteristic of the Batak people of the northern regions of the island of Sumatra in Indonesia.

In addition to protecting against the evil eye and witchcraft, defensive magic protected from other malevolent supernatural forces, including vampires and ghosts, as well as simple bad luck. The "bottle-trees" of West Africa and the American South, arrangements of empty bottles on dead trees or holders set in Earth for the purpose, were designed to trap evil spirits, including ghosts, or "haints" in bottles till sunrise, so the Sun could destroy them. Originally, the bottles were blue, a color having a powerful association with benevolent or protective magic in many traditions.

Defensive magic took the form of not only acts of violence or objects such as hamsas or witch bottles but also spoken words and gestures. These are legion. A common defense against the evil eye is to denigrate a person or object when it is praised, as in cultures where the evil eye is associated with envy praise it's often considered dangerous. The horns, a gesture made by extending the index and fourth fingers while curling the thumb, middle and index finger into the palm, is also identified as protection against the evil eye and misfortune generally, particularly in Italy and the Mediterranean region generally.

See also: Amber; Amulets and Talismans; Blood; Caul; Evil Eye; Excrement; Execution Magic; Garlic; Ghosts; Horseshoes; Incantation Bowls; Iron; Lightning; Lilith; Mandinga Pouches; Mirrors; Mistletoe; Obeah; Onions; Precious Stones; Relics; Roses; Silver; Tea; Toads; Vampires; Witchcraft

Further Reading

Briggs, Robin. 1996. *Witches and Neighbors: The Social and Cultural Contexts of European Witchcraft.* New York: Penguin; Hutton, Ronald, ed. 2016. *Physical Evidence for Ritual Acts, Sorcery and Witchcraft in Christian Britain.* New York: Palgrave Macmillan; Sabar, Shalom. 2010. "From Sacred Symbol to Key Ring: The Hamsa in Jewish and Israeli Society." In Simon J. Bronner, ed. *Jews at Home: The Domestication of Identity.* Oxford: Littman Library of Jewish Civilization. Pp. 140–162.

Devil's Mark

During the early modern European witch hunt, an increasingly prominent belief was that a witch had a mark, a small, insensitive area usually on a concealed part of the body. This mark had been left by the devil when the witch first made an agreement with him and was known as the devil's mark. This belief was not known in the Middle Ages or in the early stages of the witch hunt in the fifteenth century. It began to emerge in witch hunting in the early sixteenth century and played an important role in the judicial process in many parts of Europe, particularly in Protestant areas. The first important Protestant demonologist, the French Calvinist Lambert Daneau (c. 1530–c. 1590), placed great emphasis on it. The Catholic Inquisition, by contrast, never used it, and many Catholic demonologists and witch hunters were skeptical. One exception was the French magistrate and witch hunter Pierre de Lancre, who claimed he could see the devil's mark even in a witch's eye. Devil's marks were usually inconspicuous but not always; a Brazilian witch, Leonor Martins, had a hollow in the small of her back with a figure having a humanlike face composed of her flesh that she claimed was her familiar. English people, whose conception of witchcraft placed a great deal of emphasis on the witch's relationship with a familiar, conceptualized the devil's mark, also known as the witch's mark, as a nipple by which the witch nourished the familiar, or "imp," often on her own blood. The witch's mark was usually located around the vagina or anus. Although this process was a demonic parody of a woman nursing her baby, male witches could also have these marks and nurse imps of their own.

The devil's mark was tangible evidence of the pact the witch had made with the devil. The stripping of female witches to search for the concealed devil's mark was a ritual of humiliation that was justified because the devil often marked women, but not men, on the breasts, anus, or genitals. "Witch prickers" marketed their expertise in identifying devil's marks and showing that they did not feel pain or bleed when pierced with a sharp object. The executioner of Rocroi, Jean Minard, pricked hundreds of suspects, boasting of having been responsible for the death of over 200, before the Parlement of Paris, the most powerful court in France, put a stop to his activities by sentencing him to the galleys for life in 1601. Scotland, which saw some particularly fierce witch hunts, was particularly reliant on witch pricking. Scottish witch prickers practically constituted a profession, even taking their special expertise to the north of England for witch hunts there. Witch prickers in this region were often paid per witch identified, a practice that encouraged fraud. Fraudulent witch prickers could usually find warts or other insensitive spots on the body of the suspected witch. Fraudulent witch prickers also used special knives with a retractable blade. The end of the great Scottish witch hunt of 1661–1662 was accompanied by a series of scandals in which witch prickers were revealed as frauds. The late seventeenth and early eighteenth centuries, when witch hunting was declining in Western Europe, saw fewer judges willing to accept the devil's marks as evidence.

See also: Blood; Satanic Pact; Witchcraft

Further Reading

Levack, Brian P. 2016. *The Witch-hunt in Early Modern Europe*. Fourth Edition. London and New York: Routledge. Neill, W. N. 1922. "The Professional Pricker and His Test for Witchcraft." *Scottish Historical Review* 19: 205–213. Souza, Laura de Mella e. 2003. *The Devil and the Land of the Holy Cross: Witchcraft, Slavery and Popular Religion in Colonial Brazil.* Translated by Diane Grosklaus Whitty. Austin: University of Texas Press.

Diamonds. See Precious Stones

Diana

In the European Middle Ages, there was a widespread belief, particularly among the peasantry, in the existence of a powerful female supernatural being, often associated with fertility. This belief seems to have been a vestige of pre-Christian paganism that the Church never entirely succeeded in erasing. Learned writers identified this powerful being with the ancient Roman Goddess Diana, who was herself sometimes identified with the Greek Hecate, goddess of magic, as well as Artemis, goddess of the Moon. One reason for this identification may be that as "Diana of the Ephesians," Diana is the only classical Graeco-Roman deity to appear in the New Testament. Diana of the Ephesians, unlike the virgin who appears in mythology, was a fertility goddess with abundant breasts. The identification of the being recognized by medieval peasants as Diana goes back at least as far as the ninth century, when a piece of Church legislation called the *Canon Episcopi*, or "Canon of the Bishops," condemned as a diabolical illusion the belief in women who traveled great distances at night led by Diana. This supernatural being was also identified with the biblical Herodias, the wife of Herod who had demanded the execution of John the Baptist. She was also called Holda in Germanic areas and Lady Habundia, or abundance, in France. The queen of the fairies in Scotland and Sicily had similar qualities, and a fourteenth-century English preacher's handbook, the *Fasciculus Morum*, referred to the queen of the elves as "Diana."

Whatever her name, the goddess was believed to travel at night, particularly on Thursdays, accompanied by swarms of followers, including the living and the souls of the dead. (This procession had some similarities to the legend of the Wild Hunt.) Honoring or welcoming the goddess and her followers could bring a reward of gifts. Until the fourteenth century, the Church continued to identify this belief as a demonic illusion, again and again repeating the prohibition of the *Canon Episcopi*, forbidding Christians from honoring Diana or believing that they or others traveled with the goddess and her followers. These beliefs were however fairly minor sins, atoned for with a light penance.

This official attitude slowly changed in the later Middle Ages, when authorities began to assert that the goddess and her followers were not illusions, but real demons who devoured children. Now honoring Diana and her followers was not superstition but demon-worship, a much graver crime. In the late 1380s, two women, Sibillia Zanni and Pierina de Bugatis, were tried and condemned to death in Milan for being followers of Diana, who they called "Signora Oriente," "Lady of the East." (This title may be derived from a Roman title for the Moon goddess.) Diana herself crossed the gender barrier to be identified with the devil only rarely–usually she was conceived of as a powerful but subordinate demon. Diana and her troupe played only a minor role in the witch hunts of sixteenth- and seventeenth-century Europe, replaced by Satan and his sabbat. However, belief in this being continued in some places and was found by folklorists in Europe into the twentieth century.

See also: Fairies; Moon; Wild Hunt; Witchcraft

Further Reading

Cohn, Norman. 2000. *Europe's Inner Demons: The Demonization of Christians in Medieval Christendom*. Revised Edition. Chicago: The University of Chicago Press; Ginzburg, Carlo. 1991. *Ecstacies: Deciphering the Witches's Sabbath.* Translated by Raymond Rosenthal. New York: Pantheon Books; Russell, Jeffrey Burton. 1972. *Witchcraft in the Middle Ages*. Ithaca, NY: Cornell University Press.

Divination

Divination is a term for techniques for learning hidden knowledge, particularly but not exclusively knowledge of the future. The desire to know that which is hidden and particularly events of the future is extremely common, and numerous societies have divination practices. The varieties of divination are endless. There are fully developed bodies of learning that require a great deal of training to use such as astrology or the use of the *Yijing* or the Ifa divination of the West African Yoruba people. There are also practices as simple as reading tea leaves. Some forms of divination were explicitly religious, calling on the gods to reveal the future, and some had little religious involvement. Some divination was planned, with elaborate rituals to seek the truth, some was a matter of interpreting the signs that suddenly, without warning, might present themselves such as a comet or a monstrous birth, and some was a matter of direct contact with ghosts, gods, or spirits. Some required skilled practitioners; some could be done by anyone. All classes in many societies, from rulers to the humblest, availed themselves of divination and asked questions ranging from whether or not to go to war to who to marry. Divination always had something of the supernatural about it, although the distinction between divination and natural prognostication such as predicting rain from a cloudy sky could be fuzzy.

Many forms of divination are connected with states, and many diviners, particularly in early urban societies, were state officials. Astrology was developed in early Mesopotamian societies as a way of interpreting omens for the benefit of kings. Since astrology took a great deal of technical knowledge, it was the preserve of a skilled profession. In the ancient Mediterranean

This cuneiform tablet, from the late first millenium BCE, includes interpretations of the liver of a sacrificed sheep. (The Metropolitan Museum of Art, New York, Purchase, 1886)

world, this professionalization spread to other forms of divination as well, such as hepatoscopy, the examination of sacrificial livers. In the Bible, the king of Babylon is described as using a variety of divination techniques including liver and arrow divination—later known as "belomancy." Other ancient divinatory sciences included astragalomancy, divination by throwing specially marked dice; ornithomancy, divination by the flight of birds; haruspicy, divination by the entrails of sacrificed animals; empyromancy, divination by the observation of flames; brontoscopy, divination by thunder; and cleromancy or divination by lots. The behavior of animals revealed secrets to the diviner, as in the famous Roman practice of refusing battle if the sacred chickens carried with the army or fleet refused to eat. (Divination was considered an essential part of military decision-making by the Greeks and Romans.) For more direct contact with the gods, there were oracles where a god spoke through a human whose words were then interpreted by priests—although many believed the message was carried to the oracle by an intermediary being called a *daimon*, rather than by the god directly. (The message was always granted after a sacrifice to the god.) There were established oracles such as those at Delphi and Dodona that persisted for centuries, but there were also individuals who claimed to be carrying prophetic entities in their stomachs—the so-called

Deuteronomy on Magic and Divination

"10 There shall not be found among you any one that maketh his son or his daughter to pass through the fire, or that useth divination, or an observer of times, or an enchanter, or a witch,

11 Or a charmer, or a consulter with familiar spirits, or a wizard, or a necromancer.

12 For all that do these things are an abomination unto the LORD: and because of these abominations the LORD thy God doth drive them out from before thee."

Source: Deuteronomy 18:10–13, King James Version.

"belly-speakers." Ancient theorists distinguished between direct communications from the gods through oracles or dreams and the type of divination that required skill to interpret, viewing the former as more reliable. All forms of divination, however, were regarded as gifts of the gods.

Belief in divination's effectiveness was not universal in the ancient world; many were skeptical, including the Epicurean philosophers and the Roman politician and philosopher Marcus Tullius Cicero (106–43 BCE), whose *On Divination* is one of our principal sources for ancient divination. The Stoics, who were regarded as some of the strongest supporters of divination, believed that divination drew on the ordered structure and harmony of the universe and placed less importance on daimons than did other divination theorists. The Roman Empire viewed divining outside the control of the state, and particularly divining about the emperor, as dangerous and forbade astrologers from casting his horoscope. The first Roman Emperor, Augustus (r. 27 BCE–14 CE), had around two thousand books of magic and divining burnt. However, prophecies from diviners about future greatness became a stock feature of imperial biographies (beginning with Augustus himself).

Along with dream interpretation, divining, such as the use of oracle bones, was presented as a way in which the Shang Dynasty (c. 1600–1046 BCE) of early China communicated with the gods. Chinese imperial divination broadened over the dynasties to take in other techniques such as astrology, which became part of the state bureaucracy. As elsewhere, though, divination techniques were difficult to restrict to state officials and elite circles. Under the Zhou Dynasty (1046–771 BCE) that followed the Shang, divination began to spread more widely in Chinese society. The *Yijing*, or *Classic of Divination*, was one of the five books at the center of the Chinese Canon until the twentieth century. Confucianism, which became the dominant philosophy of the Chinese elite, accepted the *Yijing* and the divination practices that went with it. Confucian philosophers wrote commentaries on the *Yijing*. However, divination was a danger as well as an aid to rulers. The Tang Dynasty (618–907 CE), like the Romans, attempted to restrict divination from being practiced outside the court out of fear that forecasting the death of the emperor or other disasters could encourage rebellion, but they were ineffective.

By the time of the Ming Dynasty (1368–1644), divination had spread to all classes of

Chinese society, and the diviner or "fortune-teller" was a common figure in Chinese life. (Many Chinese believed that blind people were particularly powerful diviners, and divination was one economic recourse for the blind who were unable to engage in other activities.) Divination from the *Yijing* remained a preserve of the elite, but there were many other techniques. Divining from the shape of the client's body and the features of his or her face, "physiognomy," was common and had the advantage that it required no equipment. One specialized branch was palmistry. Physiognomical divination differed according to whether the person being examined was male or female; a palm reader examined a male's left hand, representing yang, but a woman's right, representing yin. A simplified form of astrology based on the year, month, day, and hour of a person's birth known as the "Four Pillars" system was also common. Simple questions could be answered by shaking a jar of specially marked incense sticks and reading the first to fall from the jar. There was also a popular divination system among women and the lower classes in which strips of paper would be marked with prophetic verses and scattered on the ground. A bird would be released, and whichever strip of paper the bird first pecked would be read and interpreted. Alternatively, snakes or turtles could pick out the piece of paper.

Chinese diviners appealed to gods and goddesses, including Zigu, the "Purple Aunt," or "Third Aunt," popular with woman diviners. Ancestral spirits were also appealed to by diviners, although "spirit mediums" who invited possession by gods or spirits were viewed unfavorably by Chinese authorities and Confucian philosophers, while diviners who communicated indirectly through the use of techniques and systems were viewed much more favorably.

Other East Asian societies drew from Chinese methods, including the *Yijing*, but also developed divination methods of their own. The Japanese divination system utaura drew from interpretations of classical Japanese poems called waka. Versions of the Four Pillars system have become very popular in Korea.

Divination presented a problem for Abrahamic religions, in that it seemed to infringe on the prerogative of God to know and determine the future. The Witch of Endor in the Hebrew Bible was a model of the impious diviner practicing forbidden arts for both Jews and Christians. Although the Hebrew Bible contains sweeping denunciations of divination as impious, it also shows some uses of divination, such as the casting of lots, that are not condemned. The *Urim* and *Thummim* included on the breastplate of the High Priest appear to have been connected with divination through lots.

Abrahamic regimes often forbade various divination practices. Christian theologians following the Church Father Augustine (354–430) argued that all forms of divination involved the power of demons—derived from the Greek *daimons*—whether or not they were implicitly invoked. Divination is condemned in the Quran, and a commonly cited hadith, or tradition, states that there is no divination after the coming of Muhammad, the founder of Islam. However, whatever the religious and political authorities said, Jews, Christians, and Muslims were all avid practitioners of divination. Divination often made use of religious texts and artifacts, such as texts of the Bible or the Quran, and Christian diviners frequently invoked the names of saints. Sacred books were particularly useful for bibliomancy, a form of divination that required opening a book at random and applying the first passage found. Use of the

Bible for bibliomancy was even regarded as a pious practice. Islamic society developed both astrology and geomancy, divination from the patterns formed by thrown rocks or sand, to a high level. The "practical Kabbalah" of Jewish magicians included divination among other magical applications.

The early modern Catholic Inquisitions frowned on divination, but it was often treated as a venial offense to be handled with a penance, not taken seriously the way witchcraft was. Astrology in particular received very little opposition from the Church, unless someone was foolish enough to predict the date of the Pope's death. Pope Sixtus V (Pope 1585–1590) forbade many divinatory techniques in a bull in 1586, but the impact of this prohibition was moderate. Protestants also condemned divination as demonic. The drawing of conclusions from monsters and prodigies, however, was generally considered religiously permissible, as these could be warnings from God.

In addition to learned practices such as astrology, medieval and early modern Europe was full of traditional divination practices, often undertaken by unlearned "cunning folk" or even by the individual querent himself or herself. Some divinations required no preparation, as it was simply the interpretation of natural occurrences as omens. In Irish legend, druids were credited with the ability to divine by the observation of clouds. In medieval England, the chattering of a jackdaw was held to herald a calamity for the same day. Other forms of divination required elaborate rituals. The Jesuit demonologist Martin del Rio (1551–1608) described and condemned a practice called "onychomancy," which meant smearing the fingernail of a virgin boy with oil and soot, followed by a spell, and then seeing what the diviner wished to know on the fingernail. Desire to know the identity of a future spouse was a common motive for divination. *Mother Bunch's Closet Newly Broke Open* (1685), an English collection of charms, suggested that sleeping with a peeled onion under the pillow and saying a short prayer addressed to St. Thomas would cause a woman to dream of her future husband, a belief that persisted into the nineteenth century.

In eighteenth-century Europe, following the Scientific Revolution and the Enlightenment, many systems of divination were discredited among the intellectual and cultural elite. Divination came to be identified with popular superstition and increasingly with women. However, divination did not entirely disappear even among elite circles. The period also saw the creation of a new and influential divination system, the Tarot.

Sub-Saharan Africa has a rich tradition of divination, reflecting a society that does not draw a sharp barrier between human society and the supernatural world. Diviners, both those who communicated directly with the ancestors or spirits and those who used techniques involving manipulating physical objects, were frequently believed to have been chosen for their profession by supernatural forces. A common method of West African divination, which has spread with the African diaspora in the Americas, is the tossing of cowrie shells, which are then interpreted as to whether they have landed with the open part of the shell up or down. Cowrie shell divination is also found in East Africa and India. The West African Ifa tradition uses Kola nuts in a similar fashion.

Divination was practiced by professions, sometimes even by institutions like the Chinese Bureau of Astronomy or the ancient Roman College of Augurs. Divinatory skills were also believed to be handed down within families. Sometimes specific ethnic groups were credited with expertise in divination. Examples include the Etruscans, who the ancient Romans believed to be the best diviners, and the Roma fortune-tellers of Europe.

See also: Astrology; Bibliomancy; Coffee; Comets; Dowsing; Dreams; Eclipses; Geomancy; Goats; Hepatoscopy; Ifa Divination; Jinn; Lettrism; Lightning; Mirrors; Necromancy; Onions; Oracle Bones; Palmistry; Physiognomy; Prodigies; Rain; Roses; Scrying; Sibyls; Tarot; Tea; Witchcraft; *Yijing*

Further Reading

Driediger-Murphy, Lindsay G. and Esther Eidinow, eds. 2019. *Ancient Divination and Experience.* Oxford: Oxford University Press; Johnston, Sarah Iles. 2008. *Ancient Greek Divination.* Chichester: Wiley-Blackwell; Opie, Iona and Moira Tatern, eds. 1989. *A Dictionary of Superstitions.* Oxford: Oxford University Press; Peek, Philip M., ed. 1991. *African Divination Systems: Ways of Knowing.* Bloomington: Indiana University Press; Smith, Richard J. 1991. *Fortune-Tellers and Philosophers: Divination in Traditional Chinese Society.* Boulder, CO: Westview Press.

Dogs

Humanity's oldest companion animals, dogs. have attracted many superstitious and legendary beliefs. In medieval France, a greyhound that had saved a baby from a snake and then been mistakenly killed under the impression that it had attacked the baby was venerated as St. Guinefort. (This legend of the unjustly killed loyal dog crops up in many places.) So popular was this completely unauthorized cult that the Church had to intervene to suppress it, and even then there are indications that it survived into the twentieth century. Peasants brought their children to the alleged tomb of St. Guinefort in hopes that they would be protected from disease or to recover changelings kidnapped by the fairies.

Belief in a race of dog-headed people, usually believed to live in remote regions, is found in a variety of cultures, including classical and medieval Europe, the Islamic world, and China. These peoples are generally referred to as cynocephali and were often believed to be ferocious, natural born warriors. The legendary St. Christopher is frequently depicted in Eastern Orthodox art with the head of a dog, associating him, as a soldier-saint, with these peoples. In some versions of his legend, submission to Christ resulted in the replacement of his dog's head with the head of a man.

In Islam and Rabbinic Judaism, dogs are considered ritually impure. This belief was under tension with the many social uses of dogs, as herders, guards, and hunters. The later belief with increasing urbanization that dogs were carriers of disease led to more hostility directed to them in the Islamic world. In Islamic dream interpretation, to dream of being transformed into a dog signified having been granted great wisdom by God and having forfeited it by abuse. The dog is one of the twelve animals of the Chinese Zodiac, associated with the positive qualities of loyalty and honesty. In the West, dogs were considered to be more subject to melancholy than any other animal.

There are many superstitions all over the world about encounters with dogs as fortunate or unfortunate omens. The howling of dogs was frequently believed to herald a death or at least be a bad omen. This was usually applied to individuals, but also could apply to public affairs, as the howling of many dogs in ancient Rome was believed to presage civil wars and seditions. Encountering a dog was believed a good omen in the Indian province of Malabar, but only if moving from right to left, whereas Westerners, other than the common interest in black dogs, did not bother to distinguish much between colors and patterns of dog's coats in considering their supernatural meaning; the Chinese developed elaborate systems of discerning different meanings from different appearances. A yellow dog with a white tail presaged the birth of future officials in every generation—a highly desirable goal in a traditional Chinese family.

In Northern England, a supernatural or spectral large black dog also known as a barghest was believed to be a herald of death or to haunt graveyards or sites of murder, execution, and sudden death. (Some versions of the barghest legend claim the barghest is a shape-shifter, and it is sometimes described as a headless man or a rabbit in addition to its dog form.) Normandy has a legend of the ghosts of those who committed suicide taking the form of black dogs known as *varous*. (Black dogs were also viewed with suspicion by Muslim Arabs, who associated them with shape-shifting evil jinn.) These supernatural dogs of ill omen frequently appear in literature, such as Sir Arthur Conan Doyle's *The Hound of the Baskervilles* (1902) or Bram Stoker's *Dracula* (1897), where the first form taken by the vampire Dracula on arriving in England was that of a large black dog. (In Eastern Europe, where the vampire legends that Stoker drew on originated, vampires were far more likely to turn into dogs and wolves than bats.) Even ordinary black dogs were believed to be an omen of death or disaster, particularly when they followed a person. Small dogs were among the animals that served early modern English witches as familiars, or "imps." During the English Civil War, the dog "Boy" belonging to the royalist general Prince Rupert of the Rhine (1619–1682) was frequently portrayed in parliamentary propaganda as a satanic imp or even a shape-changed witch. (The male dog Boy was even portrayed as being a shape-changed female witch.)

The connection between the bite of a mad dog and rabies, a disease for which there was no effective cure or treatment, led to a variety of superstitions. The transformation of a normal person into a "rabid" beast after the bite from a mad dog may have been what gave rise to the idea of lycanthropy. The belief that a dog bite could be treated by applying hair from the dog that bit the person to the wound—the "hair of the dog"—became a cliché. The belief that a dog bite could be cured or at least its bad effects mitigated by killing the dog—even if it appeared perfectly healthy—was also common. Other parts of a dog were believed to have medical uses; in eighteenth century Portugal, carrying the tooth of a male dog protected against toothache.

See also: Death; Jinn; Lycanthropy; Onions; Vampires; Wild Hunt; Witchcraft; Wolves and Coyotes

Further Reading

Collier, V. W. F. 1921. *Dogs of China and Japan in Nature and Art.* New York: Stokes; Mikhail, Alan. 2014. *The Animal in Ottoman Egypt.* Oxford and New York: Oxford University Press; Schmitt, Jean-Claude. 1983. *The Holy Greyhound: Guinefort, Healer of Children since the Thirteenth Century.* Translated by Martin Thom. Cambridge, UK: Cambridge University Press; Wang, Jessica. 2019. *Mad Dogs and Other New Yorkers: Rabies, Medicine and Society in an American Metropolis, 1840–1920.* Baltimore: Johns Hopkins University Press.

Doll Magic

One way to cause supernatural harm or otherwise manipulate a person or deity in many cultures was by performing operations on a small figurine or doll of him or her.

In the West, doll magic was used in the learned magical tradition as well as by uneducated witches and other magical practitioners. Effigies could be made of many substances, including wax, lead, parchment, clay, earth, wood, straw, or rags. Surviving magical figures from the ancient world are most often made of lead, but this may be a quirk of survival, in that metal figures were mostly likely to survive the centuries. Wax in particular offered the possibility of being melted over a fire. In ancient Egypt, wax effigies of the god Set, enemy of the good god Osiris, were ritually melted. Other ways of harming an image included sticking pins in it or breaking it. In ancient doll magic, pinning or sticking a nail into a doll was not necessarily intended to cause harm, but to "fix" the person being represented to the spell caster's will. In Christian societies, images could be prepared for magical operations by being baptized, a misuse of a sacrament that aroused the particular ire of church authorities. Images were also made more potent by incorporating a piece of the victim's body, bodily excrements such as spittle, or clothing.

Although figurines do not survive from ancient Mesopotamia, Babylonian texts speak of procedures for gaining victory in battle by manipulating effigies of enemy soldiers made from tallow. Figurines of magicians were also manipulated to protect from hostile magic. Greeks and Romans used bound and pierced dolls to protect themselves from hostile ghosts and witches. Doll magic was also used as love magic, with invocations beseeching that the person represented by the doll should burn with love for the caster as the doll itself burned. Fear of doll magic continued into the Middle Ages. The Emperor Charlemagne (d. 814) forbade the making of wax images. In early modern Europe and its colonies, doll magic like other forms of popular magic figured in witchcraft cases and was discussed by demonological writers. The sixteenth-century English cunning man John Walsh, interrogated on charges of witchcraft, described a complicated formula for making images out of the earth of a new grave, ashes from a human rib, a spider, the inner pith of an elder tree, and water in which toad familiars had been washed. Although Walsh denied ever making such an image himself, he claimed that sticking pins in particular areas of the image would afflict that part of the victim's body. A pin in the heart would cause death in nine days. "Puppets," made of rags and stuffed with goat hair, were found in the house of a poor Massachusetts Irish woman, Goody Glover, in 1688. The Puritan minister, witch hunter, and demonologist Cotton Mather (1663–1728), in his *Memorable Providences, Relating to Witchcrafts and Possessions* (1689) described how by wetting her finger with her spittle and stroking the images, Glover caused the children of the Goodwin family to writhe in pain. Some Scottish witches shot at clay images with neolithic flint arrowheads, called "elf-bolts" and believed to be the work of the fairies. Doll magic could also be used as defensive magic with the witch's victim sticking pins into a small effigy of the witch.

The greatest in the land could view themselves as vulnerable to the image magic of ordinary people, and the finding of a small image of the monarch in unusual circumstances was taken seriously by political authorities. In 1320, Pope John XXII (1244–1334), known for his fear of magic, had Galeazzo Visconti of Milan tried for attempting to assassinate him by using a doll.

In modern times, sticking pins in magical effigies to cause pain and suffering has become strongly linked in popular culture to Haitian Vodou, a west African-derived religion, as shown in the common term "voodoo doll." So ubiquitous is the term that it is even used to refer to such objects in places far removed in space and time from Haiti. The connection between Vodou and "voodoo dolls" emerged out of tourist writings and popular films in the early twentieth century. In reality, doll magic plays little role in Vodou.

See also: Curses; Death; Defensive Magic; Fairies; Love Magic

Further Reading

Armitage, Nancy. 2015. "European and African Figural Ritual Magic: The Beginnings of the Voodoo Doll Myth." In Nancy Armitage and Ceri Houlbrooke, eds. *The Materiality of Magic: An Artifactual Investigation into Ritual Practices and Popular Beliefs.* Oxford: Oxbow Books. Pp. 85–102; Faraone, Christopher A. 1991. "Binding and Burying the Forces of Evil: The Defensive Use of 'Voodoo Dolls' in Ancient Greece." *Classical Antiquity* 10: 165–220; Gibson, Marion, ed. 2000. *Early Modern Witches: Witchcraft Cases in Contemporary Writing.* London: Routledge.

Donkeys. See Horses and Donkeys

Dowsing

Dowsing, or "rabdomancy," is the supernatural practice of detecting underground phenomena. Now it is most commonly used to locate underground water deposits preparatory to drilling wells, hence the term "water witching." However, for centuries dowsing was primarily employed to locate underground mineral deposits or buried treasures rather than water. It broadened to the point where dowsers were claiming the ability to identify criminals, discover the sex of unborn babies, or verify virginity. Cultures of dowsing could vary tremendously between regions. The most common form of dowsing was the use of a "divining rod" such as a forked stick, which in the hands of a dowser reacted to underground deposits of the thing sought after. There was no requirement as to the specific kind of wood, although hazelwood was common. There was a belief found in places as distant as Western Scotland and Pomerania that a branch of hazelwood cut on St. John's Day (June 24) would make the best divining rod. The association between St. John's Day and dowsing was found in many parts of Europe, although not necessarily involving hazelwood. German writers developed the idea that different kinds of wood located different metals—hazel to silver, ash to copper, and so on. Danes believed that a willow branch made the best rod for dowsing for water. Elm was also commonly recommended. It was not even always necessary for the stick to be forked.

Some ancient representations of persons carrying sticks have been interpreted as showing dowsers, and dowsers pointed to the rods of Jacob and Moses in the Bible as providing unimpeachably orthodox precedents. The modern tradition of dowsing, however, traces its origins to miners in Central Europe in the late medieval period. The German Georgius Agricola (1494–1555), the most influential writer on mining in the Renaissance, did not entirely deny its effectiveness, but advised against it due to its unreliability in comparison to other methods. Other writers supported it, such as the early seventeenth-century French noblewoman and mining entrepreneur Martine de Bertereau, Baroness de Beausoleil, the author of the first treatise devoted to the subject, *True Declaration of the Discovery of Mines and Minerals* (1632). Some, such as the Anglo-Irish chemist Robert Boyle (1627–1691), took a moderate stance, supporting dowsing for water or minerals but regarding extensions of the practice as magical and, like all magical arts, implicitly satanic. Protestant reformer Martin Luther (1483–1546) had condemned dowsing as early as 1518 as a violation of the First commandment, although Protestant Germans continued dowsing nonetheless. The Lutheran philosophy professor Johann Sperling (1603–1658) explicitly linked dowsing with witchcraft at the height of the German witch hunt. He claimed that both dowsers and witches relied on the aid of Satan. There were also Catholic writers who took this position, but dowsers were never persecuted like witches. Religious condemnation did not end dowsing, but it tended to marginalize its more magical aspects, such as the use of particular prayers or invocations or the treatment of the dowsing rod as a magical implement, in favor of more secular dowsing practices based on natural-philosophical theories. One early theory was that dowsing was a way of detecting the "mineral vapors" supposedly given off by metallic ores.

Dowsing did not work for everybody. By the mid-seventeenth century the power of dowsing was increasingly identified as coming from the dowser himself (women dowsers were rare) rather than being inherent in the divining rod. The successful dowser was considered to possess a gift, for which there were various natural and supernatural explanations. Dowsers could be born under favorable stars for the practice of their art, or they could be "human magnets" somehow responsive to intangible forces. Seventeenth-century Cartesian philosophers suggested that the items dowsers sought emitted small corpuscles to which dowsers were sensitive—a variation on the "mineral vapors" theory. Others credited dowsers' successes to demonic aid or wrote the whole thing off as simple fraud. Among the claimed supernatural abilities, dowsing was fairly easy to test, and dowsers not only accomplished some allegedly amazing feats but also incurred some spectacular failures.

Dowsing was a controversial subject during the Enlightenment, widely practiced but often rejected by the educated elite as superstitious. The Royal Academy of Sciences in Paris issued an anti-dowsing opinion in 1701. Jacques Aymar, a peasant from the Dauphine region of France, which produced many dowsers, became a celebrity in France. Aymar's use of a divining rod to identify murderers and other criminals led to him being called in by authorities to aid in criminal cases, but many condemned him as a fraud. Dowsing in eighteenth-century France came to have a class nature, as uneducated peasant dowsers relied on the support of the common people while most, but not all, academicians and savants denounced dowsing as a fraud. By the last decades of the century, dowsing became associated with mesmerism, as dowsers were increasingly thought to be particularly sensitive to "magnetic" forces emanating from objects and people. Another Dauphinois dowser, Barthelemy Bleton, even claimed support from King Louis XVI (1754–1793) and his queen, Marie Antoinette (1755–1793). Bleton asserted that he did not even need a rod, but carried one only for the benefit of observers. The sensations produced in his body were all he needed. In the early 1780s, Bleton's powers were subjected to a series of rigorous tests, including tracing the path of an underground aqueduct specifically created for the purpose. By this time, the astrological and demonological explanations of dowsing had faded from both popular and learned culture, and the question was whether Bleton was susceptible to electric and magnetic forces or simply a fraud. Bleton's record of success under testing was mixed, and the debate ended inconclusively.

There were similar conflicts in eighteenth-century Germany, whose mining academies were establishing geology as a scientific discipline. "Enlightened" mining scientists positioned themselves against the practices of German working-class miners, among whom the divining rod was in common use to locate mineral deposits. Rather than being the product of supernatural gifts or attunement to magnetic energies, eighteenth-century German miners viewed dowsing skill as a product of mining experience. The dowser was required to be an "honest miner," pious and responsible, and not to abuse the gift. Learned Germans, however, including some practitioners of *Naturphilosophie*, began to theorize dowsing in a way influenced by French debates, as a matter of sensitivity to subterranean electric or magnetic currents. German mining authorities conducted field tests of dowsers' claims as late as 1843.

Ideas about dowsing reached England around the middle of the seventeenth century, possibly carried by German miners. Dowsing continues to be widely accepted in England. The British Society of Dowsers was founded in 1933, and continues to the present day. British local water authorities used dowsers to locate breaks in underground pipes or trace the routes of lost pipes into the twenty-first century. In colonial and nineteenth-century America, where settlers frequently sought water deposits to drill wells, water-finding dowsers were in great demand. Dowsing continues to be widely practiced in America. The American Society of Dowsers was founded in 1961.

See also: Divination; Magnetism; Mesmerism and Animal Magnetism; *Naturphilosophie*; Witchcraft

Further Reading

Besterman, Theodore. 1926. "The Folklore of Dowsing." *Folklore* 37: 113–133; Dym, Warren. 2010. *Divining Magic: Treasure Hunting and Earth Science in Early Modern Germany.* Leiden: Brill; Lynn, Michael R. 2001. "Divining the Enlightenment: Public Opinion and Popular Science in Old Regime France." *Isis* 92: 34–54; Sharman, Jon. 2017. "UK Water Companies Still Use 'Magic' Dowsing Rods to Find Leaks, Despite no Supporting Scientific Evidence." *Independent*, November 22. https://www.independent.co.uk/news/uk/home-news/uk-water-companies-magic-dowsing-rods-use-engineers-leaks-no-scientific-evidence-sally-le-page-a8069616.html

Dragons

Dragons, big scaly reptilian creatures, often with the ability to fly or breathe fire or poison, are found in the belief systems of many cultures. Although they often appear in the realm of myth, they were not limited to it, but were believed to inhabit the historical world as well. The word "dragon" is derived from the Greek *draconta* "to watch," an early appearance of the popular idea of the dragon as a hoarder or guardian of treasure. Roman historians described how a Roman army in the First Punic War had encountered and defeated a dragon by using catapults, sending its skin back to Rome. The Roman natural historian Pliny the Elder (d. 79 CE) believed that dragons could be found in India, where their natural enemy was the elephant. Dragons were big enough to encircle and constrict elephants in the manner of a python, but elephants were heavy enough that they crushed the dragon in their deaths, leaving both contenders dead. Pliny described the belief that burying the head of a dragon under the threshold of a house, and performing the proper rituals to the gods, would bring good fortune to the inhabitants. The eyes of a dragon, dried and mixed with honey, formed a liniment that protected from nightmares. Various parts of the dragon when carried on the body ensured the bearer success in lawsuits or favorable treatment from those in power. Pliny also described an elaborate formula by mixing the tail and head of a dragon with parts of other animals, which magicians claimed would turn the bearer invisible, but plainly does not believe in it.

By the Middle Ages, Westerners mostly located dragons in the remote past or distant territories, particularly hot countries such as Africa or India, or on the seas where their wings served as fins. Large fossil bones, now identified as the bones of mammoths or other large animals, were identified as the bones of dragons. Medieval dragons also took on a more malevolent quality, influenced by the Christian tradition of linking reptilian qualities to the evil or diabolical, a tradition ultimately springing from the serpent's temptation of Eve in the Book of Genesis. In the allegorical "bestiary" tradition, the dragon representing evil or Satan was the natural enemy of the panther representing Christ. Dragons appeared in the works of Renaissance natural history writers such as Konrad Gesner (1516–1565), Ulisse Aldrovandi (1522–1605), and Athanasius Kircher (1602–1680). Aldrovandi even claimed to have encountered a small, two-legged dragon himself. Kircher claimed to have examined the head of a small dragon the size of a vulture a Roman hunter had killed in 1660. He recounted several stories involving dragons from Switzerland, in the mountainous regions of which encounters with dragons seemed particularly common. Dragons also appeared in traveler's accounts. One traveler claimed to have witnessed the fights between elephants and dragons that Pliny had described, but in Africa and not in India. Belief in dragons, or at least willingness to believe in the possibility of dragons existing, did not disappear in Europe until the Enlightenment. In 1734, Carl Linnaeus (1707–1778), the great Swedish natural historian, got into a dispute in the German city of Hamburg with a man who had acquired what he claimed was the preserved body of a multiheaded dragon. ("Dragons" or dragon skeletons made artfully from the remains of other animals were commonly found in European collections.) Linnaeus was actually threatened with prosecution and had to leave Hamburg. In his *Systema Naturae* (1738), often considered the basis of the modern system of biological nomenclature, Linnaeus included the dragon in the category of "Paradoxa" along with other animals he did not believe existed, such as the unicorn and the satyr.

The other major dragon tradition is that of China, which viewed dragons in a far more ambivalent or even positive way. The dragon is one of the twelve animals of the Chinese zodiac and as such is deemed auspicious. However, dragons could also be the cause of flooding, excessive rains, or other disasters, particularly ones associated with water. Dragon eggs and body parts were also believed to have medicinal value. Dragon bones, teeth, and horns could be ground into powders that had many medicinal uses, while dragon fat could be made into an ointment that protected the skin. The boundary between human and dragon was more permeable for Chinese than for Westerners—not only was it possible for dragons to take on the appearance of humans, but there were folktales of humans who had become dragons. Japanese dragons are similar to Chinese dragons, although unlike the Chinese but like the Westerners, Japanese have legends of heroic dragon killers.

See also: Elephants; Monsters; Oracle Bones; Snakes

Further Reading

Arnold, Martin. 2018. *The Dragon: Fear and Power.* London: Reaktion Books.

Dreams

Among the most mysterious of mental phenomena, dreams have attracted a range of supernatural and superstitious beliefs. The history of dream interpretation and divination by dreams, known as oneiromancy, goes back far in recorded history. Dreams figure prominently in the Mesopotamian *Epic of Gilgamesh*, humanity's oldest surviving epic poem. In the mid-second millennium BCE, dream interpreters were part of Chinese royal courts.

Dreams were often viewed as communications from supernatural entities including saints, angels, gods, and demons. These types of dreams frequently appear in sacred texts, including the Hebrew Bible. Some individuals, such as Joseph, are depicted as inspired by God with skill in dream interpretation. Such was Joseph's skill that it raised him to the rank of the pharaoh's chief minister. In China, dreams were often interpreted as messages from ancestors directing their descendants as to how they could be properly honored. Various techniques were developed to distinguish between dreams sent by benevolent supernatural entities, those sent by demons, and "ordinary" dreams with no supernatural element. Some sought to encourage prophetic dreams. As with other divination techniques, dream divination was frequently used to identify a future spouse. In eighteenth-century England, a young single woman who fasted on Friday could be rewarded that night with dreams of a future husband. Europeans put horseshoes, leaves, and keys under pillows to facilitate prophetic dreams. The dreams of pregnant women were particularly important for diviners, given their relationship to the future of the expected child.

The meaning of a dream could be interpreted according to its plain meaning, or it could be subjected to symbolic interpretation, in which things and actions that appeared in the dream could be viewed as representing something else or even their opposites, as dreaming of death could be interpreted as an omen of good fortune. In India, dreaming of being bitten by a snake guaranteed immunity from snakebite. Dreamed subjects could also be interpreted as puns. "Dream dictionaries" with correlations of dreamed phenomena with their "real" meanings appeared in many cultures. The earliest surviving is a Babylonian work possibly derived from a lost Akkadian original, iškar dZaqīqu (book of the god Zaqīqu). Another early work is the Egyptian text referred to as the Chester Beatty papyrus.

Ancient Greeks believed that inspired dreams could play an important role in the healing process. At temples dedicated to Aesculapius, the god of medicine and healing, supplicants prayed to the god to send healing dreams. Ancient Greek philosophers were often more suspicious of the importance of dreams—Aristotle suggested that if the gods wanted to send messages, they would send them in the daytime and be more selective in who received them. Despite philosophical opposition, however, interest in dreams continued throughout the history of classical Mediterranean culture.

The second-century CE *Oneirocritica* of the Greek physician Artemidorus of Daldis distinguished between several types of meaningful dreams. Some dreams were related to the past or present, such as a hungry person dreaming of food. Others predicted future events, either explicitly or through symbolism. Artemidorus usually interpreted dream symbolism literally; for example, to dream of a knife indicated division. He also suggested that the dream interpreter (the last two books are addressed to his son, a novice dream interpreter) should familiarize himself with the life and family background of the dreamer before making an interpretation, rather than simply applying a formula. The *Oneirocritica* was translated into Arabic in the Middle Ages, and it was frequently cited in the Islamic dream literature. It received its first printed edition in Europe (a Latin translation) in 1518 and was quickly published in French, English, and Italian as well.

Early Christians believed that some dreams were sent by God, but they were suspicious of dreams generally, seeing them as a way in which Satan tempted human souls. One type of dream that was not suspected of demonic connections was the dream of a martyr, and descriptions of the dreams of martyrs in the days before their death were circulated in accounts of their passions. The tradition of the suspicion of dreams continued in the works of Christian writers; the German Protestant reformer Martin Luther (1483–1546) claimed that he had prayed to God not to send him dreams, visions, or the apparitions of angels. The suspicion of dreams on the part of Christian leaders, however, has not precluded the development of a rich dream culture in Christian societies. Dream interpretation was embodied in cheap pamphlets or "dream books" often interpreted by uneducated local men and women with a reputation for wisdom. It continued to be frowned upon by Church authorities.

The dreams and visions of Muhammad played a central role in the founding of Islam, and Islam developed a rich tradition of dream lore and dream interpretation. Although Muslims believe that some dreams are deceptive or satanic, Islam does not manifest the hostility to dreaming characteristic of many Christian cultures. Many works of dream interpretation in Islamic culture were written by religious writers, and mainstream dream interpretation encouraged Islamic piety. The Quran endorses the idea of the meaningfulness of dreams by stating that in dreaming, the dreamer is brought into the presence of God. It shows Muhammad as experiencing a prophetic dream of his triumphant return to Mecca after his exile. The hadith, or traditions ascribed to Muhammad, also endorse prophetic dreaming, and Muhammad is credited with the saying that dreams constitute 146th of prophecy. The Iraqi Muslim Ibn Sirin (d. 728) was credited with immense authority in Islamic dream interpretation, and many works on dreams, some published centuries after his death, were ascribed to him. Islamic dream interpreters distinguished between ordinary dreams, prophetic dreams sent by God, and deceptive dreams sent by Satan or evil jinn. However, the believer could be sure that dreaming of Muhammad was always prophetic, as God had forbidden evil spirits from appearing in the Prophet's shape. Stories of dreams and visions often appear in histories written by medieval and early modern Muslim historians. *The Great Book of Interpretation of Dreams* is a fifteenth-century Islamic work that assumed great prominence in Islamic culture.

A particular break in the Islamic tradition is the synthesis developed by the

Persian mystic and philosopher Yahya Ibn Habash Suhrawardi (1154–1191), drawing on Sufism, Neoplatonism, and Indigenous Persian traditions. Suhrawardi claimed that dreams and visions accessed a nonphysical world, in Aristotelian terms a world of form without substance, that actually existed. His narratives of his own experiences, some of the earliest works in Persian prose, influenced generations of Islamic mystics.

In Chinese culture, *The Duke of Zhou Interprets Dreams*, ascribed to a legendary sage, was a popular book of dream interpretation. The duke was posthumously honored as the "God of Dreams," and Confucius once complained of how long it had been since he had dreamed of the Duke of Zhou. It was believed that the duke would warn of upcoming great events in the life of an individual through dreams.

The validity of an inspired dream could be influenced by the subsequent actions of the dreamer. There are a variety of superstitions about telling another person of a dream before breakfast, which would either guarantee or negate the dream's prophetic impact. English superstitions either encouraged or warned against telling Friday night dreams on a Saturday. In India, the dreamers of auspicious dreams should stay awake after their dreams, while dreamers of unfavorable dreams were encouraged to go back to sleep.

Frightening or otherwise disturbing dreams were also given a supernatural origin and viewed as a curse. Ghosts and other malevolent supernatural entities were blamed for bad dreams. The English term "nightmare" is derived from the word for an evil spirit called a "mare," which supposedly lay on the chest of sleepers, causing evil dreams. Similar beliefs existed in other European countries and elsewhere. The Korwa people of Mirzapur in India believe in a malevolent female ghost called a Reiya, which occupies the bodies of dreamers and attacks them with evil dreams and rheumatism. There were defensive techniques to ward off nightmares and evil dreams. There were also techniques to encourage good dreams. In Heian Japan, those seeking good dreams were advised to go to bed wearing their clothes inside out.

Beginning with the publication of Sigmund Freud's *The Interpretation of Dreams* (1899), the tradition of dream interpretation was secularized with the arrival of the psychoanalytic concept that dreams were revelations of unconscious mental states.

See also: Apparitions; Divination

Further Reading

Al-Akili, Muhammad M. 1992. *Ibn Seerin's Dictionary of Dreams according to Islamic Inner Traditions.* Philadelphia: Pearl; Bulkeley, Kelly, Kate Adams and Patricia M. Davis, eds. 2009. *Dreaming in Christianity and Islam: Culture, Conflict and Creativity.* New Brunswick, NJ: Rutgers University Press; Green, Nile. 2003. "The Religious and Cultural Roles of Dreams and Visions in Islam." *Journal of the Royal Asiatic Society Third Series* 13: 287–313; Kagan, Richard L. 1990. *Lucrecia's Dreams: Politics and Prophecy in Sixteenth-Century Spain.* Berkeley and Los Angeles: University of California Press.

Dung. See Excrement

Duppy. See Ghosts

Dybbuk. See Ghosts

E

Earthquakes

Among the most terrifying and destructive of all natural events, earthquakes have accumulated a considerable body of lore. Much of the lore has to do with predicting earthquakes. It is a common belief that animals can sense earthquakes and exhibit changed behavior in the hours before an earthquake.

Earthquakes are frequently associated with the wrath of God or the gods. Some Christian writers argued that earthquakes, like volcanoes, did not exist before the fall of Adam, but were marks of the world's fallenness. Although the biblical account is ambiguous, the destruction of Sodom and Gomorrah for their sins was often ascribed to an earthquake. The Roman emperor Justinian I (r. 527–565), when outlawing male homosexual acts with extreme penalties said that they led to earthquakes among other public calamities. Earthquakes were also associated with the end of the world in apocalyptic writings. In addition to being disasters themselves, earthquakes were the signs of more disasters to come. The Roman natural historian Pliny the Elder (d. 79 CE) stated that Rome had never been struck by an earthquake without it being the sign of a forthcoming disaster. In Imperial China, earthquakes, along with other natural disasters, were a sign that the ruling dynasty had lost the mandate of heaven.

There are numerous non-supernatural explanations for earthquakes. The Roman Stoic philosopher Lucius Annaeus Seneca (d. 65 CE) in his *Natural Questions* denies that earthquakes are caused by the anger of the gods but gives a brief history of natural explanations in Greek science, couched mostly in terms of four-element theory. Thales of Miletus (c. 624–c. 548 BCE), Seneca claims, thought the Earth floated in a vast ocean, and earthquakes were analogous to waves shaking a ship. Another water-based theory was that earthquakes were caused by underground rivers overflowing their banks. Seneca describes Anaxagoras (c. 500–c. 428) as believing that earthquakes were caused by fiery explosions. Anaximenes of Miletus (c. 586–c. 526) believed that Earth itself caused earthquakes, either as waterlogged portions of the Earth that cracked as they dried or as overly wet portions of the Earth dissolved. Seneca himself leaned toward one of the many air-based theories, claiming that the Earth had an abundant supply of what the Stoics called *pneuma*, or breath, the vital air that sustained life and the cosmos. When blocked or compressed, this air, which Stoics believed to be the most powerful of the elements, forced its way out, causing earthquakes.

It is a myth that earthquake-prone Japan literally ascribed earthquakes to the movements of a giant catfish under the islands. The dominant approach to earthquakes in premodern Japan was to view them as explosions of underground yang energy. The connection between catfish and earthquakes probably emerged as a metaphor, coming into widespread use in the seventeenth century and building on previous Japanese and Chinese legends of underground earthquake-causing dragons. The connection between catfish and earthquakes in nineteenth- and twentieth-century Japan continued in the form of beliefs that catfish behaved eccentrically before an earthquake or that an earthquake would be preceded by fishermen taking in unusually large hauls of catfish. Research on the subject has not discovered a connection between catfish and earthquakes.

See also: Elemental Systems; Volcanoes; Yin/Yang

Further Reading

Seneca, Lucius Annaeus. 1910. *Physical Science in the Time of Nero; Being a Translation of the Quaestiones Naturales of Seneca.* Translated by J. Clarke. London: MacMillan; Smits, Gregory. 2012. "Conduits of Power: What the Origins of Japan's Earthquake Catfish Reveal about Religious Geography." *Japan Review* 24: 41–65.

Eclipses

The idea of eclipses as omens dates back at least as far as ancient Babylon, where the earliest astrological records are based on interpreting eclipses. Babylonian priest/astronomers worked out the cycle of lunar eclipses, although their understanding of the less common solar eclipse was not as advanced. Greek astronomers worked out the physical mechanism of eclipses and continued to view them as omens. Greek and Roman historians were fond of stories of how eclipses affected the course of historical events. These stories did not always support the idea of eclipses as meaningful; the Athenian general Nicias (c. 470–413 BCE) was blamed for delaying his departure from the siege of Syracuse because of a lunar eclipse. His decision to delay his departure for twenty-seven days on the advice of soothsayers was blamed for Athens' disastrous defeat. The sixth-century CE philosopher Olympiodorus speaks of a popular belief among Greeks that demons wandered the Earth during the darkness of an eclipse; making a racket by beating on bronze helped drive them off.

The Bible contains many references to eclipses, seen as events of great significance. The Book of Joel in the Hebrew Bible proclaims, "The sun shall be turned into darkness and the Moon into blood before the great and terrible day of the Lord." This passage connects solar and lunar eclipses with the end of the world for Jews and Christians. The most famous "eclipse" for Christians is the darkness claimed to have covered the Earth at the death of Christ, but the description of this event in the Bible does not match up with an eclipse, nor is there a record of a total solar eclipse in the Middle East around this time. The linkage of eclipses with disasters and great changes continued for Christian societies through the Middle Ages and into the early modern period. In Islamic dream interpretation, to dream of an eclipse signified the death of a leader or ruler.

The Chinese also viewed eclipses, like other celestial events, as omens. Predicting the time and duration of eclipses as well as interpreting their meaning was one of the primary duties of the Bureau of Astronomy, part of the Board of Rites. Chinese imperial records are the longest, continuous, and most detailed series of eclipse accounts. By the eighth century CE, under the Tang Dynasty, Chinese astronomer officials had developed the ability to predict eclipses. Eclipse prediction was treated as a matter of state security, and astronomers were expected to keep their methods a secret. It was the superior ability to predict an eclipse that won for a Western Jesuit the headship of the Bureau of Astronomy in the seventeenth century.

The Maya also associated eclipses with disaster. The Aztecs, who were Sun worshippers, associated eclipses with significant events in their own past. They were also among the peoples who made loud noises during a solar eclipse to scare away the demons and monsters who could appear during the Sun's temporary absence. These demons and monsters were identified with the stars and planets emerging into visibility as the sky darkened. (The idea of "primitive" peoples banging pots to drive away the dragon that had swallowed the Sun during a solar eclipse became a staple of imperialist writing during the nineteenth century.) Like the Aztecs, the Iroquois associated the founding of their confederacy with a solar eclipse.

Numerous societies have mythological interpretations of eclipses, often based on gendering the Sun as male and the Moon as female (or, less frequently, vice versa). *The Golden Wheel Dream-book and Fortuneteller*, a nineteenth-century American book of magic, states that to dream of a solar eclipse presages the death of a father and to dream of a lunar eclipse that of the death of a mother. Early Spanish chroniclers described the Inca as beating dogs during an eclipse in hopes that the Moon goddess, who they thought was fond of dogs, would relent. A common way of interpreting a solar eclipse was in terms of the Moon biting or attacking the Sun. Hindu mythology ascribes solar eclipses to the Sun being temporarily devoured by a demon called Rahu. Numerous actions, including food preparation, are forbidden during the eclipse due to the danger of such actions being polluted by the impurity of the demon, and at the end of the eclipse, a ritual bath restores purity. Norse mythology features brother and sister wolves named Skoll and Hati who chase the Moon and Sun, respectively, and devour them on the day of the final battle, Ragnarok.

In Western astrology, to have an eclipse in one's natal chart is considered malefic, although some modern astrologers connect eclipses more with crises than with evil influences. Some astrologers believe that the last eclipse preceding an individual's birth is of significance for that person. In mundane astrology concerned with the destinies of nations and societies rather than individuals, eclipses were generally viewed as negative factors and could compound the bad effects of other celestial phenomena. A lunar eclipse was believed to have heralded the fall of Constantinople to the Turks in 1453. The "Black Monday" solar eclipse of March 29, 1652, in England attracted a wide variety of astrological and apocalyptic predictions, but proved anticlimactic, both as a spectacle and a prognosticator. The failure of the eclipse to be followed by dramatic events contributed to the discrediting of both astrological and apocalyptic eclipse interpretation. Decades later, in 1715, the astronomer Edmond Halley (1656–1742) would publish a broadsheet combining a scientific explanation of a solar eclipse with an assertion that the eclipse was a purely natural phenomenon with no political meaning, particularly none calling into question the legitimacy of Britain's new king, George I.

Despite the efforts of Halley and many others, eclipses, particularly solar eclipses, continued to bear a political meaning into the nineteenth century. The American leader of a slave insurrection, Nat Turner (1800–1831), associated the beginning of his rebellion with a solar eclipse, which he interpreted as a black hand covering the Sun.

See also: Astrology; Divination; Moon; Prodigies; Sun

Further Reading

Aveni, Anthony. 2017. *In the Shadow of the Moon: The Science, Magic and Mystery of*

Solar Eclipses. New Haven, CT and London: Yale University Press; Burns, William E. 2000. "'The Terriblest Eclipse that hath been seen in our Days': Black Monday and the Debate on Astrology during the Interregnum." In Margaret J. Osler, ed. *Rethinking the Scientific Revolution.* Cambridge, UK: Cambridge University Press. Pp. 137–152.

Elemental Systems

The idea that the material world is made up of a small number of elements is one that arises in many cultures and contexts. Before the creation of the modern system of chemical elements in the nineteenth century, the best-known systems of elements divided the world into four elements, with or without a special "fifth element," but there was also the Chinese system of five elements or phases and the alchemical system of the "tria prima" or three elements.

The "four-element" theory—earth, water, air, and fire—is found in many cultures. Its origins in Greece can be traced to the pre-Socratic philosophers who explored the nature of the cosmos and sought to reduce the multiplicity of the world to a small number of categories. Thales of Miletus (c. 624–c. 548 BCE), sometimes considered the "first philosopher," believed that water was the primary element. The four-element theory in its full form is associated with the fifth-century BCE philosopher Empedocles of Acragas, who described the universe in terms of the four elements, or "roots," and the two active principles of Love, which brings the elements together, and Strife, which separates them. This cycle explains all natural occurrences. Empedocles associates each of the roots with a god or goddess, although it is not clear from the surviving fragments of his work with which deity he associates each element. The philosopher Plato (429–347 BCE) linked the elements with the regular solids—fire was a tetrahedron, sharp and penetrating; Earth a cube, stable; air an octahedron; and water a twenty-sided icosahedron. (The fifth regular solid, the twelve-sided dodecahedron, represented unity.)

Plato's student Aristotle (384–322 BCE), one of the most influential thinkers in the four elements tradition, integrated the four elements with the polarities of dry/wet and hot/cold. Earth was dry and cold, water wet and cold, air wet and hot, and fire dry and hot. The elements could be paired off by mutually exclusive qualities, with earth the opposite of air and fire of water. The four elements each had a natural place—Earth at the center of the universe, followed by water, air, and fire—and their seeking their natural place underlay much of the Aristotelian physics that would dominate scientific thinking about the cosmos in both the Christian and Islamic worlds for centuries. The things we perceive are not the elements themselves but compounds in various ratios of the four elements.

Aristotle believed that matter was capable of being transformed from one element to another. This meant that the four elements were not the fundamental reality of matter since if water was transformed into air, for example, there was an underlying matter that lost the properties of water and gained those of air. Although scholars disagree as to what Aristotle himself thought about this fundamental reality, it is sometimes referred to as the prime matter.

The four elements were associated with many other sets of four in ancient and medieval thinking. The four humors were associated as black bile with earth, yellow bile with fire, phlegm with water, and blood with air—although blood was also considered to embody all four elements. Like the macrocosm of the universe, the microcosm of the human body contained all four elements. The twelve astrological signs were divided into four sets of three, each assigned an element.

Many four-element systems, including Aristotle's, allowed for a "fifth element" that stood apart from the other four. In the case of Aristotelianism, this was the perfect, indestructible, and unchanging element of the heavens, referred to as "ether." Unlike the other four elements, this element could not be directly perceived or manipulated. (Aristotle's theory also implied that the four elements were exclusively terrestrial, not to be found in the heavens.) Other four-element systems such as that of India have adopted "Void" as the fifth element.

Although the four elements are not found in the Hebrew Bible, they were adopted into Judaism in ancient times. References to the four elements are found both in rabbinic commentaries on the Torah and in Kabbalistic literature. There are many sets of four or number divisible by four, like twelve, in the Hebrew Bible that can be aligned with the four elements. The Kabbalists, with their emphasis on microcosm/macrocosm analogies, repeated the point that humanity was composed of the same four elements as the universe. (An interpretation of the creation of man in the Book of Genesis found the four elements present at the creation.) Four-element theory also entered medieval Judaism directly from Aristotelianism in the work of the Jewish Aristotelian Moses Maimonides (1138–1204), who frequently refers to the four elements in his work.

Christians also employed four-element theory in both religious and philosophical contexts. Saint Sophronios of Jerusalem's (560–638) Greater Blessing of the Waters, a ritual still used in the Orthodox Church to bless holy water, speaks of God creating the world of four elements. Medieval scholastic philosophers drew on both the Platonic tradition of defining the elements in terms of shapes and the Aristotelian tradition of defining in terms of qualities to debate such questions as the existence of elements in "pure," unmixed form and the possibility of transformations between the elements. Another question was whether all bodies were formed from a mixture of all four elements or whether it was possible for a body to be formed of a combination of only two or three elements.

Medieval Islamic philosophers and physicians took their ideas about the four elements principally from Aristotle. Ibn Sina (970–1037), known in the West as Avicenna, was particularly interested in how the four elements joined together to form all "corruptible" things, including living things. Those things that people identify as the four elements through their senses are not the four elements in their pure form—the air as perceived by people incorporates some water, earth, and fire as well as air. Ibn Sina believed knowledge of the four elements was vital for a physician and included discussion of them in his classic *Canon of Medicine*, widely taught and read in the medieval and early modern Islamic and Christian worlds.

Classical Chinese philosophy includes the concept of "wuxing," what has been called a "five-element" system, although some scholars have argued that given the Chinese emphasis on changing from one element to another, the five should be referred to as phases rather than elements. This emphasis distinguishes the five elements from four-element systems, which tend to treat the elements as mainly static and place less emphasis on transformation. The five phases are fire, water, wood, metal (literally gold), and earth. Unlike the four elements, the order in which the phases are named matters, and there are several orders in which they can be listed, each with a different meaning, such as a "generating" order in which each element emerges out of the previous one. Like the four elements in the West, the five phases were correlated with many sets of five in Chinese thinking. The five visible planets were paired with phases as wood/Jupiter, fire/Mars, earth/Saturn, metal/Venus, and water/Mercury. In traditional Chinese medicine, the internal organs are divided into two sets of five, the yin and yang organs, and then each of the five is associated with a wuxing element. Specific organs associated with one of the five phases can be strengthened by the consumption of foods associated with that phase. The wuxing phases are also correlated with the five colors (green, yellow, black, red, and white) and the five flavors (sour, sweet, salty, bitter, and acrid) as well as many other sets of five. According to wuxing thought, a school that emerged in the Han Dynasty, to maintain the five phases in harmony, was a job for the ruler and his officials. When the five phases were disrupted and inharmonious, trouble followed. Like the Western four elements, the five wuxing elements were also used in Chinese astrology.

Late medieval and early modern alchemy saw the development of a three-element system that supplemented but did not entirely replace the four-element system. This seems to have begun with the Arab alchemist Jabir Ibn Hayyan (721–813). In addition to the four elements, Ibn Hayyan postulated two new elements, "sulfur" and "mercury" (not to be identified with the substances now known by these names), which coagulated together in various ratios to produce metals. The German physician Paracelsus (1493–1541) broadened this theory to cover materials other than metals and added salt to form the "Paracelsian triad" or "three principles." (Paracelsus also promoted the idea of "elementals," living beings associated with each of the four classical elements, as salamanders were with fire.) Paracelsus was not a clear writer, and the exact role and nature of the triad and its relations to the four elements led to many debates among his followers on issues such as whether the triad proceeded from the four elements or the other way around. Sulfur, or "oil," was identified with combustibility and bitter taste, salt with solidity and saltiness, and mercury with vapor and sourness. Like other elemental systems, the Paracelsian triad was broadened to include other phenomena. The French Paracelsian royal physician Joseph Duchesne (1544–1609), a firm supporter of the triad, introduced a new humoral system based on three humors. The most radical use of the Paracelsian triad was to identify the three principles with the Christian Trinity,

During the Western Scientific Revolution of the sixteenth and seventeenth centuries, the theory of the four elements and the three-element system was abandoned by scientists, but the idea of the four elements still retains a prominent place in culture.

See also: Alchemy; Astrology; Elementals; Humours, Theory of; Kabbalah; Macrocosm/Microcosm; Yin/Yang

Further Reading

Bakhtiar, Laleh. 2012. *Avicenna on the Four Elements: Earth, Air, Fire and Water.* Chicago: Kazi Publications; Caiazzo, Irene. 2011. "The Four Elements in the Work of William of Conches." In Barbara Obrist et Irene Caiazzo, ed. *Guillaume de Conches: Philosophie et Science au XIIème siècle.* Firenze: SISMEL-Edizioni del

Galluzzo. Pp. 3–66; Debus, Allen G. 1977. *The Chemical Philosophy: Paracelsian Science and Medicine in the Sixteenth and Seventeenth Centuries*. New York: Science History Publications; Rochat de la Vallee, Elisabeth. 2009. *Wuxing: The Five Elements in Classical Chinese Texts*. London: Monkey Press.

Elementals

Elementals or elemental spirits are spiritual beings associated with each of the four elements in the Western four-element system: earth, air, water, and fire. The idea has roots in the distant past, but emerged into prominence in sixteenth-century occultism, particularly the works of the Germans Heinrich Cornelius Agrippa (1483–1535) and Paracelsus (1493–1541). Agrippa, in *De Occulta Philosophia* (1533), referred to spirits of the four elements, but did not give them names or lengthy descriptions. Paracelsus gave the classic descriptions and names to the elementals of the four elements in *A Book on Nymphs, Sylphs, Pygmies and Salamanders* (1566). (In the book, he states that he uses these terms because they are established but prefers undine to nymph for water spirit, sylvestre to sylph for air spirit, gnome to pygmy for earth spirit, and volcani to salamanders for fire spirits.) He wrote this book toward the end of his life, but it was published only posthumously. Paracelsus claims that the elemental spirits are neither angels nor devils, but part of nature to be studied like other natural phenomena. As they are works of God, studying them is a positive good and to be ranked higher than studying thc works of man. Paracelsus's interpretation of these beings draws from multiple sources, from Greek mythology to the folklore of his German contemporaries.

Paracelsus believed that the elementals, although not descended from Adam, were closely akin to humans. They occupied an intermediary role between purely spiritual beings such as angels and humans. Like spirits, they can move through solid objects; like humans, they eat, drink, and reproduce. Although they have spirits, they do not possess souls, which humans have. In that sense, they are intermediary between humans and animals, resembling humans physically, but soulless like monkeys. Thus, although they are intelligent like humans and live in political societies, manufacture and wear clothing to protect their modesty, and even practice medicine, when they die they completely disappear, like animals. Nor, as they lack souls, are they capable of worshipping God. The four societies have nothing to do with each other, but stay in their proper element, which is natural to them. Thus, the gnomes can move through Earth as easily as humans do air. Water elementals, who are female, can marry human men, and if they do so, the offspring have souls and the undines themselves received souls as well.

Paracelsus was drawing on popular beliefs; the Italian artist and goldsmith Benvenuto Cellini (1500–1571) recounted a story of having seen a salamander in a fire when a child. The next major text dealing with elemental spirits was a French work, *The Count of Gabalis* (1670), by a French clergyman, the Abbe Nicolas-Pierre-Henri Montfaucon de Villars (1638–1673). Gabalis is a mangled version of "Kabbalah," and despite the absence of specifically Rosicrucian references, the work was widely perceived as in the Rosicrucian tradition. The picture the fictional Gabalis, a German magical adept, gives of the elemental spirits is very different from Paracelsus's. Gabalis's elementals are active students of magic (although inferior to human adepts) and worshippers of God. Gabalis accepts Paracelsus's idea that elementals who marry humans receive souls, but focuses on sylphs rather than undines. Elementals furnish idealized mates for philosophers, far superior to human women. Male elementals can also marry human women, although Gabalis places less emphasis on this. The work was quite popular, frequently reprinted and attracting translations and sequels. Although some have claimed *The Count of Gabalis* is a satire, it was taken seriously by later occultists including leaders of the nineteenth-century occult revival such as the French poet Eliphas Levi (1810–1875) and the Russian founder of Theosophy Helena Petrovna Blavatsky (1831–1891). The Hermetic Order of the Golden Dawn, an influential British occultist group founded in 1888, even developed a ritual for their human members marrying one of them. Elementals also made frequent appearances in literature and art.

See also: Elemental Systems; Kabbalah; Rosicrucianism

Further Reading

Nagel, Alexandra. 2007. "Marriage with Elementals: From 'Le Comte de Gabalis' to a Golden Dawn Ritual." MA Thesis, University of Amsterdam; Paracelsus. 1991. *Four Treatises of Theophrastus von Hohenheim, called Paracelsus.* Baltimore: Johns Hopkins University Press.

Elephants

The huge size and peculiar appearance of the elephant has attracted many superstitious and erroneous beliefs. The elephant and its ivory are associated with luck in many cultures. In India, the elephant-headed god Ganesha brings prosperity. In parts of Africa, rings or bracelets made of elephant hair bring luck. Elephants were also believed to bring good luck in China, and statues of elephants were placed in buildings to bring good fortune. Elephant statues placed near doorways are believed to protect the household. A white elephant is associated with the Buddha, whose mother dreamed of a white elephant when he was conceived. In Buddhist Thailand, ivory amulets are believed to bring good luck.

Classical and medieval Europe, where people knew about elephants but seldom saw one, evolved some bizarre elephant-related beliefs. The philosopher Aristotle (384–322 BCE) denied the popular Greek belief that elephants, who were near-legendary in Greece, had no knee joints and had to lean against trees to rest. Despite Aristotle's theory, the belief that the elephant had no knee joint persisted into the seventeenth century and may have originated from the idea that the elephant used its trunk to gather plants because it couldn't bend down and eat them. Roman natural historian Pliny the Elder (d. 79 CE) agreed with Aristotle, adducing examples of elephants who had been seen to kneel. Pliny had a very high opinion of the intelligence and character of elephants, ascribing to them virtues such as sexual modesty, honesty, and prudence and even asserting that they worshipped the Sun, stars, and Moon. Pliny believed that the rhinoceros, another creature of whom Romans had even less direct knowledge, was a natural enemy of the elephant and that its horn was adapted for piercing the elephant's unprotected stomach, an idea that eventually extended to unicorns. Pliny claimed that the dragon was another natural enemy of the elephant, being one of the few animals large enough to eat an elephant. The idea of hostility between elephants and other animals was not restricted to large animals such as rhinoceroses, unicorns, and dragons. Pliny also stated that elephants were frightened by the grunting of hogs, an idea that would persist into the seventeenth century. (According to the early third century CE Roman rhetorician Claudius Aelianus, author of *On the Nature of Animals*, the Romans used this weakness to defeat the elephant-using General Pyrrhus of Epirus.) Yet another creature to which elephants were averse was the mouse, a belief that persists in modern times in the form that elephants actually fear mice. The Talmud uses, as an example of the power of the weak over the strong, the alleged ability of the mosquito to torment the elephant by entering its trunk.

In the European Christian Middle Ages, elephants were admired for their supposed freedom from sexual desire. It was believed that elephants mated only for the purposes of procreation, as was the ideal for humans. In order to arouse themselves, both male and female elephants consumed mandrakes. Medieval people tended to see the natural world as a series of religious allegories, and the elephant represented the innocence of Adam and Eve in the Garden before the Fall and the mandrake the forbidden fruit that brought sin into the world.

Ivory, for which elephant tusks are the main source, was believed to have healing powers in traditional Chinese medicine. Emperors used ivory chopsticks as a protection against poisoning, as it was believed that ivory would change color on contact with poison. Ivory was believed to heal ulcers when ingested as a powder. In the ayurvedic medical tradition of India, ivory "bhasma," or ash, was believed to promote the growth of hair.

See also: Dragons; Luck; Mandrakes; Rats and Mice; Unicorns

Elves. See Fairies

Emeralds. See Precious Stones

Equinoxes. See Solstices and Equinoxes

Evil Eye

Belief in the evil eye, a person's ability to cause harm by their glance, was and is widespread throughout many cultures and societies. Although magicians and witches were often thought to possess it, it was also an ability of ordinary people, particularly women. People with unusual eyes, such as eyes of different colors, cross-eyes, doubled pupils, or blue eyes, in places where they are rare are often suspected of carrying the power of the evil eye. The evil eye is frequently identified with jealousy of another person's good fortune. The connection between the evil eye and envy meant that receiving a compliment could be suspicious, as the compliment could mask a secret envy leading to an attack by the evil eye. Uttering a protective formula after paying the compliment was advisable in such circumstances. The evil eye could cause a variety of conditions including sickness, accidents, and death. The person casting the evil eye was not always thought to wish harm. Southern Italians and Sicilians had a concept of an involuntary witch called a jettator whose glance caused harm and bad luck completely unbeknownst to him or her.

Evil eye belief seems to have originated in the ancient Near East. It appears in both Sumerian and Babylonian texts, although its presence in the Hebrew Bible is disputed. The evil eye was known in classical Greece and Rome. It spread to Europe and India, but is strongest in the Mediterranean region.

An eighth–tenth-century Iranian button with designs warding off the evil eye. (The Metropolitan Museum of Art, New York, Rogers Fund, 1938)

It arrived in the Americas with the Spanish and Portuguese conquests of the sixteenth century.

Infants were believed to be particularly vulnerable to the evil eye, a belief encouraged by the high rates of infant death in many traditional societies. Medical manuals treated the evil eye along with other normal hazards of childhood. Praising an infant was considered to threaten use of the evil eye, and in many cultures, it was common for mothers to speak of their babies in deprecating or negative terms, so as to avert envy and its inevitable consequence, the evil eye. Particularly fortunate people were thought to be the envious evil eye's potential victims, but animals and plants, particularly those that were the property of a successful person, were also vulnerable.

In Hindu India, the evil eye was associated with the lower castes, whose gaze polluted upper caste people, particularly Brahmans. The gaze of strangers was also associated with the evil eye and could be blamed for cows not giving milk or the sudden failure of trade. In Malabar, various dolls or images were hung up near houses to divert the passing eye of strangers and thus divert their evil eyes. Ethiopians also associated the evil eye with a particular social group, the *buda*, who as craftspeople rather than farmers stood somewhat out of settled society. The Beta Israel community of Ethiopian Jews were also considered *buda.*

There were numerous protective practices against the evil eye. In Mediterranean cultures, the evil eye could be warded off by the color blue. Blue protective amulets, known as *nazar*, representing a stylized eye date as far back as ancient Mesopotamia. The hamsa, a stylized right hand with an eye in the middle of the palm, often blue or made from silver, is a commonly worn protective amulet in the Middle East and North Africa. (Among Indian Muslims, the hand was sometimes red and painted on buildings.) Muslims call the hamsa the hand of Fatima after Muhammad's daughter, Jews the hand of Miriam after Moses's sister, and Christians the hand of Mary after Jesus's mother. European Christians were less accepting of the hamsa; in 1526, nearly four decades after the fall of the last Muslim power in Spain, a Spanish council outlawed it. Rejection of the hamsa did not mean Spanish Christians shunned protection against the evil eye; visiting foreigners remarked on how many amulets protected Spanish babies.

In the Islamic world, the belief in the evil eye is justified by a hadith, or tradition ascribed to the Prophet Muhammad, that the evil eye is real. The expression Mā shā' Allāh, what Allah wills, was used to protect against the evil eye. This expression, originating in the Quran, was also used to protect against the evil eye by non-Muslims in some Islamic-ruled areas, such as Ottoman Greece, and inscribed on buildings as well as used as a verbal formula and on protective amulets. Doorways were a particularly popular spot for inscriptions that protected those within. Christians and Jews also used sacred texts and objects to protect against the evil eye. The six-pointed star or "sign of Solomon" appeared in protective inscriptions or amulets in many places, including Portugal and Portuguese Brazil. Catholics believed that blessing and making the sign of the cross was powerful against the eye. Some practitioners, with or without the blessing of the Church, used particular prayers, combinations of prayers, or rituals to remedy conditions caused by the evil eye. In extreme cases, the ritual of exorcism could be appropriate. The evil eye also appears in Rabbinic literature, with some claiming that descendants of the biblical patriarch Joseph are immune to it.

Other cures and protective rituals made use of natural phenomena rather than religious symbols or texts. In early modern Scotland, livestock could be protected from the evil eye by braiding rowan in their tails, while a witch's evil eye could be neutralized by drawing blood from her forehead. In Latin America, scents, including orange blossom, rosemary, and vinegar, were believed to have power against the evil eye. St. John's wort was widely believed to have protective qualities.

Explanations of the evil eye varied but were frequently linked to premodern theories of vision that saw it as something that emanated from the eye, rather than the current scientific view of the eye as a recipient of light. The evil eye could be identified by scholars as a form of natural magic. It could also be identified as a miasma or contagion, according to one popular medical theory in the medieval and early modern world. One theory was that persons with the eye had a "corrupted" glance that could be caused by either moral corruption or physical conditions such as a humoral imbalance. Old women who had ceased to menstruate were sometimes believed to carry corrupted blood due to its retention within their bodies, and thus to possess the evil eye, although menstruating women were also associated with it. Some physicians linked the evil eye with the power of moonlight.

Some writers on the subject distinguished between a natural and a sorcerous evil eye. In early modern European witchcraft cases, the evil eye was a form of *maleficium*, an evil deed, arguably the purest form of *maleficium* as it operated solely by the witch's malice and the power of the devil, rather than the properties of herbs, ointments, potions, or charms. Witches were frequently accused of employing an evil eye to cast their spells both by ordinary people and by learned demonologists. Sometimes bewitched persons seemed to feel particular torment when a witch cast her gaze at them. The evil eye provided an easy explanation for how a witch caused children or livestock to die, and any interest in a particular person shown by a suspected witch would be cause for thinking that she was preparing to cast the sorcerous evil eye.

See also: Amulets and Talismans; Blood; Defensive Magic; Luck; Menstruation; Witchcraft

Further Reading

Ameen, Ahmed. 2020. "The Significance of the Qur'anic quotation 'Mā shā' Allāh' on both Ottoman and Greek Heritages in the Balkans." *Egyptian Journal of Archeological and Restoration Studies* 10: 73–85; Baynes-Rock, Marcus. 2015. "Ethiopian Buda as Hyenas: Where the Social Is More that Human." *Folklore* 126: 266–282; Dundes, Alan, ed. 1981. *The Evil Eye: A Folklore Casebook.* New York: Garland; Elliott, John H. 2016. *Beware the Evil Eye: The Evil Eye in the Bible and the Ancient World.* Eugene, OR: Cascadia; Souza, Laura de Mella e. 2003. *The Devil and the Land of the Holy Cross: Witchcraft, Slavery and Popular Religion in Colonial Brazil.* Translated by Diane Grosklaus Whitty. Austin: University of Texas Press; Thurston, Edgar. 1912. *Omens and Superstitions of Southern India.* London: T. Fisher Unwin.

Excrement

Urine and feces, human and animal, were ubiquitous in premodern societies and the subject of many superstitions and magical beliefs.

The medical use of excrement has a long history. Ancient Egyptian physicians used excrements including those of donkeys, hippos, and crocodiles as well as humans. They believed that the repulsive qualities of excrement would drive out the evil spirits that caused disease. Classical medical authorities including revered figures such as Hippocrates and Dioscorides continued to endorse the medical use of excrement, which continued into the early modern period. Roman natural historian Pliny the Elder (d. 79 CE) suggested that women suffering from sterility should apply a pessary made from the first excrement of a baby to her vagina. In India, cow dung was used in defensive magic, smeared on walls or formed into cakes and placed in areas to be defended from evil spirits or malevolent magic. In Bengal, sheep's dung was also used for this purpose. Cow dung served in Europe as a healing poultice applied to wounds, while goat dung mixed with vinegar was applied for skin conditions such as warts and boils. Pennsylvania Germans made an ointment from goose dung, elderwood, and tallow to cure the skin disease known as erysipelas.

The use of various animal excrements as medicine was common in the medical traditions of India. Ayurvedic physicians endorsed the idea that all natural substances were potentially medicines, and excrements were no exception. The urine and dung of the cow, a revered animal, appeared most commonly in the Indian pharmacopeia. The pañcagavya was a mixture of five cow products: dung, urine, clarified butter, yogurt, and milk. It had both religious and medical uses and was sometimes claimed to grant superhuman powers.

Excremental medicine also appeared in early Chinese texts, but expanded greatly in the Tang Dynasty as it spread with the influence of Buddhism and Indian medicine. Excremental medicines fit the Buddhist concept of compassion, in that they were common and cheap and hence available to the poor. (Excremental medicines generally employed the product of common domesticated animals that would be easy to acquire as opposed to exotic or wild animals.) However, some rejected these medicines as impure and defiling, and excremental medicines appear less often in the medical literature addressed to scholars and members of the elite. Human as well as animal excrement appeared in Chinese medical recipes. The term "yellow-dragon decoction" has multiple meanings, but one referred to a product caused by sealing human feces in a jar for a long period of time. The yellow-dragon decoction was the liquid that resulted, which was given to suffering patients to drink.

In parts of Europe, stepping into dung is lucky, as long as it is done accidentally rather than deliberately. A bird defecating on a person also brings luck, and some authorities recommend that the bird droppings not be wiped off or cleaned.

In the West, urine had strong associations with both witchcraft and defensive magic. Urinating through a wedding ring protected against impotence. Witches who possessed a person's urine had the ability to work harmful magic on them, although spitting into the urine made it useless for the witch's evil purpose. Urine was a key ingredient in the form of defensive magic used in the British Isles and known as the witch bottle. The urine of the bewitched person was placed in a bottle, along with a variety of ingredients often including pins and needles, and then buried in a remote corner of the house. This was believed to both protect the victim and, in some cases, to kill the witch. A variation was to thrust the bottle into a fire. When it shattered, the witch would die. Witch bottles date from the seventeenth century in England, and a few have been found in areas of English settlement in the United States. A related practice was the witch cake, a flour cake baked with a small portion of the victim's urine. Baking the witch cake would draw the witch to the house of the victim so he or she could be identified.

Feces too had magical power; burning the excrement of a bewitched person would force the witch to come to the scene, thus revealing his or her identity. Burning a person's excrement also posed dangers, though, as it could cause the person to develop colic or a painful burning sensation in the bowels. English natural philosopher Sir Kenelm Digby (1603–1665) relates a story of how he detected the family's burning a child's excrement as a cause of the child's fever and how placing the child's next bowel movements in a basin of cool water broke the fever.

There are numerous superstitions about which direction to face while urinating or defecating. These are generally organized around the idea of separating the impurity of excrement from the divine. The ancient Greek poet Hesiod warned against urinating in the face of the Sun or urinating or defecating into a stream or river as things that brought bad fortune. There was also a sailor's superstition that it was bad luck to urinate into the wind.

In Islamic dream interpretation, feces, human or animal, commonly signify money.

See also: Defensive Magic; Foxes; Goats; Pregnancy and Childbirth

Further Reading

Despeux, Catherine. 2017. "Chinese Medicinal Excrement." *Asian Medicine* 12: 139–169; Opie, Iona and Moira Tatem, eds. 1989. *A Dictionary of Superstitions.* Oxford: Oxford University Press.

Execution Magic

In medieval and early modern Europe, the executioner and his victim stood between life and death. As such, they were associated with magical power.

In the early modern period and in some places into the nineteenth century, Continental European executioners supplemented their incomes with a trade in human fat, a substance believed to have healing properties. The application of unguents made from human fat could treat wounds, scars, contusions, and other forms of external damage as well as ease the pain of gout. Although these remedies did not require the fat of executed criminals, executioners did have access to a continuing supply of fat. (As control over the bodies of executed criminals passed from executioners to the medical profession beginning in the mid-eighteenth century, physicians and medical students did not hold themselves above making a side income from the sale of human fat.)

Other pieces of an executed criminal had powers that were not shared by other corpses. Ironically, considering the fate of their original possessor, the powers of an executed criminal's body parts were mostly positive. These beliefs focused on relatively detachable parts, such as teeth or fingers. In fifteenth-century Italy, touching a sore tooth to the tooth of a hanged man would cure the toothache. Early modern Germans believed that the finger of an executed victim would bring good luck. Body parts could be stolen, but also given away or sold by the executioner. Other magical uses required more preparation. Belts of leather made from human skin, for example, were believed to ease labor pains.

Some forms of execution magic did not require the possession of body parts. In eighteenth-century France, there was a superstition that passing under the corpse of a hanged man cured barrenness in women. In much of Europe, the touch of a hanged man's hand was believed to cure epilepsy or wens, a belief that persisted into the nineteenth century. (There is some evidence for the belief that this power only worked across gender lines—women benefiting from the touch of a hanged man, men forced to avail themselves of the less common hanged woman.) The behavior of crowds at executions seeking the touch of the dead man's hand was one factor in the drive to end executions as public occasions by moving them into the restricted space of the prison in the nineteenth century. Executioners could profit from their control over access to the hanged body. The belief in the power of the touch sometimes extended to the executioner, as people sought his healing touch.

Hangings did not produce blood, but beheadings did, and the blood of executed victims also had magical powers. Drinking the blood of an executed victim could cure epilepsy, a belief that persisted in Germany into the early twentieth century. (Blood drinking was mostly a northern European phenomenon; there is little evidence for it in Mediterranean Europe.) In China, the blood of executed criminals was consumed as a cure for tuberculosis. Executioners tried to control access to the blood, but scaffolds were sometimes rushed by mobs. Even the criminal himself could have a voice, as those who wished to consume his blood obtained his permission beforehand. The blood of an executed criminal could also bring luck, as handkerchiefs dipped in criminal blood were good luck artifacts.

The magic of execution extended from the criminal's body to the executioner's tools. Nails or chips from the execution site were made into amulets bringing luck or good health. The executioner's sword was reputed to have power against werewolves and other monsters. The most popular of these artifacts was the hangman's rope, which was believed to have the power to cure several medical conditions. In sixteenth-century Spain, the hangman's rope was a frequent ingredient in love charms. Hangman's ropes were particularly popular among gamblers, who thought they brought luck. Control over the rope was an important perquisite of executioners, who sometimes sliced them into pieces as short as an inch to maximize sales. Control over the rope was zealously defended by executioners as governments attempted to suppress rope sales in the nineteenth century, in the name of combating "superstition." (The hanging rope did not bring good luck to everyone. There was a superstition in Europe and America that a rope-maker who made hanging ropes would soon go out of business.)

Executioners themselves sometimes used magic to aid in the performance of their difficult task. Sixteenth-century German executioners carried a splinter from a judge's staff to protect themselves from botching an execution or covered the convicted person's head to prevent them from using the evil eye to disrupt the proceedings.

The place where a gallows, execution block, or gibbet had stood also had supernatural power, being haunted by the ghosts of the executed or even just causing horses to shy decades after the last time it had been put to use. Executions could shape beliefs about time as well as space. In nineteenth-century America, there was a common belief that it was bad luck to start a new enterprise on a Friday because that was the day executions took place. However, the idea of Friday being a bad day to begin anything is found in many places and not necessarily related to any execution save that of Jesus Christ, whose crucifixion on a Friday was probably at the root of the superstition.

See also: Amulets and Talismans; Blood; Death; Ghosts; Hand of Glory; Luck; Mandrakes; Weapon-Salve

Further Reading

Davies, Owen and Francesca Matteoni. 2017. *Executing Magic in the Modern Era: Criminal Bodies and the Gallows in Popular Medicine.* Cham: Palgrave Macmillan; Harrington, Joel F. 2014. *The Faithful Executioner: Life and Death, Honor and Shame in the Turbulent Sixteenth Century.* New York: Picador; Peacock, Mabel. 1896. "Executed Criminals and Folk-Medicine." *Folklore* 7, 268–283.

F

Fairies

"Fairies" is a collective term for supernatural creatures resembling humans, usually employed in a European context. Belief in fairies is found in many European societies, notably Britain and Ireland, but fairy belief and the fairies themselves take a variety of forms. They are often believed to be smaller than humans—"wee people" is another designation—and usually claimed to be attractive. However, they can also be of human size and engage in love affairs with humans, or even larger. Fairies are sometimes viewed as spirits of the dead, particularly the pagan dead who died before the coming of Christianity. Another version of this legend made fairies the dead in battle whose bodies had not been recovered. Belief in fairies as spirits of the dead persisted in some areas into the twentieth century.

Fairies are frequently identified with Celtic peoples, and fairy belief hung on longer in Ireland than just about anywhere else. (In 1895, an Irishman named Michael Cleary murdered his wife Bridget because he believed she was really a fairy impostor.) However, belief in fairylike creatures such as the Sicilian "Ladies from Outside" can be found in many parts of Europe from the Mediterranean to Scandinavia. Fairy lore can be traced to the early Middle Ages, although its roots are much older. Until the fifteenth century, the creatures later referred to as fairies were generally referred to in English as elves and by analogous terms such as "aelf" in Germanic and Scandinavian languages. Elves were originally thought of as possible allies for humans against monsters, but the introduction of Christianity meant elves were more often portrayed as demonic. *Beowulf*, an Anglo-Saxon Christian epic poem of uncertain date about Scandinavian pagans, puts elves in the same category as the hero Beowulf's adversary, the monster Grendel, a descendant of Cain. Elves were also originally thought of as male, but became more closely associated with beautiful women by the eleventh century. Anglo-Saxons referred to sudden pains suffered by man or beast as "elf-shot," ascribing them to invisible arrows shot by the elves. Neolithic arrowheads were believed to be the work of elves. The idea that some illnesses were caused by fairies persisted into the sixteenth century.

The official position of the medieval church was at first to deny the existence of the fairies as mere superstition. Beginning in the thirteenth century, however, Church authorities began to identify fairies with devils and suggested that involvement with fairies led mortals to hell. A variation on this was to suggest the fairies were devils, but a variety of devil less bad than the fallen angels of hell. The idea that alongside the rebellious angels that fell with Satan there were some who were unworthy of heaven for lesser reasons such as trying to be neutral in the struggle between God and the devil went back to the third century CE, but was applied to fairies in the Middle Ages. (This association persisted in popular fairy belief into the twentieth century.) Fairies were still dangerous in the minds of the clerical establishment no matter their

Ladies from Outside

From 1579 to 1651, the Spanish Inquisition, which had jurisdiction in Sicily, dealt with a series of cases involving fifty-seven women and eight men, all poor people, who identified themselves as the "ladies from outside." Belief in the ladies from outside can be traced as far back as the mid-fifteenth century. This term had two meanings, one referring to a group of supernatural beings such as fairies, the other to their human associates. The human ladies from outside journeyed in spirit on Tuesday, Thursday, and Saturday nights to revel in meetings presided over by a goddess called the "Queen of the Fairies" or "the Greek Lady" or "the wise Sibyl" or several other names who resembled the Diana figure of European folklore. The gatherings were pleasant, with fine food, music, and enjoyable sex. The fairy ladies from outside resembled humans, except in having hands and feet like animals. The human ladies from outside claimed to be organized in companies and to have a "sweet blood" that gave them the power to heal those diseases in humans or animals that were caused by the fairies. Some functioned as magical healers or cunning folk in their communities. The cult was widely known—members discussed it openly, and the Inquisition never had trouble finding witnesses. The Inquisitors attempted to deal with the ladies from outside by explaining to them that their meetings were really sabbats, presided over by the devil, but never succeeded in convincing most Sicilians.

Source: Hennigsen, Gustave. 1990. "The Ladies from Outside: An Archaic Pattern of the Witches' Sabbath." In Bengt Ankarloo and Gustav Henningsen, eds. *Early Modern European Witchcraft: Centres and Peripheries*. Oxford: Clarendon Press.

origin; among the charges leveled at Jeanne Darc at her trial in 1431 was consorting with fairies at a "fairy tree." The prayers of mendicant friars, a thirteenth-century innovation, were portrayed as particularly powerful in banishing fairies. However, many people, including some clerics, had a far less hostile view of the fairies and even suggested that some could hope for salvation.

In the Middle Ages, both male and female fairies were believed to seek out humans for sex, although such couplings were invariably heterosexual. (The French theologian and bishop William of Auvergne (d. 1249), one of the few writers to deal with fairy homosexuality as even a theoretical possibility, thanked God that no one had ever heard of it actually happening.) Such affairs could even produce offspring. Merlin, the legendary British wizard, was sometimes described as the child of a fairy and a human woman (although Merlin's father was more commonly described as a demon), and the Plantagenet line was described as the offspring of the water fairy woman Melusine. Female fairies were viewed in masculine legend as desirable sexual partners, and necromancers sometimes claimed the ability to summon them and bind them to the caster's will. Clerical writers were eager to warn that such pairings inevitably ended badly.

The demonization of fairies culminated in the early modern witch hunt. Fairies and witches were often treated together by demonologists and were often viewed as comrades. Witches and "cunning folk" sometimes claimed to cure the afflictions sent by the fairies. Some accused witches claimed to get their power from the fairies, perhaps as an alternative to the witch hunter's belief that the source of a witch's power could only be the devil. Writers on the sabbat, the gathering of witches under the leadership of the devil, frequently compared it to fairy gatherings. Familiar spirits were also described as fairies or something very similar, and claims to consort with fairies could be viewed by witch hunters as admissions of associating with devils. In 1576, a woman named Bessie Dunlop was burned at the stake in Edinburgh for consorting with the fairies and the queen of the elves. (Scottish witchcraft was particularly strongly influenced by fairy belief.)

Fairy realms were usually located underground, but they were also sometimes identified as "high places." In Ireland, the "Sluagh Sidhe" were believed to live under hills, but only under natural hills rather than man-made ones. There were also stories of fairies who lived deep in the forest (the forest of Broceliande in Brittany was particularly strongly associated with fairies) or under bodies of water.

Fairies were sometimes believed to have the gift of changing themselves to animals. They were also believed to have the ability to summon bad weather, reflected in their being referred to as "powers of air." Winged fairies seem to have been more of a literary convention than an aspect of popular belief.

Relations between fairies and humans were frequently treated as abductions. Many fairy legends center on a person who attends the fairy court or a meeting of fairies. Fairies were usually believed to be organized into a monarchy. The fairy court is described as splendid, but a common trope was for this splendor to be revealed as an illusion and the reality as corrupt and foul. Fairies were frequently blamed for stealing human children, particularly unbaptized ones, from the cradle and substituting a fairy child or "changeling." This changeling was usually sickly and ill-mannered and sometimes no child at all, but an aged and decrepit fairy. Changeling babies were also believed to be voracious, draining the milk from multiple nurses while still failing to thrive. There were various rituals for returning the changeling to the fairies in exchange for the original baby. In the late medieval English mystery plays, even the infant Christ is referred to as a changeling, although they are villains and fools who make the reference. What happened to the original human baby was less discussed, but it was often assumed to be raised by the fairies.

Belief in changelings seems to have become more common in the early modern period. It continued among rural people well into the Enlightenment; the Swedish botanist Carl Linnaeus (1707–1778) was shown a thirteen-year-old boy believed to be a changeling in 1741 while touring remote areas of Sweden. The boy was sickly, unable to walk or to talk in anything but a mumble, and also showed feminine characteristics. Linnaeus dismissed out of hand the possibility that the child was a changeling, but suggested an incident where the mother was frightened while witnessing a knife-fight while pregnant with the child was the true cause.

"Domestic" fairies were frequently believed to be attached to families, from the Irish banshee whose wailing announced the death of a member of the family to the more prosaic English and Lowland Scottish brownie, who if treated with respect kept the house clean and assisted in the churning of butter. The brownie could be mischievous if disrespected or not properly rewarded for his work; one common bribe for brownies or other sorts of fairies was a bowl of milk or cream.

Fairies were also potential guides to buried treasure. There are many stories of people presenting themselves as intermediaries with the fairies and conning would-be treasure seekers. Victims were not always ordinary people or unlearned. The English politician Goodwin Wharton (1653–1704), an educated member of the gentry, spent many years of his life planning a marriage with the fairy queen, which would also win him enough treasure to pay off his many debts.

See also: Death; Diana; Pregnancy and Childbirth; Saffron; Wild Hunt; Witchcraft

Further Reading

Briggs, Katharine. 1967. *The Fairies in Tradition and Literature.* London: Routledge and Kegan Paul; Clark, J. Kent. 1984. *Goodwin Wharton.* Oxford and New York: Oxford University Press; Green, Richard Firth. 2016. *Elf Queens and Holy Friars: Fairy Beliefs and the Medieval Church.* Philadelphia: University of Pennsylvania Press; Hall, Alaric. 2007. *Elves in Anglo-Saxon England: Matters of Belief, Health, Gender and Identity.* Woodbridge, Suffolk; Rochester, NY: Boydell & Brewer; Kirk, Robert. 2008. *The Secret Commonwealth of Elves, Fauns and Fairies.* Garden City, NY: Dover.

Feces. See Excrement

Flat Earth

Belief in a flat Earth was common to many archaic civilizations, including those of Egypt, Mesopotamia, and archaic Greece. The pre-Socratic Greek philosopher Anaximenes of Miletus (c. 586–c. 526 BCE) endorsed a flat Earth, together with a flat Sun and Moon. However, belief in a flat Earth was a minority belief in the West following the establishment of Greek natural philosophy. Aristotle (384–322 BCE), with his picture of the universe centered on a spherical Earth, was particularly influential in this regard. Ptolemy (c. 100–c. 170 CE), the most influential astronomer and geographer among the ancients, also presented a spherical Earth. Among the ancient Greeks and Romans, the most influential champions of a flat Earth were the fourth-century CE Latin Christian writer Lactantius and the sixth-century CE Greek traveler in the Indian Ocean Cosmas Indicopleustes. Lactantius rejected the sphericity of the Earth as part of a rejection of Greek philosophy that asserted things that could not be known with certainty, such as the shape of the Earth. He went beyond this, however, to assert that a spherical Earth was self-evidently absurd, as on the other side of the world the sky would be lower than the ground and people's heads under their feet. Cosmas, the author of a work called *Christian Topography,* gave the most fully fleshed-out picture of a flat Earth in the Western tradition. Cosmas's theory was in part caused by his confusion between the whole world and the "Oikumene," or inhabited world, frequently represented by ancient geographers as a squarish territory centering on the Mediterranean. The main force driving Cosmas's flat Earth belief, however, was a literal reading of the Bible, which frequently set forth the picture of the Earth as having four corners and the sky above it being hung like a tent or a ceiling. Cosmas also believed that the Earth corresponded to the Tabernacle of Moses, a flat surface on which offerings were made. Neither Lactantius nor Cosmas, however, had much influence on subsequent Western thought about the shape of the Earth. The idea that before Columbus Western people thought the world was flat was a product of nineteenth-century anti-Catholicism and the controversies over Darwinian evolution. The idea that the Church supported a flat Earth was incorporated into the popular idea of a war between religion and science.

The sacred geography of Hindus, Jains, and Buddhists depicts a flat world with a center at the sacred mountain Sumeru. The Sun and the planets circle the mountain, with night occurring when the Sun goes behind the gigantic mountain. The world is a vast ocean surrounded by mountains with a few continents scattered in it. Traditional Chinese astronomers also held to a flat Earth view, with some debating as to whether the flat Earth was circular or square. Mesoamericans believed in a flat Earth disc, with a number of unearthly but also flat realms suspended above it.

See also: Geocentrism

Further Reading

Russell, Jeffrey Burton. 1991. *Inventing the Flat Earth: Columbus and Modern Historians.* New York: Praeger.

Fortune-Telling. See Divination

Fossils

The modern category of "fossil" emerged only slowly from its original meaning of "something dug up." Originally, the category of fossil included crystals and other unusually shaped stones. Individual types of curious stones attracted superstitions. For Roman natural historian Pliny the Elder (d. 79 CE), a type of fossil shaped like a ram's horn was called the Horn of Ammon and had the power to make dreams come true. (These fossils are now known as ammonites.) Another common fossil, shark's teeth, was called glossopetrae by Pliny after their purported resemblance to the human tongue. He recounted superstitious beliefs that they fell from the heavens during the waning of the Moon, could be used for divination, and had the power to calm gales, although he was skeptical about all of them. Interest in strangely shaped stones continued through the Middle Ages and Renaissance. The German mining engineer Georgius Agricola's *On the Nature of Fossils* (1546), for example, included many unusual stones. These objects were avidly collected as curiosities, or "jokes" of nature, and a fossil resembling a part of a living being could be considered the same kind of object as a fruit shaped like a human head.

Fossils by the modern definition—the petrified remains of what were once living things—began to emerge as a separate category in the seventeenth century. The major split in the study of fossils was between those, such as the Jesuit natural philosopher and collector Athanasius Kircher (1602–1680), who believed that fossils were formed within the Earth by a "plastic virtue" shaping stone into the forms of living things, and those such as the English polymath Robert Hooke (1635–1703), who believed that fossils were the remains of actual living things. The living things did not have to be ancient; some argued that the eggs or seeds of animals and plants could be caught in stone and develop into a stone version of the original creature. This did not explain how fossils of marine creatures could be found on mountaintops many miles from the ocean or deep within solid rock, however. The idea of fossils as possible remnants of what had been once living things also conflicted with biblical chronology. It was difficult to place all fossils in the short history of the Earth that early modern Christians accepted. When did all these fossils appear in the few thousand years from creation to the seventeenth century? Noah's Flood explained some fossils and was particularly helpful in explaining marine fossils on dry land, but it was still difficult to explain fossils of unknown creatures, as early modern science lacked the concept of extinction. Some late seventeenth-century people continued to uphold the theory that fossils were formed by the Earth itself. With the general secularization of knowledge in the eighteenth-century Enlightenment, the idea of fossils as formerly living things became more widely accepted, but the lack of a theory of extinction still posed problems into the early nineteenth century.

See also: Otoliths, Fish

Further Reading

Findlen, Paula. 1990. "Jokes of Nature and Jokes of Knowledge: The Playfulness of Scientific Discourse in Early Modern Europe." *Renaissance Quarterly* 43: 292–331; Oldroyd, D. R. 1996. *Thinking about the Earth: A History of Ideas in Geology*. Cambridge, MA: Harvard University Press.

Fountain of Youth. See Bathing

Foxes

Foxes have played a prominent cultural role in many societies, frequently identified with cunning and supernatural power. They were among the animals associated with witchcraft, as witches were believed to be able to turn themselves into foxes. In Western physiognomy, an individual who bears a resemblance to a fox is cunning.

A sudden shower of rain appearing when the Sun is shining is believed to accompany a "fox's wedding," and it is wise to stay indoors when it happens, as the foxes prefer their privacy and can punish those who violate it. This idea is found in a variety of Eurasian societies, stretching from Japan to England with numerous variations. Spanish Basques associate the double rainbow rather than the Sun shower with the fox's wedding, and in Korea, the animal getting married can be either a fox or a tiger. In India, the event is referred to as the "jackal's wedding."

The fox was a source of magical and medical remedies and charms. For medieval Europeans, the fox's internal organs, particularly the liver and lungs, were of medicinal value. In England, applying a fox's tongue cured cataracts, and the fat of a fox, rubbed on the scalp, cured baldness. A North American version of this superstition recommended pouring the liquified fat of a fox into a person's ear to cure earache. There was a Jewish legend that hanging a fox tail between a horse's eyes would protect the animal from the evil eye. In Persia, fox dung was an ingredient in aphrodisiacs. South Indians carried the nasal bone of a fox or jackal with them as a protective charm. The Han Dynasty Chinese physician Hsu Shen lists three uses for the fox: Eating its flesh cures ulcers, its liver can cause those who have suddenly died to revive, and its blood refreshes the drunken. In twelfth-century China, a self-proclaimed magician admitted under investigation from a magistrate that the source of his magical power was a jar of fox saliva in his possession.

Ancient Persian Zoroastrians classified the fox, along with the hedgehog, the otter, and several other animals as a variety of dog and held it, like other "dogs," in respect as an enemy of demonic powers. Islam was somewhat less favorable, as the fox was associated with cunning and cowardice. In the Islamic tradition of dream interpretation, the fox is a symbol of an attractive woman or a deceitful man. (In *Oneirocritica* of the second-century CE Greek pagan Artemidorus, a work of dream interpretation that greatly influenced the Islamic tradition, a fox is the symbol of a cunning enemy just as a wolf was the symbol of a violent one.) For a man to dream of a fox jumping into his shoes is a sign that his wife will be unfaithful. The nineteenth-century American dream book *The Golden Wheel Dream-book and Fortune-teller* agreed with these negative evaluations, associating foxes with thieves, lewd women, unfaithful servants, and false friends. In Brazil, the local fox, the "crab-eating" fox, is associated with bad luck.

In East Asia, the fox was a powerful supernatural being. The fox was a liminal creature, existing in the space between wild and domesticated, human and beast, and male and female. In China, the fox was believed to have an extremely long life, extending to as long as a thousand years. Chinese writers wrote of the long life span of the fox and of the powers it gained at particular milestones, the power to change to a human form after five hundred years (or in another version, fifty years) and ascension to heaven as a "celestial fox" after a thousand years.

The fox was identified with yin and sought yang to balance its powerful yin. This meant that foxes, according to some authorities male foxes as well as female, assumed the guise of attractive women to seek out men and absorb their yang energy in order to balance the yin, a necessity for living a long life. The sexual power of beautiful women over men was frequently identified with "foxes." However, foxes could also assume male guise. Foxes in male guise were also seductive, and like female foxes, their sexual power represented a threat to Chinese families. They were less likely to be portrayed sympathetically. Fox spirits were even objects of local cults, although these cults were discouraged by state authorities. Malevolent foxes could work evil magic on people by cutting off their hair, a practice Chinese people believed to be a characteristic of sorcerers generally.

East Asia also has legends of a nine-tailed fox. In ancient China, eating a nine-tailed fox protected against poison, although the appearance of a nine-tailed fox was also a good omen. The Queen Mother of the West, a prominent deity in early China, had a nine-tailed fox as one of her attendants. The fox was considered auspicious because it turned its head to its place of origin while dying, providing an example of piety. Unusual foxes of varying colors or larger size than normal were presented to Chinese courts as auspicious animals.

Beliefs about foxes in Korean culture were more uniformly hostile than in Chinese culture. Korean shape-shifting foxes were almost always portrayed as malevolent. The fox of Korean legend receives another tail for every century it lives, and when it has nine it can become human. First, however, it must eat a human organ.

Japanese beliefs about shape-shifting female foxes and their ability to seduce men were similar to those of China. To detect a shape-shifting fox, note that it could not pronounce the word "moshi," and the common use of moshi-moshi in conversation was a way of demonstrating to the interlocutor that he or she was not speaking with a fox. The belief that foxes could start fires by striking the grounds with their tails, also known in China, was common in Japan. (The association of foxes with fire, possibly deriving from the color of the red fox, is also found in the biblical story of Samson tying firebrands to foxes' tails to set the fields of the Philistines alight.) In rural Japan, some families were believed to have prospered over the generations through their alliance with fox spirits. Such families were viewed with suspicion. Malevolent fox spirits could cause mental troubles for humans by possessing them. However, the associations of the fox in Japanese culture are not purely negative. Small, coin-like metal amulets depicting Inari Okami, the god of foxes and prosperity, as a fox are believed to bring wealth.

See also: Amulets and Talismans; Defensive Magic; Dreams; Evil Eye; Physiognomy; Witchcraft: Yin/Yang

Further Reading

Blust, Robert. 1999. "The Fox's Wedding." *Anthropos* 94: 487–499; Johnson, T. W. 1974. "Far Eastern Fox Lore." *Asian Folklore Studies* 33: 35–68; Kang, Xiaofei. 2006. *The Cult of the Fox: Power, Gender and Popular Religion in Late Imperial and Modern China.* New York: Columbia University Press; Uther, Hans-Jorg. 2006. "The Fox in World Literature: Reflections on a 'Fictional Animal'." *Asian Folklore Studies* 65: 133–160.

Frogs

The many species of frogs are found throughout the world and have attracted numerous legends and superstitions.

The Roman Pliny the Elder (d. 79 CE) discussed frogs in his *Natural History.* Pliny stated that frogs were dissolved into slime for six months out of the year and then reconstituted into their original form. He also believed that frog meat was an antidote to many naturally occurring poisons, such as those of scorpions. Pliny recounts many superstitions and magical practices involving frogs, although he distances himself by ascribing these beliefs to others. He described the Magi of Persia as believing that a man could keep his wife faithful by piercing a frog from its sexual organs to its mouth with a reed and then dipping the reed in his wife's menstrual blood. Pliny ascribed to unnamed authors the belief that a large frog he called a "bramble frog" had two bones with special powers. One would instantly cool a pot of boiling water by being thrown in it, while the other would instantly bring a cool pot of water to boil the same way. The first bone would also cool fevers and diminish sexual desire, while the second bone would assuage the anger of barking dogs and act as an aphrodisiac. Such were the many uses of the frog, said Pliny in one of his rare jokes, that they should be considered more useful to the commonwealth than laws.

Frogs had a particular meaning to Jews and Christians, due to the plague of frogs with which God afflicted Egypt in the Book of Exodus. Christian attitudes to frogs are also influenced by the reference to "unclean spirits, like frogs" emerging from the mouths of the dragon, the beast, and the false prophet in Revelation 16. Frogs were associated with divine warnings. Jewish lore expanded on the description of the plague of frogs in Exodus by describing how all the frogs that plagued Egypt were descended from a single frog or emerged from a single monstrous frog. Rains of frogs continued to be a prodigy, indicating God's displeasure. When frogs covered the lawn of a magistrate who had persecuted groups of Christians dissenting from the state church in 1661, it set off a major pamphlet controversy, pitting those who saw the invasion of frogs as a divine sign against those who claimed that there were natural explanations. Muslims were more positive about frogs. In Islamic dream interpretation, frogs represented pious and holy people. However, plagues of frogs could have meaning outside the world of Abrahamic religions—the fall of the kingdom of Baekje in the Korean peninsula in 660 CE was preceded

A Love Charm

"Take a healthy, well-grown frog. Place it in a box which has been pierced all over with holes with a stout darning needle or gimlet. Then carry it in the evening twilight to a large ant-heap, place it in the midst of the heap, taking care to observe perfect silence.

After the lapse of a week, repair to the ant-heap, take out the box, and open it, when in place of the frog you will find nothing but a skeleton. Take this apart very carefully, and you will soon find among the delicate bones a scale shaped like that of a fish and a hook. You will need them both. The hook you must contrive to fasten in some way or other into the clothes of the person whose affections you wish to obtain, and if he or she has worn it, if it is only for a quarter of a minute, he will be constrained to love you, and will continue to do so until you give him or her a fillip with the scale.

This method is over three thousand years old, and it has been practised by thirty-thousand of our ancestors with the most complete success."

Source: Fontaine, Felix. 1862. *The Golden Wheel Dream-Book and Fortune Teller.* New York: Dick and Fitzgerald.

by the appearance of thousands of frogs in the treetops, according to one chronicler.

Frogs were widely believed to have medicinal value. In England, it was believed that putting a frog in a child's mouth would cure thrush. Swallowing live frogs or a soup made out of frogs was good for a variety of complaints, including stomach troubles, tuberculosis, and whooping cough. Indigenous people in Amazonia applied the poison of the highly toxic frogs of the area to their skin as a medical treatment bringing general health and fertility, as well as bringing good luck on hunters and protecting from evil spirits. (A version of this practice has spread to the modern West, where it is known as "Kambo.") In traditional Chinese medicine, the fat found near the fallopian tubes of some frog species, known as "hasma," is believed to have several uses when consumed, including treating respiratory diseases, increasing the beauty of skin, and restoring strength to women exhausted by childbirth. Hindus, relying on the affinity of frogs for water, believed frogs effective in rainmaking rituals.

See also: Prodigies; Spontaneous Generation; Toads

Further Reading

Burns, William E. 2002. *An Age of Wonders: Prodigies, Politics and Providence in England, 1657–1727.* Manchester: Manchester University Press; Schwartz, Donald Ray. 2000. *Noah's Ark: An Annotated Encyclopedia of Every Animal Species in the Hebrew Bible.* Northvale, NJ: Jason Aronson.

G

Garlic

The "stinking rose" has a long history during which it has been the subject of many superstitious beliefs, many influenced by its distinct and pungent aroma.

The Roman natural historian Pliny the Elder (d. 79 CE) asserts that garlic was believed to be effective in treating many diseases. He lists sixty-one remedies for it, one of the highest numbers for any plant he discusses. Its smell drove away poisonous snakes and scorpions. It cured the wounds made by wild animals, either consumed by the victim or administered to the wound. Garlic eaten by women who had just given birth encouraged the expulsion of the afterbirth—an opinion Pliny ascribes to the "Father of Medicine," Hippocrates. Raw garlic was a treatment for madness, and garlic was also good for respiratory conditions and expelling worms. Animals as well as humans could benefit from garlic; Pliny states that applying garlic to the genitals of a beast of burden would ease the passage of urine. Other ancient authorities state that it encouraged the menstrual flow. Greeks and Romans believed garlic enhanced courage, leading soldiers to chew it before a battle. In terms of the Aristotelian qualities, garlic was considered a hot and dry substance. In Galenic medicine, it was particularly recommended for those with cold and wet constitutions, dominated by the humor of phlegm. Those with hot, dry constitutions were to avoid it. It was also believed to protect against plague. In the early Middle Ages, Anglo-Saxon medical writers recommended a mixture of garlic, onion, and goose fat poured into the ear to treat earache. The belief that garlic is effective against the whooping cough is found as late as the twentieth century. Persians believed that garlic warded off evil and so buried their dead with garlic in their mouths to avoid putrefaction. Astrologically, garlic was identified with the planet Mars.

Garlic is widely used in traditional Chinese medicine, where it is considered a hot ingredient of a yang nature. Legend ascribed its cultivation to the Yellow Emperor, the founder of the Chinese medical tradition. It is believed to aid the digestion. Garlic was particularly beneficial for the spleen, kidney, and stomach. Like Europeans, Chinese believed in the power of garlic against the plague. Garlic is also widely used in the Indian ayurvedic tradition to treat respiratory and heart conditions, among others.

Greek writers under the Roman Empire claimed that anointing a magnet with garlic deprived it of its power to attract, an example of an "antipathy" between two natural substances. An early Byzantine source claimed that the magnet could be reactivated with goat's blood. Another medieval belief was that vinegar could restore the power of a garlic-affected magnet. A variation on the theory of garlic's effect held that anointing a magnet would cause it to repel rather than attract iron. Medieval natural philosophers tried to explain this phenomenon; the Arab physician and philosopher Ibn Rushd (1126–1198), known in the west as Averroes, claimed that contact with garlic changed the material of the magnet, rendering it unable to function. This belief in garlic's antimagnetic properties persisted into the early modern period; experimental disproof was greeted with attempts to save the belief by restricting the effect to particular types of magnet or particular types of garlic. As late as the eighteenth century, garlic was banned from ships so as not to interfere with the compass.

Christians viewed garlic with suspicion, one legend saying that when Satan was expelled from the Garden of Eden, garlic grew from his left footprint and onions from his right, a belief that drew upon the traditional hostility to the left side, and it also occurs in an Islamic context. In Islamic dream interpretation, dreaming of garlic is a bad omen of future suffering. Jews had a more positive attitude to garlic, believing that eating garlic enhanced male virility and fertility. The Talmud recommended chewing it on Sabbath eve as a preparative for marital sex. Such was the popularity of garlic among ancient Jews that Jews and Samaritans, a closely related religious group, were referred to collectively as "garlic eaters." Garlic's pungent odor, however, has contributed to the belief in many religious traditions that it is forbidden to consume it before entering a sacred place.

Perhaps due to its smell, garlic was strongly identified with defensive magic. It appears as an ingredient in defensive spells as early as sixth century BCE Egypt. In Celtic Britain, garlic at the threshold of the door of the house, whether growing in front of it or hung above it, protected the house from evil spirits. In the Mediterranean world, garlic was believed to protect against the evil eye. Braids of garlic were hung in houses as a precautionary measure. Its best-known defensive use is against vampires. Garlic was worn or hung in the house to keep vampires away, and cloves of garlic were stuffed into the mouth, nose, and eyes of corpses to prevent them from returning as vampires. If a person was suspected of being a vampire, giving them a clove of garlic to chew would reveal the truth, as a vampire would refuse. Garlic appears in Bram Stoker's novel *Dracula* (1897), which created a great deal of the modern vampire myth, as an antidote to Dracula's power. Stoker was drawing on centuries of folk-belief in garlic's antipathy to vampires and harmful magic generally. Belief in garlic's power, however, is not restricted to European traditions; it is also used as a defense against the Filipino monster known as the manananggal.

See also: Defensive Magic; Evil Eye; Herbs; Humours, Theory of; Left-Handedness; Magnetism; Vampires; Yellow Emperor

Further Reading

Hobbs, Christopher. 1992. "Garlic—The Pungent Panacea." *Pharmacy in History* 34: 152–157; Sander, Christoph. 2020. "Magnets and Garlic: An Enduring Antipathy in Early-Modern Science." *Intellectual History Review* 30: 523–560.

Gems. See Precious Stones

Gender and Sexual Difference

Although all societies have recognized the distinction between male and female, this distinction has been understood in many ways. Most male-dominated societies viewed women as inferior in some ways, but the way this was conceptualized and expressed and its implications for social relations varied widely.

As a category system, gender has also mapped on to many other category systems, particularly dualistic ones. In China, women were identified with yin energy and men with yang, although both sexes relied on both energies to flourish. (The increasingly negative association of yin has been related to social tendencies toward male domination.) In the West, women were identified as "cold and wet" and men as "dry and hot" in the system of the four "Aristotelian" qualities. Masculine and feminine have also served as categories for things that lack an obvious sexual identity. This operates on a deep level in those languages that, unlike English, gender their nouns. However, the idea of gendering nonbiological entities has been particularly influential in the magical tradition. Magicians, astrologers, and alchemists have classified the entities they deal with as masculine or feminine. Masculine entities, like the planet Mars, tend to be more associated with violence and feminine ones, like Venus, with sexuality.

Religion has had a pronounced influence on the way societies have thought about gender difference. The Babylonian creation myth centered on the struggle between a male godlike hero, Marduk, and a female dragon, Tiamat. Marduk's victory has been considered a precedent and justification for a male-dominated society. Confucianism identifies man with heaven and woman with Earth, seeing their relationship as not only complementary but also hierarchical, with heaven/man occupying the unchallenged seat of authority. In Abrahamic religions, the justification for gender hierarchy has drawn on the story of the fall of humanity, for which the first woman, Eve, was held responsible and led to the idea of both female corruption and the divinely sanctioned subordination of women to men. The Eve myth also contributed to the idea that women are more subject to the passions, including sexual passions, than "rational" males. (In Judaism, Lilith personifies the destructive force of unchained female passion.) Some Buddhist traditions hold that women cannot attain enlightenment and need to be reborn as men first, although not all Buddhists accept this. The belief that women are "impure" is found across many religious traditions and is associated with specifically female biological functions such as menstruation and childbirth. This belief is not merely characteristic of learned religious authorities but is found across all levels of many societies and among men and women alike. It contributes to the belief that women were particularly likely to be casters of the evil eye or, in some societies, witches.

Medical and philosophical theories of gender difference draw on both the anatomical differences between men and women and on differences in their sexual and reproductive functions. In the Western tradition, one basic distinction is between those models that emphasized the similarities of the sexes and those that emphasized their differences. The similarity school drew on Aristotelian ideas and the work of the ancient physician Galen (129–216), particularly his anatomical treatise *On the Use of Parts*. Similarity theorists believed women to be incomplete men, whose organs had not been fully pushed out due to insufficient heat. Male and female sexual organs were "homologous": the vagina was an inside-out penis and the ovaries were female testes. Male nipples were analogous to the female breasts. Biological processes as well as organs were common to male and female—menstruation was the equivalent of hemorrhoidal bleeding. Another consequence of similarity thinking was the emphasis placed on the female orgasm in conception. Since a man must climax to beget a child, a woman must also climax to conceive one. The similarity school believed in the possibility of spontaneous sex changes from female to male caused by a sudden excess of heat. The idea of woman as an incomplete version of the perfection of man also received support from the Bible, in Genesis 2:22, in the story of the creation of Adam from Eve's rib. (This led some to argue that men had one fewer rib than women, although the empirical evidence was strongly against this claim.)

Difference thinkers, a more intellectually varied group with less ancient textual authority, conceived of men and women as different and complementary, at least in their sexual and reproductive roles. Difference thinkers minimized the importance of structural similarities and denied the possibility of spontaneous sex changes. They emphasized those characteristics that clearly divided women from men—menstruation and the ability to bear young. This kind of thinking dominated medicine in the Latin Middle Ages, when Galen's *On the Use of Parts* was not widely circulated or read. Galenic ideas about homology had more influence among the Arabic medical writers of the Islamic world. Difference theorists did often retain the Aristotelian categories of cold and heat and wet and dry to explain sexual difference. Rather than the story of the rib, religious thinkers emphasizing the differences between men and women drew upon the alternative egalitarian passage in Genesis 1:27, "Male and Female created He them," to claim that since God created woman, and God created all things perfect, women were equally perfect as men. Cases of spontaneous sex change were of merely hermaphrodites whose male organs had been concealed and suddenly emerged or possibly women with enlarged clitorises, or simply frauds. Difference theorists rejected the idea that women were imperfect men but did not usually conclude from this that women should receive equal treatment or exercise equal power in society. It was possible for both participants in a hierarchical relationship to be equally perfect.

The Renaissance revival of ancient Greek medical writing led to a revival of the one-sex theory of human sexual difference. However, the increased attention Western physicians and anatomists gave to the clitoris after its "discovery" by anatomists in the late sixteenth century complicated this picture as the clitoris competed with the vagina for the title of "female penis." Theories emphasizing difference and those emphasizing similarity continued to be put forth through the Scientific Revolution, but the Enlightenment period saw a sharpening of the scientific concept of sexual difference, with the sexes increasingly defined as not only different but also opposite. Enlightenment male experts viewed the male as the norm and the female as a radically different kind of being. The one-sex model disappeared from scientific and learned medical discourse by the early nineteenth century (although it persisted in popular culture and medical handbooks aimed at a popular audience). The idea that women, like men, needed to climax sexually for a child to be conceived was also abandoned by many learned physicians and scientists, and women were increasingly regarded as the less sexually passionate of the two genders. This was a startling inversion of traditional Western thinking on the genders, which had identified women's sexual lusts as much greater than those of men. The Enlightenment also saw the dominant rationale for women's social subordination shift from a religious one to a scientific and medical one based on the idea of natural differences between men and women.

Ideas about the cosmic and transcendent nature of maleness and femaleness did not disappear in the Enlightenment. *Naturphilosophs* and other Romantic scientists, who saw a universe structured by cosmic dualities, identified these dualities as male and female and in the fashion of Western esoterists correlated them with other dualities. *Naturphilosophs* conceived of the animate universe as a pregnant female.

See also: Alchemy; Aphrodisiacs; Astrology; Curses; Devil's Mark; Diana; Evil Eye; Intersex Conditions; Lilith; Love Magic; Menstruation; *Naturphilosophie*; Physiognomy; Pregnancy and Childbirth; Sex Change; Syphilis; Witchcraft; Yin/Yang

Further Reading

Cadden, Joan. 1993. *Meanings of Sex Difference in the Middle Ages: Medicine, Science, Culture*. Cambridge, UK and New York: Cambridge University Press; Furth, Charlotte. 1999. *A Flourishing Yin: Gender in China's Medical History, 960–1655*. Berkeley: University of California Press; Laqueur, Thomas. 1990. *Making Sex: Body and Gender from the Greeks to Freud*. Cambridge, MA: Harvard University Press; Park, Katharine. 2010. "Cadden, Laqueur, and the 'One-Sex Body.'" *Medieval Feminist Forum* 46: 96–100; Schiebinger, Londa. 1989. *The Mind Has No Sex?: Women in the Origins of Modern Science*. Cambridge. MA: Harvard University Press; Schiebinger, Londa. 1993. *Nature's Body: Gender in the Making of Modern Science*. Boston: Beacon Press; Schleiner, Winfried. 2000. "Early Modern Controversies about the One-Sex Model." *Renaissance Quarterly* 53: 180–191.

Geocentrism

Geocentrism, the belief that an unmoving Earth is the center of the universe, is the most obvious conclusion to draw from the evidence of the senses. People see the fixed stars, the Sun, Moon, and planets all revolving around the Earth, which seems immobile. Nongeocentric cosmologies are historically rare.

Although geocentrism has existed ever since human beings have been thinking about the universe, it was incorporated into Western science by the ancient Greeks. The physics of Greek philosopher Aristotle (384–322 BCE), the dominant physical

Thomas Burnet and the Sacred Theory of the Earth

The English geologist Thomas Burnet (c. 1635–1715) is best known for his attempt to write a history of earth's geological development in harmony with the biblical narrative, *Telluris Theoria Sacra* (1681), "Sacred Theory of the Earth." Burnet was a catastrophist whose narrative of biblical geology focused on the Flood of Noah and the subsequent restoration. He described the world before the flood as perfectly smooth and featureless and without seasons as its axis was perpendicular to the line connecting it to the Sun. The flood itself included the breaking forth of subterranean waters in addition to the forty days and nights of rain. Burnet claimed that the uneven and jagged surface of the present-day earth was a result of the flood, as was the tilted axis, and that the flood waters had retreated underground. He was also a millenarian, who believed that soon there would be a massive eruption of the world's volcanoes that would burn its surface to ash. The ash would settle in a way to restore the earth's perfect roundness, and for the thousand years before the Last Judgment, Paradise would be restored on Earth.

system in the Western and Islamic world for centuries, rested on a geocentric foundation, as, for Aristotle, heavy bodies fell because they sought their natural place at the center of the universe, which was the center of the spherical Earth. Another Greek, Ptolemy (98–168 CE), mathematically worked out the implications of geocentric astronomy, describing the orbits of the Sun, Moon, and planets around the Earth. The outer stars were carried on a sphere that rotated around the Earth once a day. Not all Greek astronomical and physical systems were strictly geocentric; the atomistic physics of Epicurus (341–270 BCE) did not provide for the universe having a center as it was infinite. However, Epicurus was considered both religiously and scientifically marginal. There were a few Greek scientists such as Aristarchus of Samos (c. 310–230 BCE) who challenged the geocentric consensus in the name of heliocentrism, but they were a decided minority.

In addition to science, geocentrism also had the support of religion. It was endorsed by the Bible, the sacred text of Judaism and Christianity. A standard proof text was Joshua 10, 12–13, in which Joshua bid the Sun stand still in the sky. With the combined authority of Greek science and religion, geocentrism reigned supreme in both medieval Christian science and the science of medieval Islam whose sacred book, the Quran, also contains geocentric passages describing the paths of the Sun and Moon around the Earth. Hindu and Buddhist cosmologies, although very different from the Western biblical/Aristotelian cosmos, are also geocentric. Medieval Hindu India did produce one possibly heliocentric astronomer, Aryabhata (476–550), but this is disputed and did not lead to the founding of a nongeocentric school.

The revival of heliocentrism by Polish astronomer Nicolaus Copernicus's (1471–1543) *On the Revolutions of the Celestial Spheres* (1543) at first had little effect on the geocentric consensus. The modest advantages it had over geocentrism in terms of astronomical calculations were more than balanced by the fact that giving up geocentrism would require abandoning Aristotelian natural philosophy and starting physics over from scratch. The lack of a "stellar parallax," a shifting of the stars against the background of the sky caused by the Earth occupying different positions at different times of the year, also seemed to tell against the Copernican view. For this reason, the Copernican system was often presented as a way to make physical calculations rather than a true picture of the physical universe. Geocentrism was not usually defended on religious grounds in the sixteenth century simply because there was no need to.

The sixteenth century also saw the first radical innovation in geocentrism since the Greeks. Danish astronomer Tycho Brahe (1546–1601) modified geocentric cosmology by putting the planets in orbit around the Sun, which in turn, along with the Moon, revolved around the Earth. This system offered the mathematical advantages of Copernicanism while avoiding the problems of the Earth's apparent motionlessness and the stellar parallax.

The religious defense of geocentrism became more important in the seventeenth century as heliocentrism was attracting increasing interest from scientists such as Tycho's German disciple Johannes Kepler (1571–1630). The defense of geocentrism was particularly strong in the Catholic world where it became bound up with the post-Reformation defense of the Catholic Church's supreme authority to interpret the Bible. The Church began to enforce geocentric astronomy and physics on its intellectuals. Tycho's theory would become the most popular geocentric system in the Catholic Church, ironic because Tycho was a Protestant and Copernicus a Catholic priest. The most dramatic example of the Church's support for geocentrism was the 1633 trial of Italian astronomer Galileo Galilei (1564–1642) before the Roman Inquisition. The elderly scientist was forced under the threat of torture to publicly renounce heliocentrism. The Orthodox Church also continued to support biblical geocentrism. The patriarch of Constantinople condemned Methodios Anthrakides (1660–1736) for heliocentrism among other offenses in 1723.

Geocentrism was defended far less strongly in the Protestant world, where the fact that the Catholic Church supported it became a point against it. In 1687, English natural philosopher Isaac Newton (1642–1727) published *Mathematical Principles of Natural Philosophy*, providing a powerful system of heliocentric physics. In the eighteenth century, the invention of more accurate telescopes solved the problem of detecting the stellar parallax. Even the creation of fundamentalism in the late nineteenth century failed to revive geocentrism in the Protestant world, despite its reverence for the literal interpretation of the Bible. Geocentrism in the Protestant world became and remains a marginal position of extreme biblical literalists on the fringes of creationism. In the Catholic world, the growing scientific consensus in favor of heliocentrism made the prohibition on it increasingly a dead letter, particularly in Catholic countries such as France where the Church's institutional authority was relatively weak. The Catholic Church formally permitted the teaching of heliocentric astronomy in 1822, marking the end of the Catholic geocentric tradition. Among Catholics, as among Protestants, geocentrism is now only held by a small marginalized fringe. There is also a small minority of geocentrists among Orthodox Jews, particularly the Lubavitcher sect, and among Salafist Muslims.

In most Orthodox Christian and non-Christian societies, geocentrism held sway until the introduction of Western science, two of the places where this happened earliest being Russia and Japan where heliocentrism was introduced in the eighteenth century.

See also: Flat Earth; Sun

Further Reading

Graney, Christopher M. 2015. *Setting Aside All Authority: Giovanni Battista Riccioli and the Science against Copernicus in the Age of Galileo.* Notre Dame, IN: Notre Dame University Press; Grant, Edward. 1994. *Planets, Stars and Orbs: The Medieval Cosmos, 1200–1687.* Cambridge, UK and New York: Cambridge University Press.

Geomancy

Geomancy is a system of divination based on reading lines randomly scratched in the dirt or objects scattered with the hand. The results were reduced to an ordered set of four binaries, in a manner similar to that of the *Yijing*, although the *Yijing* generated hexagrams and geomancy tetragrams. There are sixteen types of figures in all, each of which is associated with certain meanings as well as connected with other occult or natural phenomena such as planets, astrological signs, elements, or humours. Figures can also be classified as day and night, good and bad, and male and female. Four figures, referred to as the "Mothers," were generated randomly and

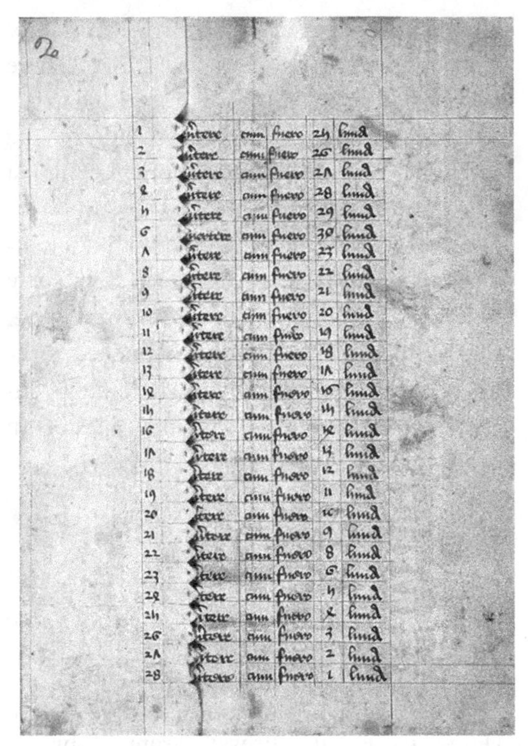

A page from a manuscript of the Experimentarius, a medieval Latin translation of an Arabic geomantic text. (Manuscripts and Archives Division, The New York Public Library. New York Public Library Digital Collections.)

then manipulated according to set rules to generate twelve more to form a full geomantic layout of sixteen four-line figures. The figures were then interpreted according to their intrinsic meaning, their relation to each other, and the nature of the query. Geomantic schools differed, so a layout could have different meanings depending on the interpreter.

Although practiced in several cultures, it was most popular in the medieval Islamic world, where it was referred to as *'ilm al-raml*, the "science of the sand." According to legend, it had been founded by Idris, an Arabic name for the prophet known in the West as Hermes Trismegistus. Another figure invoked as the founder of the science was the Jewish prophet Daniel. Some writers even identified the first geomancer as Adam, who had been taught the science by God, before it had been lost to be rediscovered by later prophets. Geomancy may have originated in North Africa, and experts from that part of the world were frequently invoked as the authorities, even as far east as Persia. Geomancy was closely linked with other Islamic occult sciences, such as lettrism and astrology. Many astrologers also practiced geomancy. The figures were linked with other elements of the cosmos, such as the four elements, the four humors, or the twelve zodiacal signs. Most Muslim jurists accepted the Islamic legitimacy of geomancy, but some regarded it as forbidden. The historian and opponent of occultism Ibn Khaldun (1332–1406) argued that as a pre-Islamic divinatory technique geomancy had been made obsolete by the Quran.

At its height in the early modern period, geomancy in Arabic and Persian writings spread through the entire Islamic world, from Central Asia to Morocco. A relatively inexpensive science, it was practiced among ordinary people and among the early modern courts of the Ottoman, Safavid Persian, and Mughal Indian rulers. Geomancy entered the Christian world along with other Arabic sciences in the twelfth century. In the early modern period, it appeared in the spurious fourth book of the German magician Heinrich Cornelius Agrippa's *On Occult Philosophy*, published in 1559. The English magician Robert Fludd (1574–1637) also wrote on the subject. European geomancers distributed the first twelve figures of the layout into astrological houses. Interest in the West faded with the Enlightenment, but geomancy was part of the nineteenth-century occult revival. It also influenced West Africa, where it provided the foundation of Ifa divination and other systems of divination. Geomancy continues to be practiced in the Islamic world, although its prestige has fallen considerably in the modern era.

See also: Astrology; Elemental Systems; Hermes Trismegistus; Humours, Theory of; Ifa Divination; Lettrism; *Yijing.*

Further Reading

Melvin-Koushki, Matthew. 2018. "Persianate Geomancy from Ṭūsī to the Millennium: A Preliminary Survey." In Nader El-Bizri and Eva Orthmann, eds. *Occult Sciences in Premodern Islamic Culture.* Beirut: Orient-Institut Beirut. Pp. 151–199.

Ghosts

Ghosts, spectral reappearances or returns of the dead, appear in many cultures. Ghosts appear for many reasons, including vengeance on those who murdered them or otherwise caused their deaths, or to carry out a task they failed to complete in life. Unlike other "undead" monsters such as the vampire, the ghost is not a corpse and is usually treated as having no material substance, thus enabling it to move through walls and closed doors. Sometimes a ghost can appear to one person but not to another accompanying them or can be perceived by animals, particularly dogs and horses, while remaining unperceived to people. The appearance of ghosts is also typically at night. The ghost also frequently appears in literature, as the genre of the ghost story appears in many cultures.

Some areas become famously haunted, often by particularly well-known historical personalities. King Henry VIII's second queen, Anne Boleyn (c. 1501–1536), is said to haunt several locations in England. Among them is the Tower of London, where she was held before her execution. The tower is reputedly among the world's most haunted locations, with several ghosts including that of Anne identified as appearing there.

Ghosts were known to the ancient Mesopotamians, who did not, however, view them as much of a threat. If a ghost interfered with the living, it was usually because the living had failed to honor it with sacrifices. Angry ghosts could cause sickness, and it was the responsibility of the Mesopotamian physician to discover what ghost or other supernatural being his patient had angered to cause his disease. Ancient Egyptians also had the idea of an angry ghost returning to punish the living or promote the redressing of a wrong.

There is only one appearance by a ghost in the Hebrew Bible, but it is a dramatic one—the summoning of the ghost of the prophet Samuel before King Saul by the Witch of Endor. Although the story of Samuel's ghost would be frequently invoked by later champions of ghosts' existence, it was also possible to interpret the ghost as a fake, a demonic spirit taking on the prophet's guise. For ancient Greeks and Romans, ghosts could appear in dreams when they wanted the living to do something. In Homer's epic *Iliad*, the ghost of the slain warrior Patroclus appears in the dreams of the champion Achilles to demand a proper burial. Romans enjoyed ghost stories and distinguished between ghosts who appeared in dreams, who were generally benevolent, and ghosts that appeared in the waking world, who were generally angry about something, whether an unavenged murder or improper burial.

Ghost belief continued into early Christianity. No ghosts appear in the New Testament, but the resurrected Christ is mistaken for a ghost, and although he denies being one, he does not deny their existence. Many of the Fathers of the Church, however, denied the existence of ghosts, instead claiming that they were demonic illusions, devils in disguise.

In the Middle Ages, the idea spread that a ghost appeared because it required something to be done before it could be eligible to enter heaven, such as the righting of a wrong or even just the saying of a Mass. This was associated with the rise of the doctrine of purgatory, an intermediate state in which souls ultimately destined for Heaven could do penance for their sins on Earth. Carrying out the instructions of a ghost was one way the living could help the dead pass through the agonies of purgatory and enter heaven. Medieval ghosts sometimes appeared in dreams, but those who appeared in the waking world were given more credit. They often appeared to relatives or other people they had known in life who might be persuaded to carry out the tasks the ghost needed to be done or at monasteries, where the monks' prayers were believed to be particularly powerful. When the ghost's request had been carried out, it could change its clothing from black to white, to indicate its passage through purgatory and attainment of heaven. Ghost stories in the Middle Ages very frequently made a moral point, to encourage the hearer to be a better Christian. Ghosts also became entangled with other Christian ideas, as in the widespread belief that a ghost could not appear on Christmas Eve.

Ghosts became a subject of theological controversy after the Protestant Reformation. Since Protestants denied the existence of purgatory, only allowing for heaven or hell, they denied the traditional Catholic interpretation of the ghost. Thus, Protestants frequently denied the existence of ghosts. The apparitions misclassified as ghosts, Protestant theologians claimed, were either illusions or the tricks of devils. These arguments were put forth in the most influential work on the subject dating from the Reformation, the Swiss theologian Ludwig Lavater's (1527–1586) *De Spectris, Lemuribus et Magnis Atque Insolitis Fragoribus* (1569), quickly translated and published in English as *Of Ghostes and Spirites Walking by Nyght* (1572). There were many other editions and translations into German, French, and Italian as well as English. Catholics responded with collections of anecdotes allegedly demonstrating the existence of ghosts and therefore establishing the doctrine of purgatory. Ordinary Protestants continued to see ghosts, however, as "official" Protestant theology had only a limited impact on ordinary believers, the inheritors of centuries-old ghost lore. In Protestant early modern Scotland, ghosts appeared as spirit guides, sometimes even introducing the guided person to the lands of the fairies.

In the late seventeenth-century and the eighteenth-century Enlightenment, Protestant interest in ghosts revived, as the existence of ghosts or malevolent spirits became a way to refute "atheism" or "Sadducism." Writers such as Joseph Glanvill (1636–1680), author of a collection of supernatural stories called *Sadducismus Triumphatus* (1681), and Richard Baxter (1615–1691), author of *The Certainty of the World of Spirits* (1691), collected accounts of ghosts to refute those who would deny the existence of a spiritual world. Ghost stories also appeared in the Scottish writer George Sinclair's (d. 1696) *Satan's Invisible World Discovered* (1685). Another Scottish writer, Robert Wodrow (1679–1734), collected stories of ghosts that returned after death specifically for the purpose of rebuking atheists. However, as the Enlightenment progressed, ghost stories were increasingly treated as popular superstition.

Ashkenazic Jewish tradition has a ghost called a dybbuk, whose main threat lies in its ability to possess people, particularly women. The dybbuk is a person condemned to wander the Earth for their sins. Some dybbuks are condemned until they perform a particular deed to make up for their sins on Earth. A dybbuk who possesses a person can be exorcised, and exorcising dybbuks is one of the powers of the wonder-working sages of Hasidic Judaism.

Unlike Judaism or Christianity, Islam offers little scope for the role of the ghost, and Islamic experts have denied that ghosts even exist. However, in many Muslim areas, pre-Islamic beliefs about ghosts continue, sometimes in an "Islamized" form. Malayan and Indonesian Muslims have the belief in a vampiric ghost called the Pontianak or the Kuntilanak. This creature is the ghost of a woman who died while being pregnant. The city of Pontianak in Borneo is named after these creatures, who were defeated by Sultan Abdurrahman Alkadrie who then went on to found the city in 1771 in the place where he had defeated the monsters. However, the area is still haunted by them.

The existence of ghosts was debated in ancient China. The philosopher Mozi (470–c. 391 BCE), often regarded as a proto-rationalist, suggested that ghost accounts should be believed if the witnesses were persons of credibility, the same procedure that should be followed in deciding whether to believe anything one had not personally experienced. The Chinese skeptical philosopher Wang Chong (25–100 CE) argued against ghosts on the grounds that the ghost of a murdered person, if as it was commonly asserted it wanted vengeance, would immediately appear before a magistrate, when this in fact never happened. Wang Chong also asserted that if, as was the common belief, more people lived in the past than in the present, ghosts should outnumber the living, when in fact ghost sightings were rare. Wang Chong's arguments, however, had little impact on popular belief in ghosts. Chinese ghosts are similar to Western ghosts in that they are spirits of the dead. Ghosts are usually portrayed negatively. "Hungry ghosts" are greatly feared in traditional Chinese culture. Being reborn as a hungry ghost is a punishment for evil deeds on Earth. There are various schemes for classifying hungry ghosts according to their crimes on Earth and their powers as ghosts. Ghost stories are a popular genre of Chinese literature.

The "Ghost Festival" is a traditional Chinese festival during which ghosts are believed to come to Earth. The seventh month of the lunar calendar is believed to be "ghost month," when the barrier that divides the world of the dead from that of the living is thinner and ghosts walk the Earth. These ghosts are generally not feared, but venerated by the members of their families. They are honored with sacrifice, the burning of incense, and theatrical performances where the front row of seats is left empty for the ghosts. The Ghost Festival has spread to other areas where Chinese have settled or have been influenced by Chinese culture, such as Vietnam and Japan.In Japan, mothers who die in childbirth or when their children are small are believed to come back as ghosts to help care for them. This is typical of Japanese ghosts, known as yurei, who are generally bound to Earth by some emotion or obligation. For example, a person who had not received proper funeral rites might appear as a ghost until the rites had been performed. However, Japanese ghosts are not always harmless or benevolent; the "hungry ghost" exists in Japan just like in China.

Ghosts known as bhoots were known and feared in India, where they could be recognized by their feet being turned backward, by dressing in white, or not casting a shadow in the moonlight (like Western ghosts, bhoots usually appear at night). Unlike ghosts in other traditions, the bhoot is not someone returning from the dead, but a dead person held on Earth before undergoing their ultimate fate, whether that be reincarnation or entrance to heaven or hell. A particularly dangerous and malevolent kind of bhoot is a chudail, the ghost of a woman who died during childbirth. Bhoots have the power to possess the living, and like Western ghosts, they can haunt particular places. Bhoots can be exorcised in such a way as to send them on their journey to the afterlife or their next incarnation. There are various techniques to prevent a dead person from returning as a bhoot. In nineteenth-century South India, some people would bury a pot full of dung, a broomstick, and a firebrand at a crossroads where three roads met to prevent a recently dead member of the family from returning as a ghost.

The blending of African and European cultures in the Americas has produced rich ghost traditions. Jamaica and the anglophone Caribbean generally is the home of a kind of malevolent ghost called a duppy. The duppy is a malevolent spirit that usually appears at night and frightens people by making noises and shaking chains. There is an extensive body of counter-magic to protect people against the duppy.

Ghosts, sometimes known as "haints," were also prominent in the folklore of African Americans, both before and after slavery. (In the post-Civil War south, Ku Klux Klansmen presented themselves as ghosts to frighten freed former slaves, although the freed men were usually not fooled.) Enslaved African Americans feared that particularly cruel masters would return as angry ghosts. The hurried burials of many enslaved people, with their masters taking little care for the performance of funeral rites, encouraged the belief that dead slaves could return as ghosts. Stories about how the ghosts of slaves took revenge on cruel masters were widely circulated. Haints were not always malevolent; however, haints could appear to aid and comfort their living descendants.

The belief that the color blue protects against haints has resulted in many blue-painted houses in the American south. Another way of defending against haints and other spirits was the "bottle tree" with bottles, preferably of blue glass, hung on the branches of a dead tree. The bottle tree can be traced to the Congo region from which many enslaved Southerners came. The haints would be trapped in the bottles until sunrise, when the Sun would dispel them.

See also: Caul; Death; Defensive Magic; Doll Magic; Execution Magic; Monsters; Vampires

Further Reading

Gorn, Elliott J. 1984. 'Black Spirits: The Ghostlore of African American Slaves." *American Quarterly* 36: 549–565; Iwasaka, Michiko and Toelken Barre. 1994. *Ghosts and the Japanese: Cultural Experience in Japanese Death Legends.* Logan: Utah State University Press; McGill, Martha. 2019. *Ghosts in Enlightenment Scotland.* Suffolk: Boydell and Brewer; Schmitt, Jean-Claude. 1999. *Ghosts in the Middle Ages: The Living and the Dead in Medieval Society.* Chicago: University of Chicago Press.

Ginseng. See Herbs

Goats

In the Western tradition, the goat has strong negative associations. This may go back to the ancient Hebrews and the idea of the scapegoat that carried the sins of the people. (Although the Hebrew example is best known, other ancient Near Eastern cultures also had scapegoat rituals.) The biblical reference to the saved sheep on the right hand and the condemned goats on the left at the last judgment also contributed to the goat's satanic associations. Another source of negative associations was the goat's association with lustfulness. The "lustful goat" image goes back to the goat/human combinations of classical mythology, fauns and satyrs, which influenced the Christian depiction of Satan. The goat was the most common animal form taken by Satan in depictions of the Sabbat, the gathering of witches described in early modern demonology and witch trials. The devil was frequently claimed to take the form of a goat at the sabbat to have sex with witches or be kissed by them. In the Basque country between France and Spain, which had a developed culture of witch belief, the place where witches gathered for the sabbat was referred to as *akelarre*, literally "he-goat meadow." Witches were also sometimes believed to be carried on flying goats to the sabbat. In Britain there was a claim that once every day, goats visited the devil to have their beards combed, and for that reason, it was impossible to keep a goat under observation for a continuous twenty-four hours.

The male goat's sexuality was also associated with power; early modern alchemists believed that steel tools were best hardened by being quenched in the blood of a he-goat. The Roman natural historian Pliny the Elder (d. 79 CE), followed by many other authorities, believed that goat's blood softened "adamas," a type of jewel that subsequent interpreters identified as diamond. The English writer on superstition Sir Thomas Browne (1605–1682), who doubts this belief, suggests that it is related to another he finds doubtful—that goat's blood is good for kidney stones. (Pennsylvania Germans treated kidney disease with goat's urine.)

The association of goats was not always negative; keeping a goat on a farm was believed throughout much of Western and Germanic Europe to keep other animals healthy, and in particular to prevent cows and mares having premature births. The rank odor of male goats was sometimes believed particularly powerful. The earliest evidence for these practices is found in the late eighteenth century, but they are probably much older. This use of the goat may be associated with the scapegoat, the farm goat carrying the diseases of the other animals as the scapegoat carried the sins of the community. Goats were also sometimes associated with human health; in the nineteenth century, a "Goat Lymph Sanitarium" in upstate New York claimed to heal "nervous diseases" with injections of goat lymph. In the twentieth century, an American quack physician, John R. Brinkley (1885–1942), capitalized on the associations between goats and sexuality by promoting the transplantation of goat testicles into men as a cure for impotence. Brinkley later claimed that goat glands could cure many other medical conditions including prostate trouble.

Goat horns were associated with plenty in the concept of the "cornucopia," the hollowed-out horn filled with the products of the harvest. The zodiac sign Capricorn is a goat above the waist and a fish below and is also associated with plenty. For that reason, the Roman Emperor Augustus claimed to be astrologically associated with it and adopted it as a symbol of his reign. The goat is also one of the twelve animals of the Chinese zodiac. Individuals born in the year of the goat are believed to be introverted and creative. In the classical Mediterranean, goats were used for divination, either through examining the entrails of a sacrificed goat or using goat or sheep anklebones, known in Greek as "astragalos," as dice to cast lots. Goats were also among the most common and highly regarded sources of bezoars.

See also: Astrology; Bezoars; Divination; Doll Magic; Witchcraft

Further Reading

Lee, R. Alton. 2002. *The Bizarre Careers of John R. Brinkley*. Lexington: University Press of Kentucky; Smith, J. B. 2004. "Goats with Cattle and Hands from Graves: Towards a Fresh Look at our Insular Superstitions." *Folk Life* 43(1): 115–120.

Gold

Gold is the most valuable of all metals economically, to such degree that in some contexts "gold" has come simply to mean "wealth." Its value combined with its beauty has given it a central cultural role.

In Indian, Western, and Mesoamerican astrology, gold is associated with the Sun. It is one of the seven alchemical metals and usually considered the "noblest" or most perfect of the group. Such was the nobility of gold that alchemists in the Paracelsian tradition argued over whether it was a "pure," unmixed substance or whether like other metals it was a compound of the three elements of salt, sulfur, and mercury. Gold was also viewed as masculine, in contrast to feminine silver.

Much of alchemy was oriented to the manufacture of gold, an effort known as "chrysopoeia." (Since the creation of gold could lead to untold wealth, it was a relatively easy project to get funded.) The philosopher's stone, the object of many an alchemical quest, was valued for, among other things, its ability to transform "base" metals into gold. However, gold was not merely produced by the philosopher's stone; it was also frequently viewed as an ingredient of the stone.

Gold was also used for medical purposes by alchemists and others. Aurum potabile, or "drinkable gold," was marketed as a cure-all by Paracelsian physicians. (There is a biblical precedent for this, as Moses had forced the children of Israel to drink water in which a ground powder made from the idol of the golden calf had been mixed.) The perfection of gold meant that it could perfect defects in the human body. In its crude form, aurum potabile consisted merely of water with some flakes of gold leaf suspended in it, but more sophisticated approaches required actually dissolving the gold. Alchemists kept their formulas for preparing drinkable gold secret, but many formulae included dissolving the gold in aqua regia, "royal water," a mixture of acids that is one of the few ways to dissolve gold. Others argued that the gold in aurum potabile was not literal gold, but a substance called "philosophical gold." On a more popular level in the West, solid gold was widely used to treat conditions of the eyes. Sties in eyes could be cured by rubbing them with gold. This was particularly potent if the gold was in the form of a ring such as a wedding ring. In Ayurvedic medicine, gold ash, or Swarna bhasma, is used for general longevity and to treat several conditions including arthritis and diabetes.

Since gold was the most perfect form of metal, and according to Aristotelian natural philosophy all things strive toward perfection, many believed that gold was naturally "ripening" in the Earth, producing itself without human assistance. Chrysopoeia could be seen as artificially hastening this process, and alchemists spoke of seeds of gold. Mesoamericans, connecting gold, the

Sun, and the heart in the human body, used gold to treat heart ailments.

The Spanish quest for gold in the Americas led to a host of legends, some of which prompted lengthy and expensive expeditions. In northern South America, the legend was of "El Dorado," literally "the man of gold." El Dorado was the name given to a line of kings of the Muisca people of Colombia who were covered by their subjects with gold dust, which was then washed away into a lake. Since gold is very dense, the dust would have simply collected at the bottom of the lake, making the Spaniards who found it very rich. This never happened, but the term "El Dorado" broadened to denote a mythical kingdom of wealth and happiness. In the far north of the Spanish American Empire, what is now the southwestern United States, the legend was of Cibola, the "Seven Cities of Gold." The expedition led by Francisco de Coronado (1510–1554) from 1540 to 1542 had the mission of finding the seven cities. Despite the expedition's failure, the legend persisted for decades.

Gold is also associated with purity in Hindu culture, where the gods are frequently depicted as adorned with gold. Wearing gold jewelry or ornaments brings positive spiritual energies manifested in health, good luck, and prosperity. Such is the respect given to gold that it is forbidden to wear gold below the waist, as that is believed disrespectful. Gold protects newborns and infants from demonic enemies and disease.

Gold also signifies prosperity in Chinese culture and benefits from the favorable associations of the color yellow. Chinese Daoist alchemists strove to transform cinnabar, a commonly found ore of mercury, into gold. Li Shaojun, the first recorded Chinese alchemist, suggested to Emperor Wu (r. 141–87 BCE) of the Han Dynasty that he emulate the legendary Yellow Emperor in summoning supernatural beings to assist in the transmutation of cinnabar into gold. Eating and drinking from gold vessels would prolong life and eventually make it possible to attain immortality, the ultimate goal of Daoist alchemists. Some Chinese alchemists distinguished between "artificial" gold made from cinnabar and "natural" gold and argued that the former was more effective in gaining immortality.

See also: Alchemy; Elemental Systems; Silver; Sun; Yellow Emperor

Further Reading

Debus. Allen G. 1977. *The Chemical Philosophy: Paracelsian Science and Medicine in the Sixteenth and Seventeenth Centuries.* New York: Science History Publications; Mehrotra, Nilika. 2004. "Gold and Gender in India: Some Observations from South Orissa." *Indian Anthropologist* 34: 27–39; Nummedal, Tara. 2019. *Anna Zieglerin and the Lion's Blood: Alchemy and End Times in Reformation Germany.* Philadelphia: University of Pennsylvania Press; Pregadio, Fabrizio. 2019. *The Way of the Golden Elixir: An Introduction to Taoist Alchemy.* Third Edition. Mountain View, CA: Golden Elixir Press.

Golem

The golem is a form of artificially living man created by a Kabbalist. The creation of a golem is considered the supreme work of Kabbalah. The idea is rooted in postbiblical Jewish tradition. (The one appearance of the term in the Bible refers to something like "embryo.") The earliest datable reference to an artificial man, which actually predates the earliest Kabbalistic texts, occurs in the sixth-century Babylonian Talmud. A sage, Rava, is depicted as creating an artificial man (not referred to as a golem) and sending it to another sage. Rav Zira.

When the artificial man does not speak, Rav Zira commands it to crumble to dust, which it does. There is no reference to the technique, magical or not, by which the man was created, and the point of the story is the limitation on human power—the sage can create a man, but not one that speaks.

The fundamental text for the later golem mythology is the mysterious *Sefer Yetzira*, or "Book of Creation," whose date is unknown, but was sometime between the third and tenth centuries CE. *Sefer Yetzira* describes the world and everything in it as created through the permutations of the twenty-two letters of the Hebrew language. This idea was applied to the creation of a golem by the Kabbalist Rabbi Eleazar of Worms (1165–1230?) who described a process by which, after purification and the study of the *Sefer Yetzira*, the would-be golem creator would make a man from soil that had never been cultivated and inscribe on his limbs combinations of the twenty-two letters. This creation is specifically referred to as a golem. What the creator later does with the golem is not discussed; the point is creation, not practical use. This is bound up with a theology by which a Jewish elite of Hasids, or "Pious," have through their perfection enabled themselves to share in God's power of creation.

A fanciful representation of a Kabbalist creating a golem. (Mikolas Ales)

Subsequent famous Ashkenazic rabbis such as Elijah of Chelm (1550–1583) were described as makers of golems. The best-known golem in Jewish mythology was allegedly created by the "Maharal," Rabbi Judah Lowe of Prague (1520–1609). In contrast to Eleazar's idea of the golem as simply demonstrating the Kabbalist's power of creation, Rabbi Lowe's golem was described as a useful servant, particularly since he lacked the power of speech. The idea that Rabbi Lowe, one of the greatest figures of early modern Jewish history, was the creator of a golem does not seem to stem from his own time, however, but as part of a body of legends that gathered around him after his death. The first appearance in print of a connection between the Maharal and a golem does not appear until the nineteenth century, although the idea may have been circulating in oral tradition much earlier. Ideas about the golem as they developed among German and Eastern European Ashkenazic Jews in the early modern period emphasized not just the golem's usefulness but its danger also. Both Rabbi Elijah and Rabbi Lowe are described as overcoming dangerous golems that they had created and later lost control over. Another later aspect of the golem legend is the idea that the golem served as a defender of the Jewish community from its Christian enemies. This idea does not appear in connection with Rabbi Lowe until the twentieth century.

See also: Kabbalah; Spontaneous Generation

Further Reading

Kieval, Hillel J. 1997. "Pursuing the Golem of Prague: Jewish Culture and the Invention of a Tradition." *Modern Judaism* 17: 1–23; Shaffer, Peter. 1995. "The Magic of the Golem: The Early Development of the Golem Legend." *Journal of Jewish Studies* 46: 249–261.

Great Chain of Being

One way of understanding the relation between different levels of existence was the idea of the great chain of being or ladder of life. The great chain was a hierarchical continuum including everything in the universe and spanning, although not really including, the two end points of nonexistence and God. This theory emerged in its full-fledged form among the Neoplatonists of the Late Antique Mediterranean and had roots in the works of both the ancient Greek philosophers Plato and Aristotle. Plato had promoted the idea of a hierarchy of forms, and Aristotle proclaimed the existence of natural continua, emphasizing intermediate types such as the marine life that seemed partway between plants and animals. Aristotelianism's lack of a sharp distinction between the living and the nonliving also contributed to the picture of the continuous universe. In the fully developed theory, the hierarchy of created things—the "Chain of Being"—stretched from rocks and stones through worms and vermin to beasts and eventually to the angels that surrounded God's throne, carefully arranged in hierarchies with no gaps. Man was around the middle of this sequence, between the brute beasts and the angels, unlike those above or below combining matter and spirit. Like much else in Neoplatonism, it was adopted into Christianity. In the European Middle Ages, there was little direct awareness of the Neoplatonic pagan philosophers, but there was enough in the Church Fathers on the great chain to make it a fundamental philosophical concept. The chain-of-being doctrine also influenced Muslim theology and philosophy in the Middle Ages.

The doctrine of the great chain is often considered politically conservative, in that by viewing the universe as hierarchically arranged it endorsed a hierarchical ordering of society. People striving for a more equal society tended to be skeptical of it, sometimes adopting more materialist philosophies. The chain of being was also opposed or substantially modified by those wishing to exalt the human potential for union with God, something requiring transcendence of humanity's place in the chain of being. However, this belief did not require denying the chain, just saying that humans, but possibly no other entities, were not entirely bound by it. Supporters of this belief include the Muslim theologian Abdul Hamid Al-Ghazali (c. 1058–1111) and the Italian Renaissance philosopher Giovanni Pico della Mirandola (1463–1494). In the Enlightenment and post-Enlightenment periods, the great chain of being influenced the idea of a continuous racial hierarchy.

An allied doctrine was the "principle of plenitude," which also originated with the late antique Neoplatonists although it has roots in Plato. The principle of plenitude holds that all things that could possibly exist must exist due to the perfection of God's power and benevolence. God's power could not be limited by the lack of existence of a thing or God's benevolence by denying to a thing that could exist the blessing of existence. The principle of plenitude was sometimes considered to require an infinite universe, or at least one much larger than that allowed by Aristotelian and Ptolemaic cosmology. The Italian heretic priest Giordano Bruno (1548–1600) used the principle of plenitude to argue for an infinite universe, although the French philosopher Rene Descartes (1596–1650) denied it as an infringement on God's freedom not to create. But the idea of plenitude, incorporated into natural theology, went on to be very popular in the eighteenth-century Enlightenment.

See also: Race

Further Reading

Lovejoy, Arthur O. 1936. *The Great Chain of Being: A Study of the History of an Idea.* Cambridge, MA: Harvard University Press; Truglia, Craig. 2010. "Al-Ghazali and Giovanni Pico della Mirandola on the Question of Human Freedom and the Chain of Being." *Philosophy East and West* 60: 143–166.

Guinefort, St. See Dogs

Haints. See Ghosts

Hamsa. See Amulets and Talismans; Defensive Magic; Evil Eye

Hand of Glory

In medieval and early modern Europe, a common form of magic used the hands or fingers of corpses, particularly the hands of unbaptized or unborn children, executed people, or murderers. The most famous example was the "hand of glory," a belief that was originally particularly strong in Germanic Europe but spread to other parts of Europe, including France and England. Different versions of the idea of a hand taken from a corpse and prepared by a witch to have magic powers appear in the *Three Books of Demonolatry* (1595) by the Lorraine magistrate and witch hunter Nicolas Remy (1530–1616), *Six Books of Magic* (1599–1600) by the Southern Netherlands Jesuit priest and demonologist Martin del Rio (1551–1608), and the *Compendium Maleficarum* (1608) of the Italian priest and exorcist Francesco Maria Guazzo (1570–?), as well as appearing in art depicting witches and witches' gatherings.

The hand of glory, as the idea eventually formed, was made from the hand of a hanged criminal that had been dried and preserved to use as a candleholder. The candle itself, it was often suggested, should be made of ingredients including the fat of a criminal hanged in a gibbet. People believed that candles held in the fingers of the dead hand would never burn out (unless extinguished with milk) and that if an intruder held one the inhabitants of a house would not wake up while it would be burning—an alternative term was "thieves' candle." (If someone in the house was not asleep while the candle was lit, one of the fingers would not burn.) In some versions of the legend, hands of glory also had the ability to open locked doors and chests. Hands of glory were also associated with poisoners, who poured poison into the throats of sleeping people while the candle prevented them from waking up. Some magical books contained elaborate formulae for the preparation of a hand of glory, but magic using the hands and fingers of corpses was also practiced, or thought to be practiced, by illiterate witches and sometimes featured in witch trials. Procurement of unborn children's hands to make a hand of glory was sometimes adduced as a motive for killing and mutilating pregnant women.

The hand of glory was prominently featured in the French grimoire, *Petit Albert*, first published in 1706 and widely reprinted into the twentieth century. The *Petit Albert* gave a recipe for the creation of the hand of glory from the hand of an executed criminal, as well as claiming that the candle should be made of the fat of an executed criminal. In addition to wide circulation in France, the *Petit Albert* also circulated among Francophone African Americans in Haiti, New Orleans, and elsewhere, spreading the idea of the hand of glory across the Atlantic.

See also: Amulets and Talismans; Death; Execution Magic; Witchcraft

Further Reading

Davidson, Jane P. 2012. *Early Modern Supernatural: The Dark Side of European Culture, 1400–1700.* Santa Barbara, CA: Praeger.

Hares and Rabbits

The hare and the closely related rabbit were frequently the subject of superstitious beliefs. In northern Europe, hares were often associated with witches. Like the cat, the hare was a form that female witches were believed to temporarily adopt for sinister purposes, although witch hares were treated as somewhat less malevolent than witch cats. A crime frequently associated with old women who transformed themselves into hares was the stealing of milk from cows. There were also stories of witches taking the form of hares that could not be caught or killed for the purpose of frustrating hunters. Rabbits and hares were among the small creatures that served English witches as malevolent "imps." Swedes believed in the "milkhare," or "carrier," a small creature created by a witch from bits of wood or pieces of old brooms to steal milk or turn it into blood and pus. The milkhare took the form of either a hare or a small gray ball. Negative associations of the hare or rabbit existed outside the witchcraft context as well. The English and Irish had the belief that it was bad luck to meet a hare on the road, particularly when first setting out in the morning. In England and Central Europe, the sight of a hare, particularly a running hare, in a town or village was a sign that there would soon be a fire. British fishermen also believed it was very bad luck to have a hare or rabbit, or even to utter the words "hare" or "rabbit," on a fishing boat.

Hares were frequently believed to be hermaphrodites or to change their sex every year. One early Church Father warned

An Irish Tale of Hares and Witches

"A tailor one time returning home very late at night from a wake, or better, very early in the morning, saw a hare sitting on the path before him, and not inclined to run away. As he approached, with his stick raised to strike her, he distinctly heard a voice saying, 'Don't kill it.' However, he struck the hare three times, and each time heard the voice say, 'Don't kill it.' But the last blow knocked the poor hare quite dead; and immediately a great big weasel sat up, and began to spit at him. This greatly frightened the tailor who, however, grabbed the hare, and ran off as fast as he could. Seeing him look so pale and frightened, his wife asked the cause, on which he told her the whole story; and they both knew he had done wrong, and offended some powerful witch, who would be avenged. However, they dug a grave for the hare and buried it; for they were afraid to eat it, and thought that now perhaps the danger was over. But next day the man became suddenly speechless, and died off before the seventh day was over, without a word evermore passing his lips; and then all the neighbours knew that the witch-woman had taken her revenge."

Source: Wilde, Jane Francesca Agnes. 1888. *Ancient Legends, Mystic Charms, and Superstitions of Ireland.* London: Ward and Downey. P. 179.

against the eating of hares because their bad sexual habits would draw the eater in a sinful direction. The association of the hare with sodomy continued into the Middle Ages when it was believed that hares grew a new anus every year for sexual purposes. Medieval medical writings also warned that eating hare meat could cause insomnia. A pregnant woman who ate a hare's head or saw a hare first thing in the morning risked having a child with a harelip. Early modern Persians would not eat hares and did not like even to hear them named, because they believed female hares menstruated like women and were therefore unclean.

The Chinese saw a hare or a rabbit in the Moon, frequently accompanied by a toad. The lunar "Jade Rabbit" made medicines for the Gods. Like the Chinese, the Aztecs saw a rabbit in the Moon, one they believed the gods had thrown there as an insult. In Islamic dream interpretation, a rabbit represents a coward. *The Golden Wheel Dream-book and Fortune-teller*, a nineteenth-century American dream book, is by contrast adamant that rabbits in a dream signify nothing but children, particularly multiple births, and are disastrous to a single woman unless she plans to marry quickly.

The parts of a hare, particularly the foot, also had protective or healing power. The Roman natural historian Pliny the Elder (d. 79 CE) claimed that carrying a hare's foot would protect against gout. Medieval and early modern Europeans continued to believe that a hare's foot had magical power. A hare's foot worn on the left arm protected from danger, or a hare's right foot would keep dogs from barking. Irish people cured the "red rash" by applying the blood of a hare with a red rag and then burying the rag. Hares had medical as well as magical uses. An early medieval Anglo-Saxon "leechbook" or medical manual recommended smearing the gall of a hare on the eyes as a cure for cataracts.

See also: Amulets and Talismans; Moon; Pregnancy and Childbirth; Sex Change; Witchcraft

Further Reading

Ni Dhuibhne, Eili. 1993. "'The Old Woman as Hare': Structure and Meaning in Irish Legend." *Folklore* 104: 77–85. Nildin-Wall, Boldil and Jan Wall. 1993. "The Witch as Hare or the Witch's Hare: Popular Legends and Beliefs in Nordic Tradition." *Folklore* 104: 67–76; Pentangelo, Joseph. 2019. "The Grant, the Hare and the Survival of a Medieval Folk Belief." *Folklore* 130: 48–59; Salisbury, Joyce E. 1994. *The Beast Within: Animals in the Middle Ages*. New York and London: Routledge.

Hepatoscopy

Hepatoscopy is a form of divination by examining the liver of a sacrificed animal. Its earliest appearance was in ancient Babylon, where the sacrificed animal was usually a sheep and the diviners formed a specialized profession. The liver was considered the central organ of life and the source of the blood. It was divided into zones that corresponded with different elements of human life. Closely allied organs such as the gall bladder also carried meaning. Markings, including creases, folds, and holes, also revealed the nature of things to the skilled diviner. Divination was available to a wide range of people, as long as they could afford a sheep to be sacrificed. The highest of the profession were interpreters of sacrificial livers for the king. The Book of Ezekiel in the Hebrew Bible portrays the king of Babylon himself as practicing hepatoscopy along with other kinds of divination. "For the king of Babylon standeth at the parting of the way, at the head of the two ways, to use divination; he shaketh the arrows to and fro, he inquireth of the teraphim [idols], he looketh in the liver" (Ezekiel 21:21) Clay models of the liver were used to teach hepatoscopy and possibly also as comparisons during the actual examination of the liver.

From Babylon, hepatoscopy spread to the classical Mediterranean including the Hittites and Greeks. The Greek dramatist Aeschylus listed the interpretation of the liver as one of the arts that the renegade Titan Prometheus had taught to humanity. One of the groups that became best known for hepatoscopy was the Etruscans, a group of peoples who lived in modern Tuscany. A bronze model of a liver, the "Liver of Piacenza," has been preserved to this day. It is marked with the names of Etruscan deities and divides the liver into sixteen parts corresponding to the sixteen parts into which the Etruscans divided the sky, indicating possible connections between hepatoscopy and astrology. The Etruscans associated hepatoscopy and other forms of divination with a legendary semidivine figure known as Tages who rose from the ground and is sometimes depicted in Etruscan art examining a liver. The "books of Tages," now lost, contained the basic precepts of Etruscan divination. The Etruscans had a more complex understanding of the parts of the liver for divination than did the Greeks.

From the Etruscans, hepatoscopy and other forms of divination were passed on to the Romans, who always viewed the Etruscans as the greatest practitioners of the art. The diviners by entrails, or *haruspices*, in Rome were usually Etruscan. Oxen and goats along with sheep provided livers and other organs for interpretation. It was theoretically possible to divine by the livers of sacrificed humans, but sacrificing humans was a crime Romans viewed with particular horror, and it was believed that the fear a human felt before death would flood the liver with bile, ruining it for divinatory purposes. Romans, like Greeks, saw divination as particularly important in a military context, portraying commanders as making decisions based on diviners' interpretations or sometimes as ignoring them. Like other forms of divination by entrails, hepatoscopy diminished in importance with the rise of astrology in the early imperial period and then was largely eliminated by Christianity. As late as 488, though, a survivor of a failed pagan-backed rebellion against the Christian East Roman Emperor spoke with bitterness of how they had been deceived by the livers that had promised victory.

See also: Divination

Further Reading

Aveni, Anthony. 2002. *Behind the Crystal Ball: Magic, Science and the Occult from Antiquity through the New Age.* Boulder: University Press of Colorado; Collins, Derek. 2008. "Mapping the Entrails: The Practice of Greek Hepatoscopy." *The American Journal of Philology* 129: 319–345.

Herbs

Used for culinary, medicinal, and recreational purposes, herbs were incorporated into a variety of belief systems. Herbs were the basis of most systems of medical treatment, including Western medicine, the Ayurvedic system of India, and traditional Chinese medicine, before the nineteenth century and the invention of synthetic drugs. In addition to these written traditions, knowledge of herbs was also passed down orally. The study of herbs, like botany generally, was originally a branch of medicine, and herbals dealt primarily with the medical properties and uses of herbs, frequently including veterinary uses. Culinary properties were usually secondary, if they were mentioned at all.

In the West, sweet-smelling herbs were widely believed to be protectors against disease, which medical theory associated with corrupt "miasmas." During plagues, people carried sachets of sweet-smelling herbs to ward off disease. Pungent herbs such as garlic were also believed to protect against plague. Herbs also had stimulative effects. Pliny the Elder (d. 79 CE), who included the discussion of herbs in his *Natural History*, identified arugula (*Eruca Vesicaria,* also known as "rocket") as an aphrodisiac. Many other herbs were classified as aphrodisiacs—the ancient Egyptians believed calamus (*Acorus Calamus* also known as "sweet flag") was an aphrodisiac. It was valuable enough to the Egyptians that it was imported from Asia. Other herbs had the opposite effect—Pliny claimed that a purple variety of lettuce was an anti-aphrodisiac.

This page from an Arabic translation of the ancient Greek herbal authority Dioscorides contains recipes for treating coughs and chest pains. (The Cleveland Museum of Art, John L. Severance Fund)

Galen (129–210), the founder of the dominant medical tradition of the West, classified herbs in the Aristotelian categories of hot/cold and dry/wet. In this, he was followed by Western and Islamic physicians and medical botanists in the Middle Ages and Renaissance. The Aristotelian qualities were important in terms of cultivating optimum health by maintaining a balance. Cold conditions would be treated by hot herbs and vice versa. Not only the quality but also the quantity was important—to cure a severe disease of cold, a herb or treatment needed to be not just hot but also to have sufficient heat; a herb lacking sufficient heat would not be strong enough to do any good. Standing at the crossroads between the herbs of many different regions, medieval Arab physicians played a particularly important role in the establishment of the Afro-Eurasian herbal pharmacopeia, still in basically Galenist terms.

Although some knowledge of the effect of herbs was based on trial and error, much of it was a result of classifying herbs according to how their characteristics appeared, in the "Doctrine of Signatures." This idea, which has roots in the ancient Greek medical and botanical writers, is that plants bear some resemblance to the part of the body they are meant to treat. In one Christian version of the doctrine, this was the work of God's benevolence in providing humans with clues on how to treat their medical conditions. This reflects a broader idea that God had created the universe for human use, and the link between the body's features and those of the natural world also resonated with the macrocosm/microcosm analogy. The doctrine of signatures was revived in the Renaissance by the German physician and medical writer Paracelsus (1493–1541), among others. It was particularly important as part of a shift endorsed by Paracelsus but extending beyond the Paracelsian movement for using single herbs, "simples," as medicines, as opposed to complex mixtures of many herbs and other ingredients such as theriac.

Not merely the appearance but other aspects of herbs also could provide clues about their medical use. Saxifrage, or "rock-breaker," was the name for a group of plants that split rocks as they grew. It was used to treat kidney or bladder stones. Location was also a consideration. Paracelsus emphasized that, by the benevolence of God, herbal remedies grew where the diseases they treated originated. For this reason, some Europeans treated the "American disease," syphilis, with the American herb, tobacco. Spaniards in the colonial Andes, arguing against eradication of the coca leaf, claimed God had put it there to energize "lazy" Natives. Other Spaniards, who saw the prominent role of coca in Native culture as an obstacle to Christianization, argued that the devil had put it there. Nineteenth-century believers in progress would secularize this argument to claim that coca was responsible for the "racial degeneration" of Native Andeans.

Although the doctrine of signatures is associated with European cultures, similar beliefs are found elsewhere. Cherokee Native Americans, who thought the stems of purslane resembled worms, used them to treat intestinal worms. Some scholars have argued that the doctrine of signatures and related ideas should be seen as mnemonics, as a way of remembering which herbs were good for which conditions, rather than as a way of discovering new treatments. The doctrine of signatures was discredited with the decline of magic and the rise of mechanical theories of nature and medicine in the seventeenth-century West.

Herbs had magical and religious as well as medical uses, and as such were caught up in larger systems. In England, vervain and dill were believed to defend against witches. Mugwort was another protective herb, particularly powerful if gathered on St. John's Day. In herbals, herbs might be classified by their astrological sign. One of the most influential and frequently reprinted herbals in English, *The English Physitian* (1652) was written by an astrological physician, Nicholas Culpeper (1616–1654). For Culpeper and other astrological physicians, diseases also had astrological influences, and pairing a disease with a herb of opposed influence pointed to a cure. Plants themselves had antipathies; Culpeper claimed that rue and basil never grew together, as rue was an antidote to poison and Culpeper believed basil to be poisonous. Herbs could also be alchemical ingredients, as moonwort was used in processes to make silver.

East Asians classified herbs in several ways, including the yin/yang system, the "Four Natures" hot, warm, cool, and cold and the "Five Flavors" acrid, sweet, bitter, sour, and salty. The earliest known work of herbal medicine in the Chinese tradition is ascribed to Shennong, the "Divine Farmer," one of the mythical rulers of China in the time before recorded history. However, it was probably first composed around the first century BCE. Herbal treatments are the most common in traditional Chinese medicine, although animal and mineral substances are also used as remedies. Like Westerners, Chinese physicians used herbs to maintain a balance. Chinese physicians would usually administer combinations of herbs, including "main" and "auxiliary," as well as herbs to correct possible excesses created by other herbs. Chinese herbs were usually delivered in dried form and administered as decoctions. The most popular herb in Chinese medicine is ginseng root. Like the mandrake, the ginseng root resembles very roughly a human being, one reason it is considered efficacious in restoring vigor and treating a wide range of diseases including difficulty in breathing and cancer.

See also: Aphrodisiacs; Astrology; Chocolate; Coffee; Defensive Magic; Garlic; Macrocosm/Microcosm; Mandrakes; Mistletoe; Natural Magic; Saffron; Sage; Tea; Theriac; Tobacco; Yin/Yang

Further Reading

Allen, Gary. 2012. *Herbs: A Global History.* London: Reaktion Books; Bennett, Bradley C. 2007. "Doctrine of Signatures: An Explanation of Medicinal Plant Discovery or Dissemination of Knowledge?" *Economic Botany* 61: 246–255; Gelles, Paul. 1985. "Coca and Andean Culture: The New Dangers of an Old Debate." *Cultural Survival Quarterly* 9: 20–23.

Hermaphrodites. See Intersex Conditions

Hermes Trismegistus

The *Corpus Hermeticum* is a collection of ancient pagan Greek texts influenced by astrology and philosophy produced in Egypt around the second century AD. They emphasized a dualism of matter and spirit and the possibility of gaining knowledge of the divine through mystical contemplation. Their alleged author was "Hermes Trismegistus," "Hermes the Thrice-Great," a figure who emerged in Egypt after its conquest by Alexander the Great as a blend of the Greek god Hermes with the Egyptian Thoth but who was thought of as a wise man as well as a god. Also thought to be authored by Hermes was a lost Greek text that survived in a Latin translation, *Asclepius*, describing magical procedures, allegedly

A 1493 German engraving of Hermes, from the workshop of Michel Wolgemut. ("Hermes Trismegistus," workshop of Michel Wolgemut, 1493. Rijksmuseum)

used by Egyptian priests, for animating statues of the gods by drawing down celestial powers. There are other fragmentary texts dating from late antiquity that were also considered Hermetic. The Hermetic texts pretended to be much older than their actual date of creation—some asserted that Hermes was a contemporary of Moses or even that he lived before the Flood. (Some authors believed Hermes designated a succession of sages with the same name.) The degree to which the ideas expressed in Hermetic writings are Egyptian as opposed to Greek is still debated by scholars.

In addition to theology and philosophy, Hermes influenced magic and technology. The earliest surviving alchemical writings, those of the early fourth-century Egyptian Greek Zosimus of Panopolis, invoke Hermes as a sage and adviser to the alchemist. Although it is not part of the Hermetic Corpus proper, Hermes was also believed to be the author of a treatise on the Egyptian decans, the division of the Zodiac into thirty-six parts used by astrologers. There are astrological references in many other places in the Hermetic writings.

Isaac Newton's translation of the Tabula Smaragdina

"Tis true without error, certain & most true.

That which is below is like that which is above & that which is above is like that which is below to do the miracles of one only thing

And as all things have been & arose from one by the mediation of one: so all things have their birth from this one thing by adaptation.

The Sun is its father, the moon its mother, the wind hath carried it in its belly, the earth is its nurse.

The father of all perfection in the whole world is here.

Its force or power is entire if it be converted into earth.

Separate thou the earth from the fire, the subtle from the gross sweetly with great industry.

It ascends from the earth to the heaven & again it descends to the earth & receives the force of things superior & inferior.

By this means you shall have the glory of the whole world & thereby all obscurity shall fly from you.

Its force is above all force. For it vanquishes every subtle thing & penetrates every solid thing.

So was the world created.

From this are & do come admirable adaptations whereof the means (or process) is here in this. Hence I am called Hermes Trismegist, having the three parts of the philosophy of the whole world

That which I have said of the operation of the Sun is accomplished & ended."

Source: Dobbs, B. J. 1988. "Newton's Commentary on the Emerald Tablet of Hermes Trismegistus." In I. Merkel and Debus A. G., eds. *Hermeticism and the Renaissance.* Washington, DC: Folger. Pp. 183–184.

Although the Hermetic writings were pagan, their philosophical monotheism and exalted conception of God meant that Christians were not necessarily hostile to them. Some Church Fathers viewed Hermeticism as anticipating Christianity in some ways and, like other ancient philosophies, helping to prepare people for the Christian message. Muslim writers in Arabic took to Hermes with greater enthusiasm, identifying him with a prophet named Idris mentioned in the Quran and associating him with the lost wisdom of the time before the Flood of Noah. Idris/Hermes was also identified with the Jewish prophet Enoch. Muslim interest in Hermes began with the Abbasid Caliphate's project of translating Greek scientific and philosophical works into Arabic, but drew more on late antique magical writing (some of which had earlier been translated into Persian or Syriac) than the Hermetic Corpus. Two successive figures following Idris were also identified as "Hermes." The first lived in Babylon after the Flood and the third in Egypt. The Arabic Hermes figures lost the divine nature of the Egyptian Hermes, but they remained wise sages and powerful magicians and alchemists. The mysterious Brethren of Purity, a ninth–tenth-century CE group of learned Islamic mystics who wrote an influential encyclopedia, the *Epistles of the Brethren of Purity*, were admirers of Hermes. Arabic writers added numerous texts to the Hermetic corpus. These works were largely practical rather than theoretical, dealing with alchemy, astrology, and the making of amulets. An important Hermetic text produced in the Arab world was the *Tabula Smaragdina* or *Emerald Tablet*, an account of the creation of the world read by many as an allegorical guide to the alchemical preparation of the Philosopher's Stone. Many Arabic Hermetic texts would be translated into Latin in the twelfth and thirteenth centuries. Hermes/Idris was also credited with the founding of geomancy and was believed to be the prophet of the Sabians, a pagan group living under Islamic rule and identified with astrology and magic. There is little evidence, however, of a connection between the Sabians and the Hermetic writings.

The Central Asian Muslim polymath Abu Rayhan al-Biruni (973–c. 1050) demonstrated that much of the Arabic Hermetic corpus was a modern forgery that had no connection to the historical Hermes, whose existence he did not deny. However, al-Biruni did not attack the authenticity of the original Greek Hermetic texts, nor did his demonstration circulate or have significant influence in the Islamicate intellectual world. Like the rest of al-Biruni's work, it was completely unknown in the medieval Christian world.

The *Corpus Hermeticum* was introduced into the West around 1460, part of the recovery of Greek literature in the Italian Renaissance. It was translated into Latin by the Florentine Platonic philosopher Marsilio Ficino (1433–1499) and circulated widely in both print and manuscript forms. Hermeticism was received in two chief ways in early modern Europe, as a system of natural theology and as an alchemical tradition. As a system of natural theology, Hermeticism was assimilated into the idea of the prisca theologica or ancient wisdom, handed down directly by God through a series of Gentile sages paralleling the Hebrew prophets and now surviving only in fragments. This approach to Hermeticism drew mostly from the *Corpus Hermeticum*, and its center was Italy. Renaissance Italian scholars knew that Egypt was the oldest ancient civilization and that the ancient Greeks themselves had viewed Egypt as a source of mystical wisdom. Zoroaster, Pythagoras, Plato, and other subsequent sages were identified as disciples and followers of Hermes. Hermetic theology was believed to be perfectly compatible with Christianity. It was often claimed that Hermes derived his knowledge from Moses and prophesied the birth of Christ, and thus, Hermetic wisdom could be seen as an expression of Christianity. Hermetic natural religion was put forward as an alternative to Catholicism and Protestantism which would end the sharp religious conflicts of the Reformation era.

The alchemical version of Hermeticism centered in Germany, and its core Hermetic text was the *Tabula Smaragdina*. Hermes was put in a succession of alchemists rather than of sages, and alchemy was even referred to as "the Hermetic art." In this form, Hermes was adopted into the German alchemical tradition of Paracelsus—some Paracelsians even honored Paracelsus by calling him "the second Hermes." In the alchemical tradition, Hermes was more likely to be viewed as the elder and teacher of Moses rather than his disciple. Like Paracelsus himself, Hermeticists in this tradition were strongly anti-Aristotelian.

Several early modern scientists were influenced by Hermeticism. Copernicus, who had been educated in Italy, quoted the *Corpus Hermeticum*, and Hermetic reverence for the Sun as a divine being may have helped inspire his heliocentrism. The Italian philosopher Giordano Bruno (1548–1600), who unlike most Hermeticists took Hermeticism to the degree of opposition to Christianity, made heliocentrism and the infinity of the universe central to his magical and religious system, hoping to revive a version of "Egyptian" religion derived from the Hermetic texts. The true dating of the *Corpus Hermeticum* was revealed in 1614 by the French Protestant humanistic scholar Isaac Casaubon (1559–1614), based on philological arguments concerning the Greek vocabulary and syntax. Although Casaubon's arguments were strong and widely accepted, as they are today, they did not end all interest in the wisdom of Hermes, particularly in the alchemical tradition and Rosicrucianism. Both Isaac Newton (1642–1727) and the Jesuit polymath Athanasius Kircher (1602–1680) wrote commentaries on the *Tabula Smaragdina*, and Hermes Trismegistus was adopted as an ancient sage into the mythology of Freemasonry.

See also: Alchemy; Amulets and Talismans; Astrology; Geomancy; Hieroglyphics; Macrocosm/Microcosm; Rosicrucianism; Sabianism

Further Reading

Ebeling, Florian. 2007. *The Secret History of Hermes Trismegistus: Hermeticism from Ancient to Modern Times.* Translated from the German by David Lorton. Ithaca, NY and London: Cornell University Press; Van Bladel, Kevin. 2009. *The Arabic Hermes: From Pagan Sage to Prophet of Science.* Oxford: Oxford University Press; Van Bladel, Kevin. 2018. "Al-Biruni on Hermetic Forgery." *Gnosis: Journal of Gnostic Studies* 3: 54–66; Yates, Frances Amelia. 1964. *Giordano Bruno and the Hermetic Tradition.* Chicago: University of Chicago Press.

Hieroglyphics

The hieroglyphic images of ancient Egyptian writing were stylized pictures originally standing either for what they represented or for the sound of the word in spoken Egyptian. Knowledge of the hieroglyphs was lost a few centuries after the conversion of Egypt along with the rest of the Roman Empire to Christianity in the fourth century. However, fascination with the pictorial script continued, with many meanings ascribed to the hieroglyphs. The hieroglyphs were associated with the belief that the ancient Egyptians were great magicians. The ancient Greeks saw the Egyptians as a source of magical wisdom, and Jews and Christians associated Egypt with the magicians at the court of the pharaoh in the Book of Exodus. The Egyptian goddess Isis was portrayed as a great magician or teacher of magicians. The legendary Egyptian sage and magician, Hermes Trismegistus, was even credited with inventing the hieroglyphs. A principal source for the idea of the hieroglyphs as allegorical symbols bearing hidden meaning was the fifth-century CE Greek treatise known as *Hieroglyphica* by Horapollo, who claimed to be an Egyptian priest.

There was little interest in the hieroglyphs in the European Middle Ages, but it revived during the Renaissance. In the Renaissance, the hieroglyphs were viewed as a divinely given language expressing mystical truths that could only be understood by initiates. *Hieroglyphica* was rediscovered in the early fifteenth century and formed the textual basis of knowledge of the hieroglyphs.

However, such was the interest in hieroglyphs that artists and writers made up new symbols that they claimed were Egyptian hieroglyphs. The complex symbolic meanings of the hieroglyphs meant that they were no longer associated with spoken Egyptian or any other language but were treated as an entirely written language.

By the seventeenth century, a new generation of scholars was distinguishing between the authentic Egyptian hieroglyphs and the "new hieroglyphs" of the Renaissance. The meaning and nature of the hieroglyphs was still uncertain, however. There were efforts to connect the hieroglyphs with Mexican writing or the runic writing of Scandinavia. Alchemists were convinced that the hieroglyphs held chemical secrets. The foremost Egyptologist of the seventeenth century, the Jesuit Athanasius Kircher (1602–1680), believed that the Egyptians had possessed an ancient mystical wisdom of ultimately divine origin through Hermes Trismegistus and had expressed this wisdom in the form of symbolic hieroglyphs. His three-volume *Oedipus Aegyptiacus* (1652–1654) set forth this theory and the interpretations of specific hieroglyphs based on it. Kircher also believed, however, that in addition to this esoteric meaning, hieroglyphs also had a plain meaning based on the sound of the word for what they represented. (He was one of the first Western students of Coptic, the language of Egyptian Christians descended from ancient Egyptian.)

In the eighteenth-century Enlightenment, the idea that the hieroglyphics were the keys to ancient mystical wisdom was abandoned and even ridiculed. There was some interest in the idea that the hieroglyphs were linked to Chinese writing, and even that China had been originally a colony of Egypt. The meaning of hieroglyphic script was still unknown and would not be revealed until the discovery of the Rosetta stone in 1799 and its subsequent decipherment by Jean-François Champollion (1790–1832) in 1822.

There was a parallel tradition of scholarly study of the hieroglyphs in Arabic. Medieval Arabic authors also associated hieroglyphs with alchemical secrets. The Arab writers did not view the hieroglyphs as simply representing ideas rather than sounds the way Renaissance Europeans did, but they drew on some of the same Hermetic sources to represent the hieroglyphs as a body of ancient wisdom.

See also: Hermes Trismegistus

Further Reading

Eco, Umberto. 1997. *The Search for the Perfect Language.* Translated by James Fentress. Malden, MA: Blackwell Publishing; El-Daly, Okasha. 2005. *The Missing Millenium: Ancient Egypt in Medieval Arabic Writings.* London: UCL Press; Iversen, Erik. 1993. *The Myth of Egypt and its Hieroglyphs.* Reprint. Princeton, NJ: Princeton University Press.

Horses and Donkeys

The equines are some of humanity's oldest servants and companions and have attracted numerous legends and superstitions. Generally, the horse and its close relative, the donkey, also known as the ass, are viewed in a positive light.

Speed is a quality particularly valued in horses, and the supernatural is frequently invoked in explaining it. For the ancient Greeks, particularly swift horses were begotten by the wind rather than the normal process of generation. Arabia has been known as the land of swift horses (modern thoroughbreds trace their ancestry there). There are legends that the swiftest horses are descended from those owned by King Solomon or the prophet Muhammad. The superstitions associated with betting on horse or chariot races are legion.

Numerous parts of a horse's or other equine's body, ranging from the hairs of its tail or mane or its teeth or hooves, were credited with the power to heal or to bring good fortune. In nineteenth- and twentieth-century England, the hair from a horse's tail was believed to have the power to cure wens. Powdered horse hoof and even the foam from a horse's mouth could be ingredients in medical/magical recipes for the curing of diseases. A fifteenth-century cunning Italian woman, Matteuccia di Francesco, recommended drinking the ashes of a female mule's hoof mixed with wine as a contraceptive.

Horses are frequently associated with luck, both good and bad. The interpretations of the omens embodied by a horse differ by culture but are usually based on colors and markings, such as the Western belief that horses with white legs are lucky. However, there is also a belief in many parts of the West that white horses are a sign of death. In India, there was a system of divination by the whorls or "hair marks" formed by a horse's hair. The shape and positioning of the hair marks could bring good or bad fortune and indicate the horse's quality. This affected decisions made in buying and breeding horses, particularly the indigenous "Marwari" breed. A particular mark of luck was the ridge known to Muslims as the "prophet's finger" and believed to be originally the mark made by Muhammad on one of the horses he had owned. Hindus also viewed the ridge as auspicious and associated it with the temple bell.

Horses had a connection to the supernatural. They were frequently credited with the ability to see ghosts or spirits. Horses were believed to shy at the presence of a gallows or gibbet, or even on passing places where one had stood years or decades ago. Horses that were found in the morning sweaty and exhausted were believed to have been ridden by witches—hence the term "hagridden"—or by elves. Various forms of defensive magic, such as the horse brass, were evolved to protect valuable horses from this phenomenon. In medieval and early modern Europe, particular individuals were believed to know the "horseman's word" which could instantly bring horses under their command. The word was believed to be a closely guarded secret.

In medieval Jewish dream interpretation, to dream of a white horse was a good omen, a red horse a bad one. Horses were generally a good omen in Islamic dream interpretation. In south India, to dream of a golden horse was a good omen. The horse is one of the twelve animals of the Chinese zodiac.

The donkey, or ass, carries out similar functions as the horse in carrying people and loads but is an animal of considerably lower status. Paradoxically, this redounded to its favor in the Christian world, where thanks to the legend of Jesus riding into Jerusalem on a donkey it became associated with humility. In Christian Europe, it was believed that the donkey was marked by a cross on its back in commemoration of its role in Christ's passion. This was part of a tradition associating donkeys with prophecy and Messiahhood going back at least as far as the book of Zechariah in the Hebrew Bible, which describes the Messiah as riding into Jerusalem on a donkey. The donkey of the prophet Balaam is one of the two animals described in the Hebrew Bible as speaking, the other being the serpent that tempted Eve. (The respect paid to donkeys in this passage does not mean they are kosher—they are considered impure in Judaism.) This tradition continued into the Islamic world, where Muhammad was depicted as riding a donkey called Ya'fur, a descendant of the donkey that carried Jesus into Jerusalem. Greco-Roman pagans charged Christians and Jews with worshipping donkeys.

Like the horse, the donkey was the subject of superstitious beliefs about body parts. In medieval and early modern Europe, people believed that the hair from a donkey's back had healing powers. Ancient Greco-Roman writers including the physician Hippocrates and the natural historian Pliny the Elder (d. 79 CE) wrote about the medicinal benefits of donkey's milk, which could cure poisoning and skin conditions. The Hellenistic queen Cleopatra supposedly maintained and enhanced the beauty of her skin by bathing in donkey's milk. Like the goat, farmers believed that keeping a donkey among a herd of cattle would prevent miscarriages. In physiognomy, possession of facial features resembling a donkey, such as a high, rounded forehead or long ears, was associated with dullness and stupidity.

The horse also served as the base for the construction of many legendary animals, such as the unicorn, the winged horse, and the longma, or "dragon-horse" of China. The Buraq, a winged equine frequently depicted with a human face, is described in Islamic tradition as having borne Muhammad to paradise.

See also: Defensive Magic; Divination; Dragons; Goats; Horseshoes; Physiognomy; Solomon; Unicorns

Further Reading

Thurston, Edgar. 1912. *Omens and Superstitions of Southern India.* London: T. Fisher Unwin.

Horseshoes

Few everyday items had more magical power than the horseshoe. In Europe and America, horseshoes were associated with both good luck and defensive magic. They were also sometimes associated with healing.

In sixteenth- and seventeenth-century England, a horseshoe nailed above the threshold of a house or on the steps leading to the door prevented a witch from entering. (Horseshoes are usually made of iron, a metal considered to have anti-magical properties.) As with many magical uses of horseshoes, it was believed that the horseshoe should be one found rather than one bought for the purpose. There were also different procedures for whether the open or closed end of the horseshoe faced upward or the number of nails to be used to attach it. Enslaved Africans in the Americas also used horseshoes for defensive magic. Jamaican slaves buried horseshoes in front of their houses to protect against duppies, ghosts of ancestral spirits who could work harm. There is some evidence of African diaspora communities in the Americas burying their dead with horseshoes in the nineteenth and twentieth centuries. In the harsh regime and material conditions of slavery, obtaining a horseshoe would have involved risk, and these actions indicate the high value placed on their spiritual power. Horseshoes not only passively protected against witches but were also used in active counter-magic. A woman in the Jamestown colony in Virginia in 1626 heated a horseshoe till it was red hot in an oven and then dipped it in urine in order to sicken a person she believed to be a witch.

The idea that horseshoes or finding a horseshoe brings good luck goes back to the Middle Ages and has become virtually a cliché, as in the Anglo-American phrase "born with a horseshoe in his pocket" to describe a lucky man. Luck horseshoes had a similar mythology to defensive magic horseshoes. Some advised nailing a horseshoe with the open end up, to keep the luck from spilling, while some advised nailing with the open end down so that good luck would spill on the people passing under it. Horseshoes found with a nail or nails still attached were particularly lucky. Although the "lucky horseshoe" tradition is most strongly associated with Europe and European and African diaspora communities in the Americas, it is also found in the Indian Vastu Shastra tradition of building and spatial organization. Horseshoe beliefs, although not indigenous to China, have also been incorporated into some modern presentations of feng shui.

Horseshoes also brought good luck at sea. They were frequently nailed to the mast of wooden ships, the most famous example being the *Victory*, the flagship of the British admiral Horatio Nelson (1758–1805).

There are a few healing superstitions associated with the horseshoe. In Germany, food served on a plate branded with a horseshoe cured whooping cough. Another German superstition, found among the "Pennsylvania German" immigrant community, is that placing a horseshoe with all the nails in it in a child's crib would keep the child healthy. Pennsylvania Germans also believed that rabies from a dog bite could be prevented by cauterizing the wound with a bar of iron made from a found horseshoe with its nails attached and that wearing a ring made from a horseshoe nail prevented rheumatism. One divinatory tradition associated with the horseshoe is that sleeping with a horseshoe under the pillow will bring prophetic dreams.

See also: Defensive Magic; Horses and Donkeys; Iron; Luck

Further Reading

Fogel, Edwin Miller. 1915. *Beliefs and Superstitions of the Pennsylvania Germans.* Philadelphia: American Germanica Press; Lawrence, Robert Means. 1898. *The Magic of the Horse-Shoe, With other Folk-Lore Notes.* Boston and New York: Houghton Mifflin; Saunders, Paula. 2014. "Charms and Spiritual Practitioners: Negotiating Power Dynamics in an Enslaved African Community in Jamaica." In Akinwumi Ogundiran and Paula Sanders, eds. *Materialities of Ritual in the Black Atlantic.* Bloomington: Indiana University Press.

Humours, Theory of

The theory of the body's health as dominated by the balance of its four "humours" originated in ancient Greece. The pagan Greek physician Galen (129–210 CE) put it forth in the form that would have the most influence in medieval Christianity and Islam. Galen analyzed the human body in terms of the four Aristotelian elements: earth, air, fire, and water. Only air was taken into the body directly. The other elements were represented by bodily fluids, or "humours"–fire by yellow bile or choler, earth by black bile, and water by phlegm. (Bile is a substance produced by the liver and stored in the gall bladder to aid in digestion.) The fourth humour was blood, although the actual blood that flowed in the body was sometimes identified as a mixture of "pure" blood and two or three other humors. In the microcosm of the human body, the four humors were the equivalent of the four elements in the macrocosm of the universe. A healthy body was one in which the four humours were in balance. Humoural medicine tended to be more devoted to maintaining health through balancing humours than curing disease. Galenic physicians recommended a balanced diet and moderate exercise to keep the humours in balance. Bloodletting, a common treatment for many forms of disease, was justified as remedying an excess of blood over the other humours. The condition as well as the quantity of humours was important, and humours could "putrefy," resulting in sickness. However, in traditional medicine, many treatments for disease had little to do with the humours.

Diet was an important factor in maintaining humoral balance and involved not just the properties of specific foods but also how they interacted with each other. Early modern Persians, great eaters of melons, believed that an excessive consumption of the spring melons called guermecs would lead to an oversupply of phlegm, which with the consumption of sweet melons later in the summer would turn into choler, inevitably leading to fevers in the fall.

Medieval physicians correlated each humour with a personality type following the domination of that humour—phlegmatic for those dominated by phlegm, sanguine for those dominated by blood, melancholic for black bile, choleric for yellow bile. This scheme was widely known in the early modern period and was often turned to comic effect, as in English dramatist Ben Jonson's plays *Every Man out of his Humour* (1600) and *Every Man in his Humour* (1601). (The "humour" play is the source of the word "humor" referring to things that are funny.) Although the theory of humours was not viewed as anti-religious and was held by many pious Muslim, Jewish, and Christian physicians, it did provide a materialist way of understanding human character without invoking a spiritual soul.

The anti-Galenist German physician Paracelsus (1493–1541) denied the humours, but they crept back into his theories in different guises, such as the four "tastes": bitter, salt, acid, and sweet, which correlated with different temperaments much as the humours had. In late seventeenth- and eighteenth-century Europe, the theory of humours was defeated by mechanistic approaches to medicine, but it lingers in some parts of India, as part of the medical system known as "Yunani" or Greek ("Ionian") medicine.

See also: Blood; Elemental Systems; Evil Eye; Geomancy; Macrocosm/Microcosm; Menstruation; Sage

Reference

Horden, Peregrine and Elisabeth Hsu, eds. 2013. *The Body in Balance: Humoural Medicines in Practice.* New York: Berghahn Books; Temkin, Owsei. 1973. *Galenism: Rise and Decline of a Medical Philosophy.* Ithaca, NY and London: Cornell University Press.

Hyenas

Hyenas are the subject of several unusual beliefs. Many of these are connected with sex, as the enlarged clitoris of the female spotted hyena resembles a penis, making it difficult to sexually classify hyenas by sight. In the ancient Mediterranean, it was believed that hyenas changed their sex every year. In the European Middle Ages, when Christians had little direct knowledge of hyenas, this became the belief that they, like hares, practiced sodomy. For this reason, the Church warned Christians against consuming the flesh of hyenas lest they be corrupted sexually. (Hyenas were also considered unclean as they were believed to dig corpses out of graves to devour them.)

Many cultures had the belief that the parts of a hyena had medicinal value. The Roman natural historian Pliny the Elder (d.

79 CE) believed ground hyena jawbone cured shivering fits and hyena teeth cured nightmares. The first vertebra of a hyena cured epilepsy. The Talmud recommends a ritual including the burial of a skin of a male hyena as a treatment for the bite of a mad dog. Medieval Arab physicians suggested making a broth out of the flesh of a female hyena and dill as a cure for sore joints. In India, it was believed that consuming the tongue of a hyena was a cure for tumors and the dried nose of a hyena, if applied to the nostrils of a sufferer, helped in cases of indigestion, boils, and difficult labor. In the Sudan, the fat of the hyena was applied to the stomach as a cure for abdominal pain. The Swiss medical writer Caspar Bauhin (1560–1624) lists several medical and magical uses for the hyena, whole or in parts. The cruelty of some of these indicates the degree to which hyenas were despised. A hyena boiled alive in oil could treat arthritis or digestive issues, while if a hyena was skinned alive by a man's right hand and the skin hung over an entranceway, it would bring good fortune. A mixture of a hyena's brain and heart could cure palpitations.

Hyena parts, particularly genitals and anuses, were associated with fertility, sexuality, and attractiveness to desired sexual partners in many cultures. This is true even in South Asia, where the only hyena species, the striped hyena, lacks the genital ambiguity of the spotted hyena. Iranians believed that the possession of a dried striped hyena skin guarantees the sexual attractiveness of the possessor. Afghans used hyena hair in love charms. These beliefs are sometimes quite elaborate; Afghans believed that an effective love charm could be made by taking the vulva of a female spotted hyena, wrapping it in seven kinds of silk, and then placing it under the arm of a mullah for seven days. In Punjab, it was supposedly believed that a woman could cure her infertility by riding a hyena around in a circle seven times while naked and then feeding it with bread and butter. It is highly doubtful that this was ever tried. Among the Pashtun of Afghanistan and Pakistan, a culture with a wide acceptance of male homosexuality, the rectum of a striped hyena was believed to be a charm that enabled a man to win another's affections. Not all uses for hyena parts were sexual—Pliny describes a stone, the hyaenia, found in the eye or head of a hyena. If placed under a person's tongue, the hyaenia imparted the gift of prophecy.

Hyenas are frequently viewed as magical animals. Pliny stated that the hyena could imitate the human voice, an ability it used to lure humans to where they could be killed and eaten, and that any animal a hyena looked at three times would be unable to move. The belief that witches rode hyenas was found in both east and west Africa as well as India. There are also African and Middle Eastern legends of hyena lycanthropes, whose human form can be recognized through "hyena-like" qualities, such as hairy bodies and nasal voices. Ethiopians believed *budas*, a class of people feared for their magical abilities, were able to both change to hyenas and ride hyenas. In medieval Delhi, the hyena lycanthrope was called a kaftar and believed to attack and kill children. The Beng people of West Africa viewed hyena dung as magically contaminating; when a hyena defecated on the grounds of a village, it necessitated the village to be abandoned to prevent mass death of its inhabitants. In Islamic dream interpretation, hyenas represent perfidious enemies and treacherous women.

See also: Aphrodisiacs; Intersex Conditions; Lycanthropy; Sex Change; Witchcraft

Further Reading

Baynes-Rock, Marcus. 2015. "Ethiopian Buda as Hyenas: Where the Social Is More that Human." *Folklore* 126: 266–282; Brottman, Mikita. 2012. *Hyena.* London: Reaktion Books; Frembgen, Jurgen W. 1998. "The Magicality of the Hyena: Beliefs and Practices in West and South Asia." *Asian Folklore Studies* 57: 331–344; Long, Kathleen P. 2006. *Hermaphrodites in Renaissance Europe.* Aldershot: Ashgate.

I

I Ching. See *Yijing*

Ifa Divination

Ifa is a form of divination practiced among West African peoples, notably the Yoruba, and among much of the African diaspora in the Americas. Although the ultimate source of the Ifa diviners' knowledge is the god Ifa, also known as Orunmila, the god of wisdom and divination, Ifa divination is not based on direct contact with the god or the spiritual world. Instead, it is based on the application of mathematical formulas and memorization of complex texts known as the odu. Divination is a male profession, and the diviner is referred to as "babalawo" or "Father of secrets." Ifa is a basically geomantic system based on 16 primary and 256 derivative figures. (like the *Yijing* and geomancy, Ifa operates on a base 2 mathematical system.) Ifa divination is a religious practice that requires prayers and sacrifices to Ifa. The palm nuts and other paraphernalia of an Ifa diviner are considered sacred to Ifa. However, Muslims and Christians also consult Ifa diviners. The figures are generated either by grasping a collection of sixteen palm nuts and observing whether one or two remain outside the grasp or by tossing a "divination chain" of half-seed pods or metal objects in their shape and observing if they fall with the closed or open side up. Divination chains are believed to be less reliable and prestigious than palm nuts, but they are often preferred in that they give a faster result than casting the palm nuts multiple times. Two primary figures generate a derivative figure, of which there are 256 (16×16). Both the primary and the derivative figures have a rank order.

A vessel used by an Ifa diviner to hold palm nuts used in divination. The work of a nineteenth-century Yoruba carpenter. (The Cleveland Museum of Art, Leonard C. Hanna, Jr. Fund)

The key to Ifa divination and the sacred status of the diviner is the odu, thousands of verses, each connected to a particular derivative figure. A babalawo memorizes some of these. Four verses per derivative figure are considered a minimum, coming to over a thousand verses. The individual verses are called ese and cover an enormous amount of Yoruba history, culture, religion, and mythology. The odu are sometimes treated as the scripture of the traditional Yoruba religion. Since many eses are connected to each figure, the babalawo repeats the ese associated with the figure until the client identifies one relevant to the matter on which he or she sought divinatory assistance. (The client does not generally reveal the subject of his or her enquiry to the babalawo.) Unlike the *Yijing* and bibliomantic systems based on written texts, the odu is a constantly changing and evolving oral tradition. Following the initial cast, the client could have more specific questions, asked in a binary yes/no fashion, which are resolved by casting figures for each alternative and determining which is the highest ranked. The recommended course of action could frequently involve a sacrifice, with the babalawo advising the client which sacrifice to make and to which deity. Some of the sacrifices, particularly if money is to be sacrificed, could be kept by the diviner as payment. Muslim and Christian clients, forbidden to sacrifice to Indigenous African deities, would be advised to give alms or have a feast for their neighbors, with certain foods recommended. The diviner could also suggest the preparation and consumption of a medicine.

The earliest written evidence of Ifa divination comes from the writing of Europeans in Africa in the early eighteenth century. Related systems of divination are practiced throughout much of Africa by Muslim and polytheistic diviners. Many systems are less elaborate—a more informal system practiced by the Yoruba involves tossing a number of cowrie shells and observing if they fall with the open side up or down without arranging them in a sequence or figure as done by Ifa diviners. Babalawos employing Ifa have the most prestige of any diviners in Yoruba society and are sometimes called on for other divination tasks such as dream interpretation.

Although the full complexity of Ifa tradition did not survive the Middle Passage and the harsh conditions of slavery, it did persist in simplified form in the African American tradition, particularly in Brazil and the Caribbean. African immigration in the twentieth and twenty-first centuries has also brought Ifa divination to many places in Europe and the Americas. Some groups in the African diaspora allow for female practitioners of Ifa divination; they are referred to as "iyalawo" or "Mothers of secrets."

See also: Divination; Geomancy; *Yijing*

Further Reading

Bascom, William. 1991. *Ifa Divination: Communication between Gods and Men in West Africa.* Bloomington and Indianapolis: Indiana University Press.

Ifrit. See Jinn

Incantation Bowls

Incantation bowls were a form of predominantly defensive magic found in Mesopotamia and western Iran and dating from the fifth to the eighth century CE. They were earthenware bowls of the kind commonly used in households but marked with written incantations, usually in dialects of Aramaic. There were some written in what appear to be invented alphabets and scripts (which may have been simply scribblings to deceive illiterate buyers) and a few bowls in Persian or Arabic, the latter presumably dating from the period after the Arab conquest of Mesopotamia in 636. The largest number of surviving bowls are in Jewish Babylonian Aramaic, the next largest in the Mandaean dialect of Aramaic used by the Mandaean religious community, and the next in Syriac, a dialect of

Aramaic used by the general community, including Christians. Written material of Jewish derivation, from the Hebrew Bible or the Mishna, was common. The demons that the bowls protected against included legendary Jewish figures such as Lilith or Bagdana, king of demons, and the protectors invoked included King Solomon. However, the use of Jewish vocabulary does not necessarily mean that the persons for whom the bowls were made were Jewish—in the religiously mixed society of late antique Mesopotamia, it was not uncommon to consult a magical practitioner of a different religion. There are many non-Jewish references, including a few to Jesus Christ and to a mixture of Semitic, Mesopotamian, and Iranian gods and demons.

The cheapness of the bowls indicates that their principal role was to provide a space for the incantation rather than carrying magical power themselves. The incantations were presented in a variety of ways, including a spiral beginning in the center of the interior of the bowl and extending to its outer edge. They were usually written in black ink. Texts were often accompanied by images such as those of bound demons. The bowls supposedly had the power to protect against malicious supernatural entities and prevent ills such as headaches, sickness, or difficulty in childbirth. A few also conveyed curses to enemies of the bowl's owner. Bowls were buried upside down in courtyards and under thresholds, thus "trapping" the demon in the bowl. Sometimes two bowls were buried together with their rims joined by a glue made out of bitumen. They were always made for a specific, named individual, sometimes including their family and possessions. These individuals were often women, and there is some evidence that women were prominent among the magical specialists who wrote the incantations. The period that the bowls were created in was also the time of the compilation of the Babylonian Talmud, the major guide to Jewish law and life, but there is no reference to the bowls in the Talmud. Some scholars believe that this is because the rabbis who created the Talmud viewed magical practitioners as competition.

See also: Amulets and Talismans; Defensive Magic; Lilith; Solomon

Further Reading

Hunter, Erica C. D. 1996. "Incantation Bowls: A Mesopotamian Phenomenon." *Orientalia Nova Series* 65: 220–233.

Incubi/Succubi

The idea that it was possible for demons to have sex with human beings goes back to ancient Mesopotamia and was expressed in the legend of Lilith, Adam's demonic first wife, for Jews and in ancient Greek mythology as the Lamia. Augustine of Hippo, a Father of the Church, identified fauns and satyrs, creatures of classical mythology, as lustful incubi seeking sex with human women. According to some medieval writers, the father of the legendary British wizard Merlin was an incubus-demon,

Medieval and early modern Christians believed that demons, while not usually themselves viewed as having a sex, could take on a sexed body. Sexually male demonic bodies were called incubi and sexually female ones succubi, both terms closely relating to fairy lore and suggesting that the belief in sex between humans and fairies was another contributor to the concept of incubi and succubi. (The term "incubus" originated in the Middle Ages and did not originally refer to a sexual demon but a demon or fairy who "pressed down," *incubitus* on a sleeper.) The concept of sex between humans and demons created the theological problem of how a material human body could have sex with an immaterial spirit. (The tales of sex between gods and mortals in classical mythology did not raise the same issue, as the gods were believed to have material bodies, just of a finer material than those of mortals.) Christian theologians in the Middle Ages, elaborating on the natures of angels and devils, had claimed that devils, while ultimately nonmaterial spirits, could create a temporary "body" out of compacted air. The "sons of God" in Genesis 6:1–4 who copulated with the "daughters of men" to beget "men of renown" were frequently identified as angelic spirits, although not all demonologists accepted this identification. (The mainstream rabbinical tradition rejected this interpretation in favor of seeing the "sons of God" as human, although some Jews saw them as angels.) For uneducated people like the witches who confessed to sex with the devil at the sabbat, the relationship of spiritual devils to carnal intercourse was not a problem as devils were thought of as material. Incubi and succubi were frequently discussed in the context of demonstrating the physical reality of demons, since if demons could engage in sex they could not be denied to be real.

The artificiality of the demonic body was particularly useful in explaining how women could conceive children by demon fathers. The standard explanation of interfertility between human women and demons (the question of whether a human male could father a child on a demoness was not often discussed as male demonologists were usually far more interested in intercourse between female humans and incubi than the reverse) went back to the Scholastic theologian Thomas Aquinas (1224?–1274). Aquinas claimed that the demon first formed and inhabited a female body as a succubus, had intercourse with a human male, and preserved the semen. (It was also widely believed that demons collected semen emitted in wet dreams.) It then formed and inhabited a male body as an incubus, ejaculating the preserved semen into a human woman's vagina. Others argued that the semen was in fact that of the incubus. Many argued that children conceived by demonic intercourse were more likely to become witches themselves. Some legendary figures, such as the British wizard Merlin, were believed to be the children of devils, and some Catholics claimed that Martin Luther, the leader of the Protestant Reformation, was a child of a devil. Some theologians even asserted that the Antichrist would be born of demonic parentage. Demons were seldom linked with same-sex sexuality, however; indeed, some demonologists argued that "sodomy" was so abhorrent even demons shunned it.

After the Protestant Reformation, it was mainly Catholics who continued to believe in and think about human-demon sex as Protestants de-emphasized both angels and devils. The Italian Franciscan priest Ludovico Sinistrari (1622–1701) was the author of *Demoniality*, an exhaustive study of the subject published only after his death. Sinistrari distinguished between incubi, who he believed had some material substance, and demons proper and believed that intercourse with incubi or succubi, although highly sinful, was much less so than sex with actual demons.

See also: Fairies; Lilith; Witchcraft

Further Reading

Green, Richard Firth. 2016. *Elf Queens and Holy Friars: Fairy Beliefs and the Medieval Church.* Philadelphia: University of Pennsylvania Press; Sinistrari, Lodovico Maria. 1989.

Demoniality. Translated with an Introduction and Notes by Montague Summers. New York: Dover; Stephens, Walter. 2002. *Demon Lovers: Witchcraft, Sex, and the Crisis of Belief.* Chicago and London: University of Chicago Press.

Intersex Conditions

People whose genitalia were ambiguous or who displayed both masculine and feminine physical characteristics were often considered to have a particular spiritual or magical status. In the West, such persons were referred to as "hermaphrodites" after Hermaphroditus, the legendary child of the Greek deities Hermes and Aphrodite who was sometimes represented as the god of the intersex. Intersex people were often viewed as monstrous, and the birth of an intersex child was frequently considered a prodigy and a sign of approaching evil. Such was the horror the early Romans felt for intersex births that they were ritually murdered by being drowned in the sea. Intersex conditions were also problematic in animals—in Europe, intersex chickens, usually hens that developed male characteristics, were killed, and the purportedly intersex natures of hyenas and hares made them unclean.

The Greek philosopher Aristotle described the generation of the hermaphrodite as caused by insufficient "heat." The male heat transformed the passive female matrix into a male, but if the heat was sufficient to change the matrix only partway, an intersex person was the result. Other biological theorists, who accorded the male and female a more equal role in conception, ascribed intersexuality to a conception where neither male nor female matter clearly predominated. Another medical theory divided the uterus into seven chambers, three on the left that nurtured females, three on the right that nurtured males, and one in the middle that nurtured intersex people. The Arab physician Ibn Sina ascribed the birth of hermaphrodites to conception in the "wrong" phase of the menstrual cycle. Astrologically, the birth of an intersex person could be ascribed to the influence of Gemini, the "Twins," a zodiacal sign identified with doubleness, or to a conjunction of Venus and Mercury. Some located intersex people in the middle of a continuum, with very masculine men and very feminine women at the extremes and feminine men and masculine women closer to the intersex person.

Persons who behaved outside the gender norm for their assigned sex were sometimes believed to be intersex, although in many cases, their bodies were typical of their sex. In the West, Hermaphrodite was used as an insult for gender-nonconforming individuals. Other cultures, however, have viewed intersex people as particularly close to the divine.

The Western and Islamic traditions have generally tried to eliminate the category of the intersex person by defining intersex people according to which sex "prevails," either physically or behaviorally. Adam Kadmon, the originally created human according to Kabbalistic Judaism, was sometimes described as intersex, but this referred more to the non-applicability of gender to a purely spiritual creation than to a combination of male and female characteristics. The cosmic androgyne, expressing both male and female qualities as it mirrored the universe, would frequently appear in magical and alchemical traditions. For early Christians, intersex people, like other "monsters," were ultimately signs of God's power to create how He willed. In the Roman and Christian worlds, the tendency to impose a gender binary on intersex bodies was reinforced by the refusal of Roman law to recognize an intersex category. In the Middle Ages, intersex people identified as predominantly male were in theory eligible to fill male religious roles such as the priesthood but were in practice denied due to their "monstrous" natures. Islamic societies have placed great emphasis on how a person urinates to assign them to a male or female category. Christian writers put a greater emphasis on procreative and sexual difference, eventually dividing intersex people into three categories: "male hermaphrodites," capable of reproducing as males but not as females, "female hermaphrodites," capable of reproducing as females but not as males, and "true hermaphrodites," capable of reproducing as neither.

Polytheistic societies have generally been more accepting of a separate category that is neither male nor female, and some deities have even been represented as intersex. Lan Cai, one of the legendary group of Daoist demigods known as the Eight Immortals, is frequently referred to and represented as "neither man nor woman" although this does not necessarily refer to a physically intersex condition.

See also: Astrology; Gender and Sexual Difference; Hares and Rabbits; Hyenas; Kabbalah; Monsters; Sex Change

Further Reading

Long, Kathleen P. 2006. *Hermaphrodites in Renaissance Europe.* Aldershot: Ashgate.

Iron

Iron's hardness and durability make it a useful metal for both tools and weapons. Its involvement in so many basic human activities gives it a range of cultural meanings. In Western astrology, iron is associated with Mars, the redness of Mars linking the planet and the metal when hot with blood and war. In Indian astrology, iron is associated with Saturn, as both Saturn and cold iron are black. Iron is one of the seven alchemical metals. Iron's association with violence led many to proclaim its superiority over the more obviously precious gold and silver, as the wielder of iron could take gold and silver from their defenseless possessors. Iron's practicality was also contrasted with the merely decorative qualities of gold and silver. The Quran speaks of iron as having been "sent down" from heaven for its usefulness, possibly a reference to iron meteorites that were one of the earliest sources of iron.

Iron's usefulness and the difficulty of working it meant that the blacksmith was often considered a person set apart from the community in a supernatural or religious sense. Indian blacksmiths had a religious ritual before beginning work that included the sacrifice of a rooster whose head was struck off with an iron hammer. The ritual ensured a safe and productive day in the dangerous world of the smithy. Blacksmiths had a particular spiritual status in many African and Celtic cultures that revered iron and were often linked with kings. However, the associations of blacksmiths were not always positive; in Ethiopia, blacksmiths were particularly likely to be identified as the magically powerful but harmful people known as *buda.*

In the Western tradition, iron is frequently associated with defense against potentially malevolent supernatural powers, a belief that may be as old as ancient Egypt. The Roman natural historian Pliny the Elder (d. 79 CE) claimed that a circle traced with iron would protect those within. The idea that fairies and other supernatural beings are deterred by the presence of cold iron is an ancient one in the West. Since the failure of butter to churn was often blamed on fairies or other supernatural beings, plunging a hot iron into the churn was one solution. Iron objects were frequently employed in defensive magic, such as a horseshoe nailed above the entrance of a house to keep out witches. In Celtic communities, keeping iron near a woman late in pregnancy or a newborn child kept the child from being kidnapped by the fairies. A "cunning man" or local magical practitioner in early modern Lorraine poured the urine of a person suspected of being bewitched on a hot iron. If it boiled away, the affliction was natural, but if it remained, it was caused by bewitchment.

Despite, or perhaps because of, its association with violence, iron possessed healing powers. Pliny wrote that water in which white-hot iron had been plunged was useful for curing many diseases. He also claimed that rust could be made into a salve. Nicholas Monardes (1508–1588), a French physician, wrote *Dialogue on the Nobility of Iron which Excels all other Metals* (1574) that discusses a number of medical uses for the metal, including the treatment of hemorrhoids, bleeding, vaginal discharges, gout, wounds, and diarrhea. Irish peasants wore an iron ring on the fourth finger to protect against rheumatism. Iron is also used in Ayurvedic medicine. It is an ingredient in a preparation called lohasava. Modern medicine does agree that iron deficiency causes a range of conditions such as anemia. In Islamic dream interpretation, iron represented longevity and prosperity.

Such is iron's power that there are occasions when its use is forbidden. The Hebrew Bible states that no iron tool was used in the building of Solomon's Temple. In the Western tradition, iron tools are often forbidden in the collection of magical or medicinal terms. Among the most famous examples is the supposed prohibition of the druids from gathering mistletoe with an iron sickle, a golden one being used in its place. (although, since all the evidence we have on druidic religious practices comes from their Roman enemies, this claim may not be accurate.) Iron also has a negative religious association for Christians in that the nails binding Christ to the cross were made of iron; in early modern England, blacksmiths would not work iron on Good Friday for this reason.

See also: Alchemy; Defensive Magic; Horseshoes; Magnetism; Mandrakes; Mistletoe; Pigs; Solomon; Solstices and Equinoxes

Further Reading

Opie, Iona and Moira Tatem, eds. 1989. *A Dictionary of Superstitions.* Oxford: Oxford University Press.

Ivory. See Elephants

J

Jade. See Precious Stones

Jettator. See Evil Eye

Jewels. See Precious Stones

Jinn

The jinn are spiritual beings of Arab origin. Belief in the jinn in Arabia predates the founding of Islam, and they were worshipped by polytheistic Arabs. The jinn are mentioned in the Quran—in fact there is an entire *sura*, or chapter, titled "Jinn," that describes how a group of jinn heard the preaching of Muhammad and became Muslims. The jinn are considered by some authorities to be an integral part of the Islamic faith due to their inclusion in the Quran. Belief in the jinn has spread to a variety of cultures along with the spread of Islam from Arabia.

The jinn are considered "intermediate beings" between humans and angels. Jinn were created before humans out of fire and air, or "smokeless fire," and humans from water and earth. Under ordinary circumstances, jinn, like angels, are invisible to humans unless they choose otherwise. In Islamic theology, the jinn are capable of accepting Islam and being saved, and the message of Muhammad was considered as directed to both humans and jinn. Unlike angels, who are immortal, jinn eventually die, although they may live for many thousands of years. They also differ from angels in that they eat food. Jinn were believed capable of falling in love with, marrying, and having children with humans, although most authorities agreed such unions were forbidden in Islamic law. Although an otherwise good jinn might fall in love with a human woman, he would still have to be banished through religious or magical ritual. Drawing on earlier Christian thought about mortal spiritual beings, some Islamic theorists claimed that although the jinn were capable of having knowledge, they were not capable of learning knowledge like humans. Instead, the knowledge they had was innate.

Like humans, jinn would be judged by God, and some would dwell in heaven and some in hell. (Some Muslim writers, however, claimed that in heaven jinn would have an inferior status to humans.) Iblis, the Satan figure in Islam, is usually considered to have been originally a jinn, although some authorities assert that he was originally an angel. Jinn were described in classical Islamic texts as organized into political societies like those of humans with kingship and hierarchy. They were also viewed as religiously similar to humans, with Jewish, Christian, Sabian, and Muslim jinn, and Muslim jinn further divided into sects like those among Muslim humans.

The jinn are employed by diviners. In pre-Islamic Arabia, seers were believed to have jinn as informants or even lovers. Muhammad, the Quran, and the early Muslim community condemned divination and contact with the jinn for divinatory purposes. However, diviners in Islamic communities to the

Peris

In Persian and Armenian folklore and mythology, the peri were a race of spirits. They are usually depicted as beautiful women, but there are some references to male peris. The origins of the idea of the peri are disputed by scholars, but their roots go back to the Zoroastrian era in Persia before the introduction of Islam. The earliest representations of peris show them as malevolent, but they ultimately became good, or at worst mischievous. Peris were often represented as winged and as living on perfume. They were assimilated into Islam after the Islamic invasions of Persia, often considered a benevolent subclass of the jinn. The home of the peris is the mythical Mount Qaf. Belief in the peris is found in Turkey and among Muslims in India as well as in Iran.

present still claim knowledge of things hidden through their contacts with the jinn. Orthodox Islamic interpreters claim that since the coming of Muhammad, the jinn were barred from overhearing what went on in heaven, which had previously been their way of acquiring knowledge of the future. Shooting stars serve to bar the jinn from access to heaven. The jinn were also believed to inspire poets.

Jinn were viewed as physically powerful and credited with the construction of great works like the city of Palmyra. King Solomon is particularly strongly associated with the jinn. In the Quran, he is described as the king of the jinn as well as of humans and assigns them tasks such as building temples and creating statues. One particularly powerful jinn offers to fetch the throne of the Queen of Sheba virtually instantaneously. Powerful magicians, holy saints, or those possessed of a hereditary gift may also control jinn, to a variety of ends good or evil.

Jinn are particularly strongly associated with wild and desolate places. They can take the form of numerous animals. Ostriches, deer, snakes, and dogs are the most common. Jinn that take the form of black dogs are the most dangerous. Jinn can also take monstrous forms, particularly those with deformed extremities, such as a human body with the feet of a dog.

Although jinn could be both benevolent and malevolent, the evil jinn were more commonly referred to in popular literature. Malevolent jinn, identified as nonbelievers in Islam, could cause madness and disease. According to some Muslims, they were particularly powerful at thresholds, and lingering at a threshold was risky. Thresholds were adorned with talismans to protect against the power of the jinn. During plagues such as the Black Death, some Muslim authorities attributed the spread of the disease to an army of invisible jinn who stabbed the victims. Some malevolent jinn were drinkers of blood or raped women.

One common way that malicious jinn harmed people was through possession, and insanity or epilepsy were frequently understood as the result of jinn possession. Some claimed that for a human being to be possessed by a jinn, there had to be an underlying weakness that allowed the jinn entrance. Exorcism involves repetition of the names of God and Quranic verses, commanding the jinn to leave, and in some cases beating of the possessed person with the belief that the pain will be suffered by the possessing jinn. Many religious authorities frown on the last method as exhibiting a lack of faith in God. Exorcising or banishing jinn is a power associated with Muslim holy figures like the marabouts of Morocco or with holy places such as the shrines of saints.

There were numerous forms of defensive magic to protect people from malevolent jinn. Some of these were religious, such as the recitation or writing of Quranic verses or the Names of God. Writing could be on amulets or on the body of the person to be protected. Even when not formed into words, Arabic letters had the power to protect from the malice of the jinn. Talismans such as the hamsa could also protect against harmful jinn. Salting food was one way of protecting it from jinn, who preferred unsalted food. Unsalted food that decayed was said to have been eaten by the jinn.

The jinn were divided into several categories. Ifrits and Marids each have one reference in the Quran, although it is not clear whether these refer to individuals or to types. Ifrits are particularly powerful and usually malevolent jinn, while marids are unruly and constantly seek to know the future by means of astrology. The ghula appears in jinn lore as a being particularly adept at shape-shifting and hard to kill, eventually undergoing a complex metamorphosis into the Westernized corpse-eating monster known as the ghoul.

As Islam spread beyond its Arab origin, ideas about the jinn also spread. The jinn were frequently identified with figures already existing in pre-Islamic mythologies, although they did not always completely subsume them. Supernatural creatures of non-Arab origin, such as the Persian peri, retained a separate identity from that of the Arabic jinn, although they were also described as a subclass of the jinn. The jinn also figured in folk tales, like the ones collected in the *Book of a Thousand Nights and a Night.*

See also: Amulets and Talismans; Dogs; Possession and Exorcism; Pre-Adamites; Snakes; Solomon

Further Reading

El-Zein, Amira. 2009. *Islam, Arabs and the Intelligent World of the Jinn.* Syracuse, NY: Syracuse University Press; Maarouf, Mohammed. 2007. *Jinn Eviction as a Discourse of Power: A Multidisciplinary Approach to Moroccan Magical Beliefs and Practices.* Leiden: Brill.

K

Kabbalah

In medieval and early modern Judaism, the Kabbalah, literally "tradition," was both a system of mystical religion and a variety of popular magic. The founding document of Kabbalah, the *Zohar*, or *Book of Splendor*, was a product of thirteenth-century Spain, although Kabbalists claimed that it was far more ancient. With roots in Neoplatonism and earlier Jewish mysticism, the *Zohar* described a process by which an unknowable God, the *Eyn Sof*, or One, had created the universe through a process of successive emanations rather than the picture of creation out of nothing given in the Book of Genesis. Some Kabbalists treated the Yahweh of the Bible not as the ultimate God, but as the first emanation of the *Eyn Sof*. They also endorsed the idea of a female aspect of God known as the *Shekinah*. (The *Shekinah* as a positive image of the feminine was counterposed to the demoness Lilith, the ultimate bad woman in Jewish mythology.) Kabbalists opposed treating matter and spirit as absolutely separate. Kabbalistic theology rested on the idea of the sefiroth, the ten successive emanations between God and the material universe. The ten sefiroth are arranged in a pattern known as the "tree of life." The arrangement is hierarchical, with the highest sefiroth, Keter, the crown, the closest to God and the most difficult, if not impossible, for humans to comprehend, and the lowest, Malkuth, kingship. the closest to humanity and the farthest from God. (Some rabbis criticized Kabbalah as a threat to Jewish monotheism.) The sefiroth can be viewed as a mystic pathway of ascent to God. Twenty-two paths connect the ten sefiroth, corresponding with the twenty-two letters of the Hebrew alphabet. Gematria, the interpretation of Hebrew words and phrases according to the numerical values assigned to the letters, was frequently considered a Kabbalistic science although its origins predate the Kabbalah by centuries.

The expulsion of the Jews from Spain in 1492 helped spread Kabbalah in the Jewish world in the sixteenth century, as did the invention of printing—Kabbalistic works were being published in Italy, the center of Jewish publishing, by the second half of the century. The town of Safed in Ottoman Palestine produced an influential group of Kabbalists who saw the Jewish condition of "exile" as conditioned by the "exile of the Shekinah." The Jewish people were identified with the Shekinah exiled from the divinity as a result of human sin. Rabbi Isaac ben Solomon Luria (1534–1572) of Safed, known as the "Ari," or Lion, claimed the ability to diagnose troubled souls by reading the Hebrew letters visible to him on their foreheads. Luria put forth an influential interpretation of the Kabbalah that viewed divine creation as a catastrophic "breaking of the vessels." The Book of Genesis was an encoded description of this process. "Lurianic" Kabbalah focused on redeeming those fragments of God or divine light that had become "entangled" in matter through prayers and pious and magical acts, a process known as Tikkun or restoration. This restoration had Messianic associations.

Although Luria himself wrote almost nothing and only spent three years in Safed before he died, he had numerous disciples and interpreters. (The degree to which Luria's system is the work of Luria himself or of his disciples is disputed by scholars.) Lurianic Kabbalah was introduced to Italy, an important center of Jewish intellectual life, shortly after its formulation. It underlay the Jewish messianic movement of Sabbatai Zvi (1626–1676), which portrayed the Messiah as descending to redeem the lost fragments. Even Sabbatai's conversion to Islam under the threats of the Ottoman sultan could be portrayed in Kabbalistic terms as a descent in order to redeem.

Kabbalah began to be studied by Christians in late fifteenth-century Italy, led by the Florentine philosopher Giovanni Pico della Mirandola (1463–1494). This was part of a larger interest in postbiblical Judaism on the part of Christian intellectuals. Like the writings of Hermes Trismegistus, Kabbalah was considered to be a remnant of the ancient holy wisdom transmitted by God to Moses, an attitude also held by some Jews. Kabbalah had even more authority than Hermeticism as it was written in Hebrew, which many considered to be the original and holy language whose very words and letters had magical power. Christians increasingly studied Hebrew in the Renaissance. Kabbalah's Jewish origin gave it prestige among those Christians who saw the ancient Jews as the source of all wisdom, and the Greeks and Romans, and even the Egyptians merely as disciples. Christian magicians and philosophers such as Giordano Bruno (1548–1600), Annc Conway (1631–1679), and Gottfried Wilhelm Leibniz (1646–1716) incorporated Kabbalistic ideas into their syncretic intellectual systems, Bruno going so far as to publish a Kabbalistic work, *Cabala del cavallo*

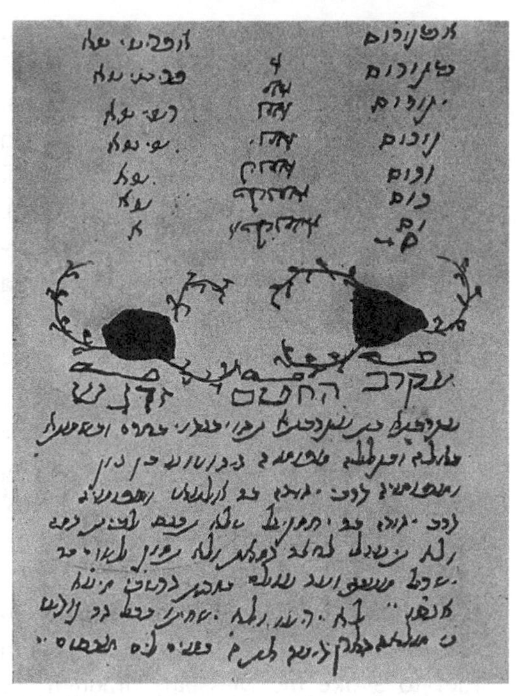

A kabbalistic charm from Morocco. (Singer, Isadore, ed. *The Jewish Encyclopedia*, 1901)

pegaseo (Kabbalah of the Pegasean Horse) (1585). This phenomenon is known as "Christian Cabala." Christian apologists also used evidence from the Kabbalah to argue for the distinctively Christian, non-Jewish doctrine of the Trinity, using what they claimed was Jewish evidence against Judaism.

The Kabbalah was also the basis of magic in the Jewish community as practiced by both lay and rabbinic figures. The tradition of Jewish magic predates the Kabbalah, but in the early modern period, it became partially identified with it in the tradition known as "practical Kabbalah." Some Jews, particularly in North Africa, bclicvcd that the *Zohar* itself had magical powers so that displaying or merely possessing a copy or reciting a passage from it protected from harm. Kabbalistic words and names appeared on the amulets carried by many Jews to ward off bad luck, illness, or demonic influence. Representations of the Tree of Life were also believed to have magical powers and appeared on amulets. The making of amulets was discussed in works of leading Kabbalists such as Rabbi Moses Cordovero of Safed (1522–1570), who recommended that the Kabbalist carrying out a magical operation concerning a particular sefiroth dress in the color associated with it. Magical recipes were presented as Kabbalistic.

Some Kabbalistic literature discusses how the Kabbalist can summon, question, and even master angels and demons. According to legend, in the 1470s the Spanish Kabbalist Joseph della Reina (c. 1418–c. 1472) brought down and bound the two chief demons Samael and Ammon of No in order to evoke the Messiah, although the operation was botched and della Reina punished. The culmination of the career of a Kabbalistic magician was the creation of an artificial life form known as a golem. The power of the Jewish Kabbalistic sage to create artificial life was evidence of Jewish superiority over the Gentile sages for Jewish writers.

The magical Kabbalah was presented to European Christians in the work of the German wizard Heinrich Cornelius Agrippa (1486–1535), *Three Books of Occult Philosophy* (1533). The association of the Kabbalah with other magical systems, including Hermeticism and astrology, also led to schemes to associate the sephiroth with the astrological planets or angels. Kabbalah was also increasingly associated with alchemy, with which it originally had little connection.

Opposition to the Kabbalah existed among Jewish rationalists who distrusted its mystical and magical elements, although some Kabbalists believed the Kabbalah reconcilable with the new developments of the Scientific Revolution of the sixteenth and seventeenth centuries. Both the magical and the mystical Kabbalah were increasingly marginalized among both Jews and Christians during the eighteenth-century Enlightenment. Among Jews, the Kabbalah was discredited by its association with Sabbatai Zvi and other fringe messianic figures, while among Christians, it suffered from both anti-Semitism and the rejection of magic and the occult by the mainstream Enlightenment. The Jewish Enlightenment, or "Haskalah," also rejected Kabbalah, which it saw as outmoded superstition that was holding the Jewish community back. Kabbalah did not entirely disappear from elite culture—historians discussed it as a philosophical theology, and elements of Kabbalah were incorporated into the mythology of Freemasonry. Kabbalistic ideas, including those of miracle workers, appeared in the Hasidic movement among eighteenth-century European Jews and continued among modern Hasidic groups such as Chabad.

Kabbalah has undergone a modern revival, associated with New Age and occultist movements. The "Kabbalah" of modern occultists often has little connection with the original Jewish Kabbalah, however, as Kabbalistic concepts are merged with other traditions such as those of the tarot or even Hinduism. The concept of Tikkun remains influential in contemporary Judaism, often divorced from its Kabbalistic roots.

See also: Alchemy; Amulets and Talismans; Astrology; Golem; Hermes Trismegistus; Lilith; Rosicrucianism; Solomon; Tarot

Further Reading

Dan, Joseph. 2007. *Kabbalah: A Very Short Introduction.* Oxford: Oxford University Press; de Leon-Jones, Karen Silvia. 1997.

Giordano Bruno and the Kabbalah: Prophets, Magicians, and Rabbis. New Haven, CT: Yale University Press; Fine, Lawrence. 2017. "Dimensions of Kabbalah from the Spanish Expulsion to the Dawn of Hasidism." In Jonathan Karp and Adam Sutcliffe, eds. *The Cambridge History of Judaism Volume VII: The Early Modern World, 1500–1815*. Cambridge, UK: Cambridge University Press. Pp. 437–474; Idel, Moshe. 1988. *Kabbalah: New Perspectives*. New Haven, CT: Yale University Press.

Kaftar. See Hyenas

Kappa. See Yokai

Karkadann. See Unicorn

L

Lapis de Goa. See Ambergris; Bezoars

Lecanomancy. See Scrying

Left-handedness

Left-handedness, like other deviations from the norm, was viewed with suspicion in many cultures. The idea of the left side being ill-fortune is embodied in the common Latin-derived word "sinister," literally meaning left and is perhaps most dramatically presented in Western culture in depictions of the last judgment, with God's chosen on His right hand and the rejected on his left. (Another influential stigmatization of the left is the story of the two thieves between whom Jesus was crucified, of which the one on His right goes to heaven and the one on His left to hell.) In Indian tantra, the term "left-hand path" refers to heterodox practices that challenge social norms, a distinction adopted by modern Western theosophists and ritual magicians.

This preference for the right is expressed in many superstitions, such as the common belief that when setting out on a journey or entering a building the right foot must take the first step. In the rite of confirmation, it was believed that being blessed with the bishop's right hand was lucky, the left hand unlucky. In early modern England, it was bad luck to be baptized by a left-hander. It is a common superstition among card players not to pick up the cards with the left hand.

Given the general disfavoring of the left, it is unsurprising that left-handed persons are viewed as unlucky or supernaturally disfavored. In Islamic cultures, which strongly discourage left-handedness, eating with the left hand is associated with demons. (The traditional division of labor between the hands in Islamic societies is that food is taken with the right hand and hygienic activities with the left, so the left is considered defiling. The left hand should not be used to touch sacred objects including the Quran. Extending the left hand in social situations can be considered an insult.) Christian societies have also shown the devil as left-handed or in other ways linked left-handedness with the demonic. There is a folk-belief in the American south that left-handers owe the devil a day's work a year. In the Black Mass, the demonic inversion or parody of the Catholic mass, the gestures and offerings to the devil were performed with the left hand, inverting the religious practice of bestowing blessings with the right hand. Even if not demonic, left-handers were unlucky; nurses in Britain watched babies carefully to see which hand they first reached out with—right-handers were lucky, left-handers unlucky.

Left-handedness is strongly stigmatized in China, a society in which authorities try to eradicate left-handedness in children. There is a Chinese belief that left-handers cannot write Chinese characters. In some traditional African societies, a candidate for leadership can be turned down on the grounds of being left-handed. Traditional Hindus also discourage left-handedness, as the left hand is viewed as impure. Religious offerings should be made only with the right hand.

See also: Black Mass; Garlic; Witchcraft

Further Reading

Coren, Stanley. 2012. *The Left-Hander Syndrome: The Causes and Consequences of Left-Handedness.* New York: Simon and Schuster.

Lettrism

Lettrism is the English term for an Islamic occult tradition based on the letters of the Arabic alphabet, the *ilm al-huruf.* Lettrism is similar to the Hebrew tradition of gematria in Kabbalah, but while Kabbalah treats the twenty-two letters of the Hebrew alphabet as the building blocks of the world, lettrism does the same with the twenty-eight letters of the Arabic alphabet. Each is assigned a numerical value, and those values are added to find the numerical value of a word. Words with identical numerical totals were linked whatever their explicit meanings. Divine and angelic names and the words of the Quran are prominent in lettrist exegesis. Like other occult sciences, it is attributed to ancient sages, but the first evidence of it is in Shia Muslim circles in the eighth century. It appears in such early Arabic sources as the writings attributed to the alchemist Jabir ibn Hayyan and the *Epistles of the Brethren of Purity.* Lettrism was particularly common in Sufi orders, but also appeared increasingly among non-Sufis, and over the centuries, it lost many of its esoteric elements to become a science that was widely known and publicly taught. In later classifications of the sciences, lettrism was categorized as a natural or mathematical science. Not all Muslim intellectuals valued lettrism; it was vigorously opposed by such luminaries as the historian and anti-occultist Ibn Khaldun and the theologian Ibn Tamiyya.

The twenty-eight letters were divided into four groups of seven. As in other forms of occultism, these groups were identified with other groups of four, such as the four seasons and the four Aristotelian qualities: heat, cold, moisture, and dryness. The evidence for the identification of the four groups with the classical four elements—earth, water, air, and fire, comes from the anti-lettrist writer Ibn Khaldun, but may reflect lettrist practice. In seven groups of four, the letters can be correlated with the seven days of the week and the seven astrological planets, and astrology and lettrism would become closely linked among Islamic occultists. The dominant system assigned to the first nine numbers the values 1–9, to the next nine 10–10, to the next nine 100–900, and to the last 1000. Certain suras of the Quran, particularly the first sura, were particularly important and were analyzed in great depth by lettrists. The single letters that appear at the head of some of the suras with no explanation are also highly significant. Lettrism had religious uses; for example, it could offer guidance on how many times to repeat a prayer. It also influenced the popular practice of the creation of talismans. The numerical values of the talismans were adorned with letters, numbers, and symbols designed to call upon the powers of jinn, angels, and even God himself. Talismans were used by members of the elite to advance their political careers or even bring success in war as well as by ordinary people for many uses, including love charms.

Like astrology, lettrism was considered relevant to politics, and numerous Islamic rulers patronized lettrists and other occultists. This was particularly the case in the Timurid period and in the early modern "Gunpowder Empires" of the Ottomans, the Iranian Safavids, and the Moguls of India.

In these empires, the ruler presented himself as a sacred or even Messianic figure. Amir Timur (1336–1405), also known as Timur the Lame or Tamerlane, the conqueror and founder of the Timurid Dynasty, was not only an astrological "Lord of the Conjunction" but also a lettrist "Lord of the Letters."

See also: Amulets and Talismans; Astrology; Divination; Elemental Systems; Kabbalah

Further Reading

Gardiner, Noah. 2020. *Ibn Khaldun versus the Occultists at Barquq's Court: The Critique of Lettrism in al-Muqaddimah.* Ulrich Haarmann Memorial Lecture no. 18. Berlin: EB Verlag; Melvin-Koushki, Matthew and James Pickett. 2016. "Mobilizing Magic: Occultism in Central Asia and the Continuity of High Persianate Culture under Russian Rule." *Studia Islamica* 111: 231–284; Schimmel, Annemarie. 1993. *The Mystery of Numbers.* New York: Oxford University Press.

Lightning

The drama and mystery of lightning has made it the subject of a wide range of supernatural beliefs. Destructive strikes from the heavens were frequently associated with the gods, with Zeus of the Greek pantheon, Thor of the Norse, Shango of the Yoruba, and Indra of the Hindus among the many gods described as "Thunderers." Lightning's suddenness, its destructive power, and its descent from the heavens to Earth identified it with divine punishment or vengeance. Its capriciousness, the fact that lightning could strike one person while leaving another nearby unharmed, contributed to the idea that there was something divine about it. The idea that lightning manifests the power and anger of divinity is virtually universal across religions and cultures. For traditional Chinese, lightning manifested the revenge of Heaven. In the Quran, lightning was an expression of the power of Allah. The ancient Romans viewed the places lightning had struck the Earth as holy; they also had inherited from the Etruscans the now lost "Libri Fulgurales" or Books of Lightning, which described how to use lighting for divination.

Early Christians ascribed lightning sometimes to God and sometimes to demons. The sulfurous smell that sometimes accompanies lightning was evidence that it was literally hellfire. However, the idea of lightning as divine punishment persisted. Some even claimed that the particular sins for which a person was being punished were revealed by which parts of the body were damaged by the lightning. This association between lightning and deserved punishment for sin was a reason many resisted the introduction of lightning rods in the eighteenth century—they viewed them as an illegitimate way of escaping the wrath of God.

Although lightning was often viewed as supernatural, it was also sometimes treated as a phenomenon subject to natural explanations. The pre-Socratic Greek philosophers had numerous theories. Anaximenes of Miletus, who believed all things were made from air, believed that lightning was the result of wind breaking free from clouds. The Roman philosopher Lucius Annaeus Seneca (c. 4 BCE–65 CE), who was skeptical about divination by lightning, viewed lightning as fire created by the friction of the clouds just as a spark could be evoked by rubbing two sticks together. Seneca's contemporary, the Chinese philosopher Wang Chong (25–100 CE), also saw lightning as a form of fire.

Various items were believed to protect from lightning. The Romans believed laurel protected from lightning, one reason for wearing laurel crowns. Lightning was particularly associated with the oak tree, which was believed to be struck by lightning more than any other. In the Middle Ages, boughs of oak were believed to protect houses from lightning. The Yule log, traditionally oaken, was also a lightning protector, and some households in Germany saved the Yule log throughout the year by burning it only during thunderstorms. A common belief in Christian Europe was that the ringing of church bells protected from lightning. There was a physical as well as a religious argument for this; in addition to provoking prayer and repentance, ringing church bells also disturbed the air and broke up the concentrations of hot air before they could issue in thunder and lightning. In fact, church towers and church bells were frequently struck by lightning, and bell ringers often died in lightning storms.

See also: Defensive Magic; Divination

Further Reading

Friedman, John S. 2008. *Out of the Blue: A History of Lightning: Science, Superstition and Amazing Stories of Survival.* New York: Delacorte Press.

Lilith

Lilith is the principal female figure in Jewish demonology, possibly originally derived from ancient Mesopotamian legend. "Lilith" referring to a female demon appears as early as ancient Sumer and is identified as an adversary of the hero Gilgamesh in the short narrative *Gilgamesh and the Huluppu Tree*. She makes only one minor, disputed appearance in the Bible as a spirit haunting the desert, but appears in many later Jewish texts, including the Dead Sea Scrolls and the Talmud, as a malevolent female spirit who snatches babies. She also appears beginning in late antique Judaism in amulets, Mesopotamian incantation bowls, and other forms of Jewish magical practice as a threatening demon connected with sexuality and reproduction. The names of three angels, Snvi, Snsvi, and Smnglof, were considered particularly potent in defense against Lilith. It is a mark of the complexity of the Lilith figure that, according to some legends, she herself provided the information about the names of the angels in order that humans could defend their children against her. Male children were vulnerable to Lilith for the first eight days of their lives. The eighth day was the traditional date for the child's circumcision, after which he was protected. Girls were vulnerable until the twentieth day.

Lilith by the nineteenth-century Pre-Raphaelite Dante Gabriel Rosetti. (Library of Congress Prints and Photographs Division, LC-DIG-det-4a26044)

The legend of Adam's first wife, with whom Lilith became identified, emerged independently, in rabbinical commentaries

An Islamic Lilith: Anaq bint Adam

In Islamic legend, Adam and Eve had a monstrous daughter Anaq. Some descriptions show her as having two heads or twenty fingers. She was either born alone or was the twin sister of Cain. She was believed to be the first to commit fornication and was killed by God. Her son Uj was a giant so tall he survived the Flood and was later killed by Moses.

on the Book of Genesis. Between the eighth and tenth centuries CE, the hitherto unnamed first wife of Adam became identified with the demonic Lilith, and the legend assumed its standard form. Lilith rejected the supine posture in sex as demeaning and was replaced with the more submissive Eve. Lilith was particularly powerful and malevolent in areas relating to sex and reproduction. In addition to baby snatching, she seduced men and also collected semen from nocturnal emissions in order to bring forth demon babies in the manner of the later succubus. Lilith also appears in Kabbalistic writings and Jewish folklore as the wife of Samael, a satanic figure, and the leader of a host of devils. Lilith appears in the *Zohar*, the most important Kabbalistic text, as a seductress who ultimately serves God by revealing weakness in men. According to later legends, the man who Lilith cannot seduce will be the Messiah. Some Kabbalists even presented Lilith as a concubine or seducer of God.

Lilith was particularly a threat to babies and infants at a time when infant mortality was high. Amulets and charms to protect children from Lilith proliferated, particularly among Sephardic Jews in the period after the expulsion from Spain in 1492. The hamsa, the hand with an eye, a charm used against the evil eye, was also used by North African Jews to protect against Lilith. This custom spread to the Muslim population. Red strings tied around infant's wrists represented blood, in the hope that Lilith would not shed the child's actual blood. Lilith was also portrayed as being among the demons subject to King Solomon and bound by the seal of Solomon.

See also: Amulets and Talismans; Blood; Defensive Magic; Evil Eye; Incantation Bowls; Incubi/Succubi; Kabbalah; Solomon

Further Reading

Frederick, Sharona. 2016. "Disarticulating Lilith: Notions of God's Evil in Jewish Folklore." In Ian Frederick Moulton, ed. *Eroticism in the Middle Ages and the Renaissance: Magic, Marriage and Midwifery.* Turnhout: Brepols. Pp. 59–82; Patai, Raphael. 1964. "Lilith." *Journal of American Folklore* 77: 295–314.

Love Magic

One of the most common uses of magic is to win the love or sexual desire of another. Such forms of magic exist across a broad range of cultures. Love magic can be aimed at a sexual encounter or at forming a lasting bond through marriage. The former type of magic is more likely to be used by men, the latter by women, although there are plenty of exceptions to both generalizations. Love magic was sometimes considered an explanation for relationships that otherwise did not make sense; a person could claim to have been the victim of love magic to leave a relationship of which they had grown tired of or if for another reason they did not wish to continue. Counteracting love magic to rid people of passions or connections they no longer, or never had, desired was another branch of magic.

As magic intended to invoke favorable feelings toward the caster, love magic had similarities to magic designed to win the favor of judges or hierarchical superiors. Another closely related form of magic is divination performed to ascertain another's feelings toward the caster. Love magic is also akin to "anti-love" magic, magic designed to destroy the relationship between two people, possibly because the caster hopes to avail himself or herself of the opportunity to form a bond with one. Rivals in love were a frequent target of curses. This can take the form of turning love to hate or of procuring impotence in the male—a very common charge against witches in Continental Europe.

Love magic could take nearly as many forms as magic itself. It could involve the preparation of potions or philters, the performance of rituals, the creation of written texts (often involving special inks), or even the invocation of demons or gods. A common practice in the ancient world was the making of "love-dolls," small representations of the desired woman who would then be the target of love spells. In Denmark, if a man catches a swallow, cuts out its tongue, and then kisses a desired woman with the tongue in his mouth, she will fall in love with him. The Aguaruna people of the Amazon use powerful magical songs, called *anen*, to win and to maintain love. Parts of a lover's or potential lover's body, such as pubic hair, could also be incorporated into love charms. In medieval Europe and elsewhere, women were believed to use their own menstrual blood to make love potions. The parts of animals believed particularly lustful, such as goats or hyenas, were frequently incorporated into love magic. Among the Tamilians of South India, the eye of a nocturnal primate called the loris was a love charm.

Love magic is influenced by gender and culture. Ancient Greco-Roman love magic, much of which survives in the form of written charms, used strongly erotic, sexual, and even violent language. Most of the written charms from the ancient Mediterranean are directed by men to desired women, and many use violent and coercive language to describe the sufferings of the woman until she accepts the love of the man. This language is similar to the language of curse spells. Other ancient love magic is designed to produce a more mutual relationship and has a different vocabulary, one more akin to that of healing magic. This magic was more likely to be used by women, frequently in the context of a marriage or other preexisting relationship. Ancient Jewish love magic, while often aimed at similar ends as the love magic of their pagan contemporaries, avoided explicit language in favor of circumlocutions or euphemisms. Despite the condemnation of all sorts of magic by Islamic religious authorities, verses of the Quran were incorporated into love charms and amulets by Muslim practitioners. A popular choice is Sura 12 verse 30, "Indeed he has smitten her to the heart with love." Chinese distinguished between the "arts of the bedchamber," a branch of medicine designed to facilitate intercourse between willing partners through overcoming physical impediments and employed principally by men, and the "way of seduction," magical formulae with which women seduced initially uninterested men. The former was praised as a healing art, the latter viewed with suspicion.

Love magic can be both carried out by the would-be lover himself or herself or be the work of a third party or hired specialist. Female magical specialists were particularly likely to practice love magic. In the European Middle Ages, the clerics and theologians who saw love magic as a problem to be combated identified women as the main practitioners, reflecting the common attitude that the domain of love and sex was particularly women's domain. Although Heinrich Kramer and Jakob Sprenger's *Malleus Maleficarum* (1486), an influential early witch hunting tract, linked female love magicians with witches, this connection was not always made. The Roman Inquisition of sixteenth-century Italy, for example, distinguished clearly between makers of love spells, usually let off with a light penalty and an admonition to desist, and actual witches, punished much more harshly. In Latin Christian culture, men who performed love magic were more likely to be seen as necromancers or performers of learned magic who performed rituals or commanded demons to win women's love or destroy existing relationships. However, in other societies, people, such as the Muslim marabouts of West Africa, were able to combine a religious role with the purveyance of love (and other forms of) magic. Marabouts are males practicing for male clients.

Although the vast majority of love magic for which we have evidence is heterosexual, there is also some where the caster desires to attract persons of the same sex. Pashtun men carried hyena rectums to attract adolescent boys. There are surviving lead tablets and papyri from Roman Egypt in which women evoke or compel gods and ghosts to make other women fall in love with them, sometimes using the harsh, violent language of male love magic.

See also: Amulets and Talismans; Aphrodisiacs; Divination; Hyenas; Mandinga Pouches; Mandrakes; Menstruation; Natural Magic; Necromancy; Obeah; Witchcraft

Further Reading

Brown, Michael F. 1986. *Tsewa's Gift: Magic and Meaning in an Amazonian Society.* Washington, DC: Smithsonian Institution; Faraone, Christopher A. 2001. *Ancient Greek Love Magic.* Cambridge, MA: Harvard University Press; Mommersteeg, Geert. 1988. "'He has Smitten her to the Heart with Love': The Fabrication of an Islamic Love-Amulet in West Africa." *Anthropos* 83: 501–510; Ruggiero, Guido. 1993. *Binding Passions: Tales of Magic, Marriage and Power at the End of the Renaissance.* New York: Oxford University Press; Steavu, Dominic. 2017. "Buddhism, Medicine, and the Affairs of the Heart: Āyurvedic Potency Therapy (Vājīkarana) and the Reappraisal of Aphrodisiacs and Love Philters in Medieval Chinese Sources." *East Asian Science, Technology and Medicine* 45: 9–48.

Luck

The concept of luck goes by many names, but good luck is one of the most frequent goals of superstitious and magical practice. Some commonly worn or displayed luck amulets, such as the rabbit's foot, the four-leafed clover, or the horseshoe, have become iconic. Luck superstitions can be widely held or individual—when a person associates good or bad luck with a particular action, that can become a personal superstition as that action is repeated or avoided for luck's sake. Sports fans are notorious for repeating actions they associate with their team winning, such as wearing a particular pair of socks. Personal amulets or wearing a particular item of clothing may be good luck practices to an individual, while being meaningless to anyone else.

There is a long-standing tension inherent in the concept of luck itself, whether luck is a random chance, a quality that inheres in certain people, something that can be manipulated magically, or actually the work of God or the gods. The ancient Greeks embodied luck in the form of the goddess Tycho, and rituals for good luck could be viewed as attempts to propitiate the goddess. The Roman equivalent, the goddess Fortuna, has become synonymous with luck itself. The Abrahamic religions, with their emphasis on God's omniscience and omnipotence in a basically deterministic universe, have by contrast always been suspicious of the concept of luck. Paganism could supply the defects of monotheism—Fortuna was invoked into the Middle Ages and Renaissance by Christians looking for luck. The image of the wheel of fortune, with roots in antiquity, was popular in medieval literature and art.

Luck superstitions can be passive, in that they involve things over which the person has no real control, or active in that the person can undertake certain actions to encourage good luck or at least not encourage bad. Carrying an amulet is one "active" practice; rain on a wedding day, believed in some cultures to herald good luck, is usually beyond a person's control.

Perhaps even more vital than maintaining good luck is avoiding bad, and some of the best-known superstitions are about the danger of bad luck. These include having a black cat cross a person's path (although in Britain black cats are good luck) or walking under a ladder. Numbers can be associated with bad luck, as thirteen is associated with bad luck in the West and four in China. Four has the same sound as "death" in Chinese, but the reason thirteen is associated with bad luck in the West remains obscure. Friday is associated with bad luck in the West because it was the day Jesus was crucified, and Friday the thirteenth is the archetypal bad luck day, a day on which no enterprise should be begun. Another, now forgotten superstition related to days referred to December 28, the Day of the Holy Innocents commemorating the biblical story of the massacre of male children in Bethlehem following the birth of Jesus. Not only was the day itself unfortunate, but serious observers of this belief, such as King Louis XI of France (1423–1483), declared the day of the week on which the observance fell as a day of ill omen throughout the year. Japan in the Heian period has a system of lucky and unlucky directions, the premier permanently unlucky direction being the northeast. Other directions were temporarily unlucky based on the movements of deities or unlucky for some but not for all depending on a person's age and gender.

Luck beliefs also affect occult sciences such as astrology, much of which involves identifying fortunate and unfortunate occasions for carrying out certain actions. Entire bureaucracies, such as the Chinese Bureau of Astronomy or the Japanese Bureau of Divination, could be devoted to determining the favorability of specific days and times. Governmental decisions or military expeditions might be timed to seek good or at least avoid bad fortune.

No community is more fixated on luck than gamblers, and gambling subcultures are rich in luck-based superstitions. Live theater is also known as a particularly superstitious milieu. Although the origins of the belief that saying the name of William Shakespeare's play *Macbeth* in a theater will bring disaster are unknown, it is one of the best-known theatrical superstitions.

See also: Amulets and Talismans; Astrology; Cats; Defensive Magic; Divination; Evil Eye; Execution Magic; Hares and Rabbits; Sage

Further Reading

Menjivar, Matt. 2015. *The Luck Archive: Exploring Belief, Superstition, and Tradition.* San Antonio: Trinity University Press; Morris, Ivan. 1964. *The World of the Shining Prince: Court Life in Ancient Japan.* Oxford: Oxford University Press.

Lycanthropy

The belief that some people intermittently transform into beasts of prey exists in many cultures. This can be seen as a literal shape-changing or as a kind of madness. It can also be seen as a power voluntarily exercised, usually but not always for malicious purposes, or as a curse borne by the unwilling. Lycanthropes differ from creatures like the Japanese kitsune, who are conceived of as animals with the power to become human while lycanthropes are humans with the power to become animals. Unlike other forms of shape-changing into small, harmless animals, shape-changing into large predators posed a danger to the community.

Some Native American peoples, notably the Navajo, have a belief in an evil witch called a "skinwalker," who can take the shape of an animal, frequently a wolf or coyote, to wreak evil. In Africa and India, there were legends of hyena shape-shifters, who took on that shape to perform evil deeds like killing and digging up corpses to eat them. Ethiopians identified artisans and the Beta Israel community of Ethiopian Jews as budas, or were-hyenas. According to one account, budas transformed themselves into hyenas by rolling around in the ashes of a hearth. Another popular animal for transformation was the leopard, associated with political power in many African societies. Societies of "leopard men" took advantage of this myth to dress themselves in leopard skins or spotted cloth resembling the leopard's skin to commit acts of violence. Jaguars and jaguar shape-shifters, known as nahuals, occupied a similar role in Mesoamerica. China and India had legends of tiger shape-shifters, who were generally perceived not as social leaders or elite warriors like their African and Mesoamerican big cat equivalents but as a menace to both people and livestock. Turkic peoples of Central Asia, who revered the wolf, had legends of shamans who transformed themselves into wolves.

By far the most common type of lycanthrope in Europe was the werewolf. Wolves were particularly suited for this mythologizing for two reasons—one of which was the persistence of wolf populations in many parts of preindustrial Europe such as Germany's Black Forest. In heavily forested and mountainous regions, wolf predation on livestock and, much more rarely, people could be a serious problem. (Britain, where deforestation began relatively early and the wolf population had been hunted to extinction by the sixteenth century, never developed an indigenous werewolf mythology, although medieval British writers do claim that there are werewolves in Ireland.) The other reason was the Christian metaphor of society as a flock of sheep under the guidance of a shepherd, which made conceptualizing the enemies of society as wolves a logical next step. It was believed that sometimes devils possessed even ordinary wolves for the purpose of wreaking destruction on humans or livestock.

The roots of the werewolf idea go back to ancient Indo-European culture including Germanic, Scandinavian, and Slavic paganism as well as Mediterranean folk-belief. The idea of a human changed to a wolf occurs in Greek mythology as in the story of Lykaon, a cruel king who served human flesh to the god Zeus and was punished by being transformed into a wolf. The ancient Greek historian Herodotus (c. 484–c. 425 BCE) describes the Neuri, a Central Asian people who became wolves temporarily for a few days each year. Herodotus stated that he did not believe the story, but claimed others do. Roman natural historian Pliny the Elder (d. 79 CE) describes a tribe called the Anthids in which one young man every year was chosen to be transformed into a wolf for nine years. If the time passed without him eating human flesh, he could be transformed into a man. Like Herodotus, Pliny disassociates himself from this belief and even ridicules the credulity of believers. The term "lycanthropy" emerged only in the late ancient world and usually referred to a form of madness in which a person believed they had been transformed into a wolf or manifested wolf-like traits such as hunger or cruelty to an extreme degree. This psychological condition could be the result of a curse or a natural condition such as an excess of melancholy.

The idea of the lycanthrope was adopted by early Christian writers, although they denied that a human could actually be transformed into an animal. Augustine of Hippo (354–430), denying the reality of transformation but puzzled by the phenomenon of the werewolf, suggested that in some cases a figment of the imagination could become so vivid that it could be perceived by others, who would then believe, along with the original imaginer, that he had been transformed into a wolf—although Augustine admitted that he had no idea how this could happen. The *Canon Episcopi*, a document originating in early tenth century France and eventually incorporated into the Church's canon law, specifically condemned the belief that a being can be physically transformed from one species to another. The term "werewolf" is derived from an old English term appearing around 1000 CE, "wer-wulf." Scholars dispute whether the original meaning is "man-wolf" or "stranger-wolf."

A sixteenth-century German depiction, by Lucas Cranach the elder, of the werewolf as an insane cannibalistic man rather than a shape-changer. (The Metropolitan Museum of Art, New York, Harris Brisbane Dick Fund, 1942)

Scandinavian lycanthrope lore often focused on the idea of a magic pelt that gave the wearer the ability to transform into whatever animal the pelt was from, two common examples being wolf and bear pelts. However, a pelt was not always necessary, and Scandinavian tales also featured people who shape-shifted through magical words.

Werewolf trials emerged in the early modern period as part of the European witch hunt, but the process by which werewolves became identified as part of the "witch problem" was a slow one. The German Dominican Heinrich Kramer's influential witch hunter's manual *Malleus Maleficarum* (1486) and most other early writings on witches take little or no notice of lycanthropes. By the late sixteenth century, however, discussion of werewolves was appearing in the work of major demonologists, including Henri Boguet (1550–1619) and Pierre de Lancre (1553–1631). All major early writers on the subject, with the exception of the French jurist and demonologist Jean Bodin (1530–1596), agreed that human beings were not actually transformed into wolves, as this would require a combination of an animal body and human soul, something beyond the power of Satan. Most ascribed "lycanthropy" to an excess of melancholy or delusions caused by demons who could create an illusion so convincing as to fool both the "werewolf" and any onlookers that a human being had actually become a wolf. Satan could even intervene to heighten the melancholic state. Artificial aids could also induce lycanthropy. The natural magician Giambattista Della Porta (1535–1615) along with many demonologists ascribed the delusions of the werewolf to the use of drugged ointments. Ordinary people seemed to have been more willing to believe in at least partial transformations of human to wolf, often triggered by items like magic wolfskins. The dominant modern form of the werewolf myth—that the transformation is a condition triggered involuntarily by the full Moon and spread by the werewolf bite—was slow to emerge, and even in the early modern period, there are very few references to it. Defensive magic to protect against werewolves included charms and the repetition of Bible verses. In late sixteenth-century Germany, the Panacea Amwaldina, an alchemical drug, was even advertised with the claim that, among its other virtues, it prevented people from becoming werewolves.

There were a series of well-publicized werewolf cases in France and Germany in the late sixteenth and early seventeenth centuries, beginning with Gilles Garnier of Franche-Comte. Garnier, a hermit, confessed to killing and eating several children and was sentenced to be burned alive in 1574. The next spectacular and widely publicized case occurred in 1589 in the territory of the Elector of Cologne, with the capture of Peter Stubbe, who confessed to incest with his daughter and sister and to killing and eating his son along with many others. Stubbe's case, the only werewolf case in the Cologne witch hunt, showed how werewolves were increasingly viewed as witches. He confessed to having received a magic belt from Satan that enabled him to turn to a wolf. (The belt was never found.) Stubbe was sentenced to a slow and painful death involving having his flesh torn with red-hot pincers and his arms and legs broken before beheading. The French case of a thirteen-year-old boy, Jean Grenier, charged with being a werewolf in 1603, displays some similar features. Grenier's confession includes such details as the Satanic Pact and the acquisition of a magical pelt allowing transformation into a wolf, as well as the crimes for which he was condemned, murdering small children and eating their flesh. He was not executed but sentenced to be confined to a monastery for life, where he was interviewed by De Lancre. Although these cases were male, women were also charged or lynched as werewolves.

Interest in werewolves varied greatly by region. Heavily wooded Franche-Comte was full of werewolves, whereas neighboring Lorraine saw very few. Livonia on the Baltic had an unusual mythology of good werewolves, similar to the non-lycanthropic Friulian benandanti, who fought devils and witches to ensure a good harvest. Many

Slavic peoples, among whom werewolf mythology was very strong, believed that persons born with a caul would become werewolves, another parallel to the benandanti. Slavs distinguished between evil werewolves who were witches or sorcerers and those to whom the condition of being a werewolf had come as a curse. Sorcerers used various rituals to transform themselves into wolves; one werewolf legend along the Pripyat river in Belarus was that sorcerers jumped or stepped over stakes or knives set in the ground to transform. The Baltic region, where Slavic and Scandinavian werewolf mythology met, was particularly rich in werewolf lore. Estonia, particularly the island of Saaremaa, was notable for the predominance of female werewolves. A female werewolf appeared in a court case in Estonia in the late nineteenth century.

Interest in werewolves was dying out among the educated population in western Europe in the seventeenth century with the decline of witch hunting and the campaigns against "popular superstition" in the early Enlightenment. After 1650, werewolf cases and books on werewolves were found only in the Holy Roman Empire, Eastern Europe, and French Canada, which had a very active culture of fear of the werewolf, or *loup-garou*. There was a furor over a werewolf allegedly stalking and killing people outside Quebec City as late as 1767. Even in the most "Enlightened" parts of Europe, however, ordinary people continued to fear the werewolf.

Many aspects of the modern werewolf myth, such as the connection with the full Moon, transmission by biting, and the belief that silver is particularly inimical to werewolves, are more the work of modern filmmakers and authors of popular fiction than of traditional believers. In many parts of the world, the "Hollywood werewolf" has displaced the lycanthropes of indigenous folklore.

See also: Benandanti; Cauls; Defensive Magic; Humours, Theory of; Hyenas; Monsters; Satanic Pact; Silver; Witchcraft; Wolves and Coyotes

Further Reading

Baynes-Rock, Marcus. 2015. "Ethiopian Buda as Hyenas: Where the Social Is More that Human." *Folklore* 126: 266–282; Ginzburg, Carlo. 2004. *Ecstacies: Deciphering the Witches's Sabbath.* Translated by Raymond Rosenthal. Chicago: University of Chicago Press; Ginzburg, Carlo and Bruce Lincoln. 2020. *Old Theiss a Livonian Werewolf: A Classic Case in Comparative Perspective.* Chicago: University of Chicago Press; Metsvahi, Merili. 2015. "Estonian Werewolf Legends collected from the Island of Saaremaa." Translated by Ene-Reet Soovik. In Priest, Hannah, ed. *She-Wolf: A Cultural History of Female Werewolves.* Manchester: Manchester University Press. Pp. 24–40; Ogden, Daniel. 2021. *The Werewolf in the Ancient World.* Oxford and New York: Oxford University Press; Otten, Charlotte F., ed. 1986. *A Lycanthropy Reader: Werewolves in Western Culture.* Syracuse, NY: Syracuse University Press.

Lychnomancy. See Scrying

Macrocosm/Microcosm

The analogy between the universe, a macrocosm, and a selected part of it, a "microcosm," takes several forms. Either the Earth or the human body can be seen as the microcosm. The term "cosmos" in Greek referred to a well-ordered arrangement of parts in an overall system. The idea of a macrocosm/microcosm analogy appears in many aspects of thought, particularly the Western magical tradition including alchemy and astrology. However, macrocosm/microcosm analogies appear in other cultural traditions, such as Chinese cosmology.

Macrocosm/microcosm analogies can be found in classical Greek philosophy, particularly Plato (428–348 BCE) and the Stoics, although their origin may be much earlier. Macrocosm/microcosm analogies often implied that the universe was in some way alive and that its functions were like those of a living being rather than a machine. For theorists in this tradition, both the individual human and the universe possessed both a physical body and a soul. By contemplating and emulating the harmonious universe, the individual may cultivate harmony within. By the late antique period, the idea was extended by Neoplatonic philosophers and magicians to the belief that the correspondences between the macrocosm and microcosm could be manipulated magically. This idea appears in the writings ascribed to Hermes Trismegistus.

The idea of a relationship between a microcosm and a macrocosm was also found in pre-Islamic Iran and early Jewish literature. The idea that God created man "in his image" in the Book of Genesis was considered a biblical endorsement of this idea. From both philosophical and Jewish sources, the idea entered Christianity, although Christians were often leery of the idea of a world soul, which seemed too like a pagan god. The macrocosm/microcosm relationship is also found in the Kabbalistic model of the universe as created by successive emanations of which the first is Adam Kadmon, the cosmic man who is a pattern both for the cosmos and for the human body. Muslims also adopted the idea of correspondence between humanity and the cosmos; it is found in the tenth-century *Epistles of the Brethren of Purity*, from which it made its way to medicine in the works of Ibn Sina (980–1037) and to mysticism in the tradition of Sufi writers. Greek Christians had a slightly different version of the analogy by which humanity, as the only entity combining a physical body and a spiritual soul, was the only being representing a universe both spiritually and physically while occupying the median point on the Great Chain of being. With the twelfth-century revival of Platonism in the West, the macrocosm/microcosm analogy shows up in more mystically inclined medieval Christian writers such as the German nun Hildegarde of Bingen (1098–1179).

Macrocosm/microcosm analogies extended to a series of correspondences between individual aspects of the body and aspects of the universe. The Sun was considered analogous to the heart, as both dispersed light and blood throughout the cosmos. (Organs are often classified in groups of seven to correspond with the seven astrological planets.) The four bodily humors—blood, phlegm, black bile, and yellow bile—correspond to the four elements. Macrocosm/microcosm analogies were also important in the connection between astrology and medicine. The connection between the movements of the universe and changes in the human body meant that in the Christian and Islamic Middle Ages and early modern periods physicians were expected to be astrologers as well.

In the European Renaissance, the rediscovery of Plato and the Hermetic writings helped spread the macrocosm/microcosm analogy. It was frequently used by magical and alchemical writers such as Paracelsus (1493–1541) and Robert Fludd (1574–1637), author of *History of the Microcosm and Macrocosm* (1617–1621), and by occultists such as the Rosicrucians. The Scientific Revolution of seventeenth-century Europe, which substituted the picture of the universe as a machine for the picture of the universe as a living being, pushed the macrocosm/microcosm connection out of mainstream natural philosophy, although it continued to be a source of literary and artistic metaphors and influenced Romanticism and German *Naturphilosophie.* Magical writers of the nineteenth-century occult revival such as the founder of Theosophy Helena Blavatsky (1831–1891) invoked the idea of the macrocosm/microcosm correspondence, and it is found in New Age circles today. Indeed, even the "scientific" idea of the human body as a machine governed by the same laws that govern the universe can be considered a successor to the premodern idea of the macrocosm-microcosm analogy.

Macrocosm/microcosm correspondences are also found in the Chinese tradition of "correlative cosmology." *The Yellow Emperor's Inner Classic*, the textual foundation of traditional Chinese medicine, contains a list of correspondences between the cosmos—"Heaven and Earth"—and the human body. Heaven is round and Earth is square; the human body corresponds with a round head and square feet. The Sun and Moon correspond with the two eyes, the nine regions of Earth with the nine orifices of the human body. The idea of a macrocosm/microcosm correspondence continued in the tradition of religious Daoism. The five viscera—liver, heart, spleen, lungs, and kidneys—corresponded with the five elements in Chinese elemental theory—Wood, Fire, Earth, Metal, and Water.

See also: Alchemy; Astrology; Blood; Elemental Systems; Great Chain of Being; Hermes Trismegistus; Humours, Theory of; Kabbalah; *Naturphilosophie*; Physiognomy; Rosicrucianism; Yellow Emperor; Yin/Yang

Further Reading

Barkan, Leonard. 1975. *Nature's Work of Art: The Human Body as Image of the World.* New Haven, CT: Yale University Press; Conger, George Perrigo. 1967. *Theories of Macrocosms and Microcosms in the History of Philosophy.* New York: Russell & Russell.

Magnetism

Magnetism is the ability of "magnetized" pieces of iron to attract unmagnetized iron and to both attract and repel other pieces of magnetized iron as well as align itself with the Earth's poles. It is one of the most mysterious physical phenomena involving as it does action at a distance and has attracted many beliefs and theories. Naturally occurring magnets, or "lodestones," were discovered in several cultures.

Ancient Romans believed that rubbing a magnet with garlic took away its power. Diamond was also believed to be a shield against the magnet's attraction. Roman natural historian Pliny the Elder (d. 79 CE) divided magnets into two categories, male and female, of which only the male possessed attractive power. He also believed that magnets varied by color, the strongest magnets being the blue magnets from Ethiopia. Pliny claimed that magnets were useful in treating diseases of the eyes. The physician Galen (129–c. 216 CE) recommended magnets as a purgative. The ability of magnets to attract iron from a distance made them a natural ingredient in love charms or, in powdered form, in love potions, but they had other magical uses as well. Magnets were placed inside magical figurines, and a magnet placed in the bed of a wife whose chastity was suspect would eject her if she was really unfaithful.

Interest in the magnet increased with the invention of the magnetic compass by the Chinese in the twelfth century and its spread to Islamic and European mariners. (Muslims were particularly interested in adapting the compass to finding the direction of the qibla, the direction of the Kaaba for Muslim prayer.) Magnets were not always the mariner's friend, however; there is a legend dating to the ancient world and found in the Middle Ages among Western and Islamic sailors of a magnetic mountain that wrecked ships by pulling the iron nails out of them. Magnetism figured in a curious myth that medieval and early modern Christians had about the tomb of Muhammad, Islam's founder. They claimed that the iron coffin of Muhammad was suspended in midair and that credulous Muslims believed this to be a miracle, when it was actually a fraud accomplished with magnets. The idea of an iron statue levitating with the power of magnets appears as early as in the work of Pliny the Elder, but it only became perceived as a "fraudulent miracle" in the work of early Christian writers such as Augustine of Hippo (354–430). The idea also appears in postbiblical Jewish literature. The first evidence of its application to Muhammad's tomb is from the early twelfth century. Muslims themselves told similar stories of fraudulent magnetic miracles, ascribing them to the "idolaters" of India.

The German physician and alchemist Paracelsus (1493–1541) claimed that the magnet's power could be taken away by immersing it in mercury. He also describes a process according to which by being repeatedly heated and cooled a magnet might become strong enough to pull a nail from a wall. Sir Thomas Browne (1605–1682) in his examination of superstition and popular beliefs in *Pseudodoxia Epidemica* (1646) denies the validity of both. The mysterious power of magnets made them a logical subject for writers about natural magic such as Giambattista Della Porta. In the sixteenth and seventeenth centuries, cosmological systems took magnetism as a fundamental force responsible for the structure and organization of the universe. William Gilbert, the author of the most important work on the science of magnetism at the time, *De Magnete* (1600), thought that the world was held together by a divine magnetic power. Following Gilbert, the early heliocentric astronomer Johannes Kepler ascribed the movement of the planets around the Sun to a "magnetic force." The "magnetic cosmology" was later worked out in great detail by the Rome-based German Jesuit Athanasius Kircher in his *The Magnetic Kingdom of Nature* (1667), one of several works on magnetism from Kircher's hand. Magnets also had numerous medical uses, purportedly effective in neutralizing the agonizing pains of childbirth or gout. The eighteenth-century French Royal Society of Medicine endorsed the treatment of toothache by magnet after a lengthy investigation. By contrast, wounds made with a magnetized knife or sword were believed deadly.

In the eighteenth century, the Scottish quack James Graham (1745–1794) sought to capitalize on the associations of magnets with vitality and fertility. He also drew on scientific awareness of the connection of magnetism and electricity. His "Celestial Bed," unveiled in 1781, was designed to encourage conception. It employed magnetized rods in addition to electrical, pneumatic, and acoustic apparatus. The idea of magnetism being essential to health later resulted in the creation of "animal magnetism."

See also: Garlic; Iron; Mesmerism and Animal Magnetism

Further Reading

Fara, Patricia. 1996. *Sympathetic Attractions: Magnetic Practices, Beliefs and Symbolism in Eighteenth-Century England.* Princeton, NJ: Princeton University Press; Lowe, Dunstan. 2016. "Suspending Disbelief: Magnetic and Miraculous Levitation from Antiquity to the Middle Ages." *Classical Antiquity* 35: 247–278; Sander, Christoph. 2020. "Magnets and Garlic: An Enduring Antipathy in Early-Modern Science." *Intellectual History Review* 30: 523–560.

Mandinga Pouches

Mandinga pouches (*Bolsas Mandinga*) were a type of amulet carried in late seventeenth- and eighteenth-century Brazil and Portugal primarily by slaves and African-descended people, although Native- and European-descended people also carried them. They were usually worn around the neck and waist and were almost always carried by men rather than women. Mandinga pouches were small bags, usually made of white cloth, containing a variety of objects, drawn from African and European religious and magical traditions. The Mandinka after whom the bags were named were a predominantly Muslim African community to which many Brazilian slaves traced their origin, and they were thought to be powerful magicians. One origin of the pouches may be the Mandinka practice of wearing bags containing papers with Quranic passages.

Many of the objects carried in the bags had a Catholic sacred meaning, particularly connected with the Eucharist. Fragments of altar stones and purificators—pieces of cloth used to wipe the chalice—were frequently found in pouches. Pieces of paper with prayers—one commonly used prayer was a prayer to Our Lady of Montserrat—written on them were also found in most bags, along with the "sign of Solomon," a six-pointed star believed to have the ability to protect from the evil eye and other forms of harmful magic. Crosses and sacred hearts also appeared on the papers. Some held that written prayers left under altar stones were more powerful, particularly if Mass was said over them or if the prayers were written in the blood of a white or black chicken or even the writer's own blood. The same text could include both prayers to God and the saints and invocations of the devil. Physical objects contained in the bags included consecrated hosts, fragments of altar candles, roots, and bones. Roots originating in Brazil were held to be the most powerful. Bags designed to protect against gunshots could include flints, bullets, and small amounts of gunpowder. Pouches could be strengthened by being buried for short periods of time or being exposed to the bonfires on St. John's night (June 23).

The bags were objects of exchange, and people made bags for sale. In Portugal, bags of Brazilian origin were thought the most powerful, and slaves arriving from Brazil were pressured to make bags for local people. Many of the purposes of the bags were defensive. In addition to protecting from harmful magic, bags were claimed to protect from bodily harm and the dangers of the sea. Bags could also instill courage and win the love of women, although there is also a case of a person who carried a bag in the hopes of being enabled to resist sexual temptation.

The Inquisition frowned on the use of Mandinga pouches, particularly those that employed sacred objects such as altar stones and consecrated waters. Some people were condemned to imprisonment or even death for manufacturing or carrying them, but this seems to have had little effect on their popularity.

See also: Amulets and Talismans; Blood; Defensive Magic; Solomon

Further Reading

Rarey, Matthew Francis. 2018. "Assemblage, Occlusion and the Art of Survival in the Black Atlantic." *African Arts* 51, 20–33; Souza, Laura de Mella e. 2003. *The Devil and the Land of the Holy Cross: Witchcraft, Slavery and Popular Religion in Colonial Brazil.* Translated by Diane Grosklaus Whitty. Austin: University of Texas Press.

Mandrakes

The mandrake is the root of the mandragora plant. It has a shape vaguely resembling that of a human. The plant also contains hallucinogenic chemicals. These qualities have caused many legends to be attached to it.

The Hebrew Bible, in recounting the story of Jacob, Leah, and Rachel, endorses the idea that mandrakes promoted fertility in women. This may be a result of translation, however, as the Hebrew word is literally "love plant," and although mandrake is the standard translation following the Septuagint, the first translation from Hebrew to Greek, not all translators agree. The love plant also appears in the Song of Solomon as an aphrodisiac, and that too is usually translated as mandrake. The idea of the mandrake as promoting fertility is at the center of the Florentine philosopher Niccolo Machiavelli's comedy *Mandragola* (1524), although he treats the idea as a superstition.

Ancient Greek physicians and botanists described the use of the mandrake as a drug, an anesthetic in preparation for surgery or a sedative for insomniacs, or as an abortifacient. Greeks also believed in the aphrodisiac qualities of the mandrake and associated it with the goddess Aphrodite. Belief in the mandrake's aphrodisiac powers continued into the nineteenth century.

The first-century CE Jewish historian Flavius Josephus describes a plant commonly identified as the mandrake to be effective against demonic possession. This idea also appears in the late antique *Herbarium of Apuleius*, which appeared in the Middle Ages both in its original Latin and translated into Old English. This may have been connected to the pungent smell of the mandrake. Mandrake legends proliferated endlessly in the Middle Ages. In the *Physiologus,* a late antique collection of animal stories widely circulated and expanded in the Middle Ages, elephants were described as consuming mandrakes as an aphrodisiac. Mandrakes were held to cure all diseases. *The Distaff Gospels*, a late medieval French collection of the beliefs of women, suggested that finding a mandrake, putting it on a bed, and feeding it meat and drink twice a day would make the owner rich, even though the mandrake neither ate nor drank.

This belief was found among French and German peasants as late as the eighteenth century. Carrying a mandrake was a recognized mark of the medieval or early modern sorcerer or witch; at the trial of Joan of Arc in 1431, she was charged with carrying a mandrake, a charge she denied. The mandrake was listed as an ingredient in the "flying ointments" allegedly made by witches. The plant's hallucinogenic properties may have played a role in deluding people into thinking they were flying. Many of the mandrakes circulated for their magical powers in the medieval and early modern period were fakes; roots such as bryony or turnips were carved into a roughly human form and passed off as true mandrakes.

The mandrake was viewed almost as a conscious being, and some distinguished between "male" and "female" mandrakes depending on whether its root resembled more a man or a woman. Its harvesting was described as dangerous. Josephus spoke of the danger of harvesting a plant he called Baaras, which if not harvested in the proper way could kill the one harvesting it. He recommended sprinkling it with the urine or menstrual blood of a woman before harvesting. He also suggested using a dog to dig it up, although the dog would die. This plant was not the mandrake, which Josephus discusses elsewhere. However, this legend would later be attached to the mandrake.

The full legend of the mandrake hunt would be developed in the Middle Ages. Some advised that the mandrake would run away from the mandrake hunter so it needed to be surrounded by iron, which it could not pass. It was also believed that the mandrake emitted a shriek on being pulled from the Earth, which killed all who heard it, an idea that first appears in a French bestiary written around 1120. Medieval mandrake hunters were advised to use a dog to harvest the

The idea of the mandrake as possessing powers to heal remained popular into the nineteenth century, as shown in this patent medicine advertisement. (Library of Congress, Prints and Photographs Collection, LC-DIG-ppmsca-43649)

mandrake. After the soil was loosened with an ivory stake, a hungry dog was tied to the part of the mandrake above the Earth. Then the mandrake hunter, his ears plugged, held out a piece of bread. The dog running toward the bread would pull the mandrake out of the Earth, hear the shriek, and die while the hunter remained safe. William Shakespeare, who frequently refers to mandrakes in his plays, suggested in *Romeo and Juliet* that hearing the shriek of a mandrake causes insanity.

The belief that mandrakes grew under gallows, nourished or formed from the semen of hanged men, developed in the early modern period. Its earliest recorded appearance is in 1532 in the work of the German botanist Otto Brunfels (1488–1534). Brunfels like most early modern botanists dealing with the mandrake does not endorse this idea, which he finds superstitious. By the seventeenth century, the idea that the mandrake grew under gallows appeared fairly often in Continental European medical literature, although not often in Britain. An Icelandic variation was that the mandrake sprung from the froth of the mouth of a hanged man. Executioners sometimes engaged in the sale of mandrakes, real or fake.

See also: Aphrodisiacs; Elephants; Execution Magic; Herbs; Iron; Witchcraft

Further Reading

Carter A. J. 2003. "Myths and mandrakes." *Journal of the Royal Society of Medicine* 96: 144–147; Davies, Owen and Francesca Matteoni. 2017. *Executing Magic in the Modern Era: Criminal Bodies and the Gallows in Popular Medicine.* Cham: Palgrave Macmillan; Van Arsdall, Anne, Helmut W. Klug and Paul Blanz. 2009. "The Mandrake Plant and Its Legend." In Peter Bierbaumer and Helmut W. Klug, eds. *Old Names-New Growth: Proceedings of the 2nd ASPNS Conference, University of Graz, 6–10 June 2007 and Related Essays.* New York: Peter Lang. Pp. 285–345.

Marid. See Jinn

Mass of Saint Secaire. See Black Mass

Menstruation

No human physiological process is as mysterious—or as hedged with superstition—as menstruation. Although the dominant approach has been to define it as unclean, there are ambiguities and even positive views.

The idea that menstruating women and menstrual fluids are impure, unclean, and contaminating is very old and found in an immense range of cultures. Menstruating women, or even women of an age to be menstruating, are believed to be too impure to enter temples or sacred spaces or participate in some religious rituals in many religions.

The Roman natural historian Pliny the Elder (d. 79 CE) viewed menstruating women as extremely powerful for both good or ill, although it is not always clear if he is describing his own beliefs or attributing them to others. A menstruating woman could stop a storm simply by exposing her body. Intercourse with a menstruating woman was very dangerous, particularly if the menstrual period coincided with either a lunar or solar eclipse. The touch or even the sight of a menstruating woman could cause a variety of disasters—her mere touch would kill plants and cause pregnant mares to miscarry. Some societies, Pliny claimed, had managed to turn this power to their advantage. In Cappadocia, menstruating women walked around wheat fields naked in order to kill insects and other pests. Care had to be taken, however, that the women did not walk in the fields at sunrise, which would harm the grain itself.

For Pliny and his ancient contemporaries, menstruation also had healing power. Menstrual fluid was an excellent topical application for gout, and a menstruating woman could cure scrofulous sores by touching them. Dogs became mad on contact with menstrual fluid, but for that very reason, cloth dipped in it could help cure rabies in humans. Most of the medical experts Pliny invoked are men like himself, but he also states that the midwife Sotira, the author of a now lost treatise on menstruation, believed that rubbing menstrual blood on a patient's feet was a good treatment for fever, particularly if done by the woman herself without the patient being aware of it. Menstrual blood had magical as well as medical powers for Pliny and his contemporaries. Application of menstrual fluids to doorposts would neutralize the spells of magicians.

The belief that menstrual blood goes into the making of a child as the female complement to the male semen was common in the ancient Mediterranean, as is the belief that it is transformed into breast milk. Both Aristotle and Pliny endorsed the idea that an infant was fashioned in the womb through a combination of the semen, that provided the form, and the menstrual blood, that provided the matter.

The concept of menstrual uncleanness is also found in the Hebrew Bible, most notably in the presentation of Jewish law in the Book of Leviticus. Leviticus views menstrual uncleanness as highly contagious, affecting anyone or anything that comes into contact with menstruating women or even comes into contact with things that have come into contact with them. The apocryphal Second Book of Esdras associated menstruation with monstrosity by claiming that menstruating women would bear monsters as a sign that the Last Days were near. This later became the common belief that sexual intercourse during a woman's menstrual period could engender a monster.

Christians and Muslims took over Jewish and pagan hostility to menstruation. The "Curse of Eve," which in the Book of Genesis refers to pains in childbirth, was expanded to include menstruation. There is a common Christian belief that the Virgin Mary never menstruated. (There is an analogous belief among some Muslims that Fatimah the daughter of Muhammad did not menstruate.) The idea that a menstruating woman was dangerous also continued into the Middle Ages. The influential late medieval tract *On the Secrets of Women*, falsely attributed to Albertus Magnus, endorsed the idea that the glance of a menstruating woman was dangerous. It also suggested that it was possible to distinguish between the menstrual flows of a virtuous and a corrupt woman in that, first, she would have menses the color of blood and that, second, she would have darker menses the color of lead. Menstrual blood continued to have power and was an ingredient in love potions.

During the Renaissance, many of the more extreme beliefs about the power and danger of menstruating women or menstrual fluid were quietly abandoned or viewed as superstitious. The German Swiss alchemical physician Paracelsus (1493–1541) is a partial exception, endorsing ideas connecting the glance of menstruating women to the evil eye and suggesting that the basilisk, a monster that kills through its glance, is engendered through menstrual blood. Menstrual blood continued to be used as an ingredient in alchemical formulae and love magic.

Ancient Greek and Roman physicians following the humoral system believed that menstruation helped keep the humors in balance in women, and even that it protected women against certain diseases caused by humoral imbalance to which men were vulnerable. In the European Middle Ages and Renaissance, menstruation was understood medically as the purging of corrupted blood. This was sometimes seen as an advantage that women had over men, who could not purge blood this way. Even more than men, however, postmenopausal women were considered to be victims of corrupted blood they could not release naturally. (One theory of the evil eye was that postmenopausal women, unable to menstruate, released their corruption through their gaze.) Bloodletting was a possible solution to the lack of a natural outlet. Bloodletting could also be recommended for women whose periods had stopped for no obvious reason.

There were attempts to find a male parallel to menstruation. In the West, hemorrhoidal bleeding was sometimes viewed as male menstruation. Anti-Semites claimed that God had cursed Jewish men with menstruation. The "Blood Libel," the myth that Jews ritually killed Christian children, sometimes involved the belief that Jewish men needed Christian blood to replenish that lost in menstruation. In Islamic dream interpretation, if a man dreamed he menstruated, it meant that he would commit a crime or a lie.

In Chinese culture, menstrual blood was both polluting and an ingredient in Daoist alchemical formulae. Menstrual blood was distinguished from other blood as being "unclean" blood. (It was not the only type of unclean blood; the blood of certain animals, such as black dogs, was also considered unclean.) As unclean, menstrual blood had the ability to counter magic—a belief Chinese shared with Pliny's ancient Romans. The failure of the Boxer Rebellion, a 1900 Chinese rebellion against foreign domination, was attributed to the foreigners' use of menstrual blood to negate Boxer magic. Menstrual blood also blocked communication from the gods; spirit mediums who came into contact with it lost their powers. Menstruating women were forbidden to enter temples lest they offend the gods, although they were expected to continue domestic ancestor worship. There was a religious belief that menstruation so offended the gods that women were tortured after their death in a hell called the "lake of blood," from which they could be liberated by the actions of filial children.

In Chinese alchemy, menstrual blood was referred to as "red lead" and used to make pills and elixirs to promote long life and youthful appearance. (Women alchemists, however, were expected to be able to make their periods cease, a process known as "beheading the red dragon.") It was also an ingredient in aphrodisiacs. Menstrual blood is extremely yin, and during their menstrual periods, women are considered unbalanced toward yin. Therefore during her period, a woman should avoid anything yin. This includes a prohibition of bathing, as water is yin. Exposure to yin at that time could lead to cramps or irregular periods. Excessive flow caused by the consumption of yin foods could weaken a woman by drawing clean blood from her body after the supply of dirty blood had been exhausted. Conversely, heat and yang foods were recommended both during menstruation and immediately afterward to promote recovery and the replenishment of the blood supply.

Sexual intercourse during menstruation was considered dangerous to both men and women. For men, it led to overexposure to the highly concentrated yin energy of menstrual fluid. For women, intercourse during menstruation blocked the flow, possibly resulting in the dangerous retention of toxic blood in the body.

The rough correspondence of the menstrual cycle and the lunar cycle has led to many superstitions and has possibly also contributed to the widespread, but not universal, identification of the Moon as feminine or ruled by a goddess.

See also: Alchemy; Bathing; Blood; Blood Libel; Curses; Evil Eye; Gender and Sexual

Difference; Humours, Theory of; Love Magic; Mirrors; Monsters; Moon; Pregnancy and Childbirth; Yin/Yang

Further Reading

Camporesi, Piero. 1995. *Juice of Life: The Symbolic and Magic Significance of Blood.* Translated by Robert R. Barr. New York: Continuum; Chu, Cordia Ming-Yeuk. 1980. "Menstrual Beliefs and Practices of Chinese Women." *Journal of the Folklore Institute* 17: 38–55; Hindson, Bethan. 2009. "Attitudes towards Menstruation and Menstrual Blood in Elizabethan England." *Journal of Social History* 43: 89–114.

Merlin. See Fairies

Mesmerism and Animal Magnetism

The Austrian physician Franz-Anton Mesmer (1734–1815) invented the influential doctrine and practice of animal magnetism, later known as "mesmerism." Animal magnetism was a method of healing physical illnesses employing the flows of an invisible fluid associated with living things. However, Mesmer quickly lost control of his movement, and "animal magnetism" included a wide variety of doctrines and practices in the late eighteenth and nineteenth centuries.

In his Viennese medical practice, Mesmer encountered the idea of the curative powers of iron magnets applied to the human body. He cured a woman suffering from seizures producing convulsions and vomiting by placing magnets on key areas of her body. He extended the idea of the "magnetic cure" to postulating a "magnetic fluid" permeating the universe. In Mesmer's time, there was a great deal of excitement over electricity, also conceived of as a fluid, and electricity had some influence on Mesmer's development of the magnetic fluid concept. "Animal magnetism" was not the

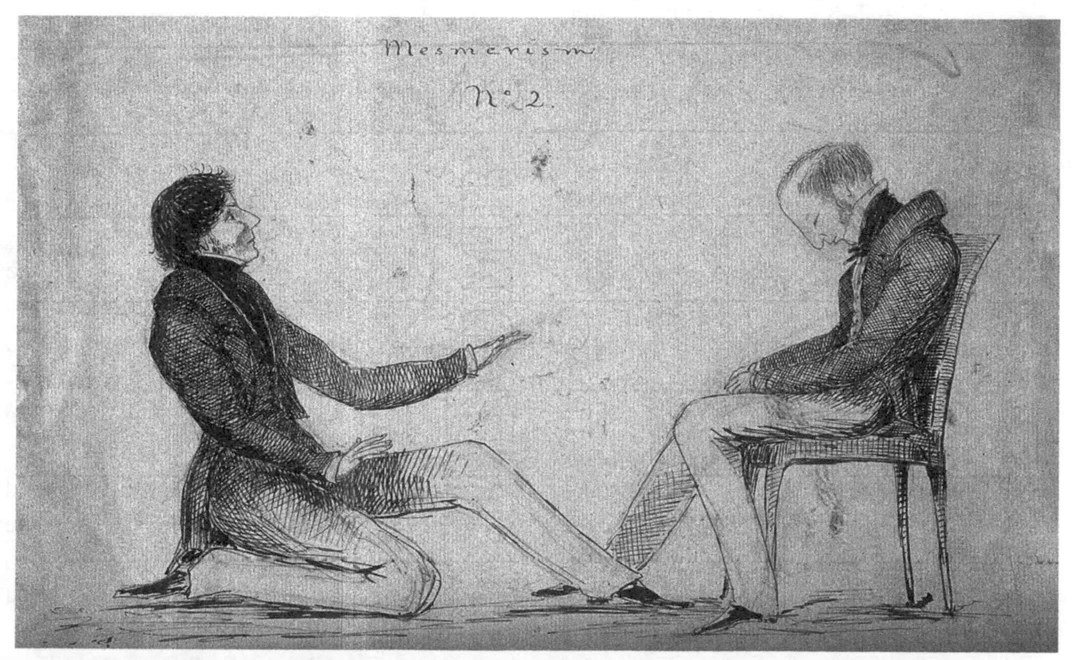

A representation of a mesmeric session, without the sensationalism of many anti-mesmeric or satirical images. ("Lt. Col. Ashburnham performing animal magnetism on Lt. Col. Forbes." Pen and ink drawing, 183. Wellcome Collection.)

same thing as magnetism, and Mesmer used substances and processes that were not "magnetic" in the conventional sense. The human body itself, Mesmer claimed, was the supreme animal magnet. The manipulation of its magnetic flows by direct physical contact between the "magnetizer" and the patient could cure many ailments, including hemorrhoids, irregular menstruation, and epilepsy. (Some have compared Mesmer's "magnetic fluid" to the Chinese concept of life energy, or qi.)

In 1778, Mesmer relocated to the center of European science, Paris. Mesmer and mesmerism took Paris by storm. He worked dramatic cures and was invited to speak before France's leading scientific body, the Royal Academy of Sciences. Demand for Mesmer's services was so great that he expanded beyond one-on-one sessions with the invention of the magnetic tub, a complicated device with bottles of" magnetized" water positioned inside an oaken tub large enough for several people. Holding hands, they created a "magnetic circle." Despite Mesmer's claims to a monopoly, other people started practicing mesmerism. The contact between mesmeric practitioners and their female patients caused many to suspect or mock them as seducers rather than physicians. Most reputable male mesmerizers insisted on having a witness present during mesmerizing sessions. Some argued that women were too fragile and emotional to be involved in animal magnetism either as practitioners or patients, but there were women practitioners.

The official test of animal magnetism was done by the Royal Academy of Sciences in 1784. Although Mesmer himself refused to cooperate, an ingenious series of tests performed on another magnetizer showed that patients could not distinguish between magnetized and unmagnetized objects if they were not told which was which. This led the academy's commission, which included the American ambassador Benjamin Franklin (1706–1790), to conclude that animal magnetism did not exist. This did not end the controversy, particularly as those opposed to the hegemony of the "scientific establishment" could now see mesmerism as another of its victims. Many future political radicals during the French Revolution were mesmerist supporters and pamphleteers. Mesmer himself became quite wealthy. Disheartened by splits in the mesmerist movement and no longer attracting large numbers of patients, Mesmer left Paris in 1785.

The most important of Mesmer's followers was the Marquis de Puseygur (1751–1825). In 1784, while mesmerizing Victor Race, a peasant suffering from congestion of the lungs, Puseygur discovered that Race had entered a condition he called "magnetic sleep," and what would now be called a hypnotized state (the word "hypnotism" was introduced in 1842). Amazed by the change between Race's normal condition and his condition under magnetic sleep, Puysegur ascribed various paranormal abilities, such as clairvoyance and telepathy, to the "magnetized" patient. Puysegur turned the doctrine of Mesmer based on flows of the magnetic fluid to one emphasizing the mental conditions of both the magnetizer and the patient and the mental and spiritual rapport between them. Magnetic sleep became a craze and the subject of several treatises.

Mesmerism had evolved far beyond Mesmer. It was now practiced to heal the patient's mind, rather than the body. It also became allied to esoteric and occultist movements. Mesmerism spread to Germany in a spiritualistic form oriented to magnetic sleep. By the 1810s, mesmerists were denying both Mesmer's magnetic fluid and Puysegur's emphasis on the spiritual rapport of magnetizer and patient and treating magnetic sleep purely as a psychological state attainable by the application of particular techniques. Mesmeric sleep was thus applicable to many medical problems, including those curable by suggestion, and also alleviated pain, such as that caused by surgery or dental extraction. Mesmerism also influenced America, where it contributed to the development of "positive thinking" and eventually the Christian Science movement of Mary Baker Eddy (1821–1910). Mesmerism appears frequently in nineteenth-century literature, including the works of Mary Shelley, Edgar Allan Poe, Honore de Balzac, Robert Browning, Alexandre Dumas (who included Mesmer as a character in his historical novel of the French Revolution, *The Countess of Charny* (1852)), and Charles Dickens.

See also: Magnetism

Further Reading

Crabtree, Adam. 1993. *From Mesmer to Freud: Magnetic Sleep and the Roots of Psychological Healing.* New Haven, CT and London: Yale University Press; Darnton, Robert. 1968. *Mesmerism and the End of the Enlightenment in France.* Cambridge, MA and London: Harvard University Press; Horowitz, Mitch. 2014. *One Simple Idea: How Positive Thinking Reshaped Modern Life.* New York: Crown.

Mice. See Rats and Mice

Mirrors

Mirrors are the object of many superstitions, of which among the best known is the association of breaking a mirror with seven years of bad luck. There are many variations on the idea of a broken mirror bringing bad luck, including that someone will die in the household within a year. Among Hindus in India, a cracked mirror or one that broke by itself was believed to be bad luck and needed to be hurriedly disposed of. (To see one's face in a mirror on rising, by contrast, was good luck.) There were also Western superstitions about it being bad luck for a baby or infant to look into a mirror or that looking too long into a mirror would cause one to see the devil—generally a rebuke against vanity, often associated with prolonged gazing into a mirror. In nineteenth-century Britain, where mirrors were becoming more common, it was frequently believed that it was dangerous to leave them uncovered in a sick room or a house where someone had recently died. Menstruating women reversed the normal process, whereby mirrors

A fifth–sixth-century CE Maya wooden mirror holder in the form of a court dwarf. The mirror would have been used for divination. (The Metropolitan Museum of Art, New York, The Michael C. Rockefeller Memorial Collection, Bequest of Nelson A. Rockefeller, 1979)

endangered people by having the power to bring corruption or foulness to a mirror, either by looking in it or being reflected in it.

Magicians used mirrors for scrying and other operations, and in the Middle Ages and Renaissance, when mirrors were rare and expensive, they were frequently seen in depictions of wizard's workshops. The English magus John Dee (1527–1608) used an obsidian mirror that originated in Aztec Mexico to summon spirits. The idea that mirrors could reveal spread from magicians to ordinary people, and a variety of practices developed for using mirrors to see one's future spouse, one's fortune, one's enemy, the thief who had taken something from the viewer, and many other persons and things. The Scottish poet Robert Burns (1759–1796) described a ritual for seeing one's future spouse by taking a candle, going alone before a mirror, eating an apple, and combing one's hair to see the face of the person in the mirror, looking over one's shoulder. Mirrors also had the protective power of deflecting the influence of the evil eye.

In China, home of some of the earliest bronze mirrors, mirrors played a central role in Feng Shui, redirecting the flow of energy, qi, in the environment. Mirrors also had a protective function; an adept Daoist wore a mirror on his back to protect against demons. Some demons had the ability to change their forms in the eyes of beholders, but a mirror would reveal their true nature. Chinese used mirrors to generate fires, "sun mirrors," and other mirrors to collect dew and condensation "moon mirrors." The sun mirror was believed to represent yang, the moon mirror yin. Holding a moon mirror to the Moon would generate rain.

See also: Death; Divination; Evil Eye; Menstruation; Moon; Natural Magic; Scrying; Yin/Yang

Further Reading

Campbell, Stuart, Elizabeth Healey, Yaroslav Kuzmin and Michael D. Glascock. 2021. "The Mirror, the Magus and More: Reflections on John Dee's Obsidian Mirror." *Antiquity* 95: 1547–1564; Opie, Iona and Moira Tatem, eds. 1989. *A Dictionary of Superstitions.* Oxford: Oxford University Press.

Mistletoe

Mistletoe, a plant parasite of the oak and other trees indigenous mostly to Europe, was surrounded by superstition. It was commonly associated with bird dung, particularly the thrush's, and some believed that it was either spontaneously generated from dung or that it had to mature in the stomach of a bird to grow. The Italian naturalist Ulisse Aldrovandi (1522–1605) denied this, pointing out that if this were true mistletoe would grow on all species of trees, which it does not. He suggested that it formed out of the tree itself, like an oak gall.

The ancient Druids of what is now France held mistletoe as sacred and gathered it with holy rites and a golden sickle. The source for the connection of the Druids and mistletoe, however, is the Roman natural historian Pliny the Elder (d. 79 CE), who wrote while Druidicism was being suppressed by the Romans, and some have argued that it is fictional. Pliny also claimed that the Druids believed that mixing mistletoe into a drink given to a farm animal would ensure its fecundity. Outside the Druid connection, Pliny also claimed that women who carried mistletoe on them would find it easier to get pregnant. The association of the mistletoe with fertility of both people and animals may be connected to its phallic-shaped leaves and white sticky sap, resembling male semen. The association would persist into modern times, leading to the custom of kissing under the mistletoe. Pliny ascribed many other powers to mistletoe. He believed that mistletoe was an antidote against all poisons and protected from injury by fire and water. He also described the belief that it could cure epilepsy if it was kept from touching the ground. The idea of using mistletoe as a preventative against epilepsy persisted into the early modern world, when it was believed that wearing it around the neck or taking it in a drink protected one from the "falling sickness." Endorsed by leading seventeenth-century natural philosophers such as Sir Kenelm Digby (1603–1665) and Robert Boyle (1627–1691), this belief was the subject of an eighteenth-century medical treatise, Sir John Colbatch's *A Dissertation concerning Mistletoe, a Remedy in Convulsive Distempers* (1720). (Like the Druids, Colbatch believed that the mistletoe found growing upon oak trees was the most powerful.) The work went through several editions. Mistletoe protected against supernatural ills as well as disease; if hung around the neck or in the house, it kept the wearer or inhabitants safe from witchcraft and evil spirits.

Mistletoe's ambiguity crops up in many legends. In the Roman legend of the hero Aeneas's visit to the land of the dead, the "golden bough" that he breaks to enter is commonly believed to be mistletoe. In Norse mythology, it was the plant made into a dart that killed the hero Balder. Despite its association with protection from evil, mistletoe never lost its associations with sexuality and paganism and was usually barred from churches.

See also: Defensive Magic; Iron

Further Reading

Opie, Iona and Moira Tatem, eds. 1989. *A Dictionary of Superstitions.* Oxford: Oxford University Press.

Mithridatium. See Theriac

Monsters

As embodiments of the "unnatural," monsters take many forms but are virtually ubiquitous in human cultures. The term has multiple meanings stretching from vast and powerful entities to deformed fetuses. Much early literature involves conflict with monsters, from the combat of Gilgamesh, the hero of the oldest surviving epic, with the monster Humbaba to the slaying of the demonic monster Kabandha by Rama in the Hindu epic *Ramayana.* Not just heroes but gods also were portrayed as being in conflict with monsters, such as the Babylonian god Marduk in a struggle with the monstrous female dragon Tiamat, a conflict that represented the conflict between order and disorder in society and the universe. Monsters continued to define the human and the normal by opposition, and many societies from the West to China identified the foreign as essentially monstrous.

Monsters can be seen as individuals or as "monstrous races," entire societies of monsters. In premodern societies, the areas of the world largely unknown to literate developed societies such as central Asia or the far north of Scandinavia were frequently identified as the lands of monstrous variations on humans such as headless, one-legged, or dog-headed people or people who were simply much larger or much smaller than normal humans. (Ancient Westerners, Iranians, and Chinese all believed that Central Asia was a land of dog-headed people.) There were several sources for the idea of monstrous races in classical antiquity, including the fifth-century BCE Greek historian Herodotus, the travelers Megasthenes and Ctesias, and the Roman natural historian Pliny the Elder (d. 79 CE). Early Iranian literature also talks about

The Transformations of the Ghoul

Ghouls were monsters originating in pre-Islamic Arabian mythology and spreading with Arabic culture in the Islamic period. The term "ghul" had many meanings, but its central one was a monster that waylaid, diverted, killed, and often ate travelers. Ghouls could be male or female and were often associated with the desert. They were usually described as ugly and often with the feet of an ass. Ghouls were often viewed as a subclass of the jinn.

The ghoul was introduced to Western literature through Antoine Galland's twelve-volume compilation of Arabic tales in French, *The Thousand and One Nights*. In a story whose Arabic original has never been identified, Galland described the ghoul as haunting graveyards and eating corpses. There is no Arabic precedent for this idea, but corpse-eating became the defining feature of the ghoul in Western literature. Galland may have been influenced by the eighteenth-century fascination with vampires to make the ghoul a corpse-eater.

Source Al Rawi, Ahmed K. 2009. "The Arabic Ghoul and its Western Transformations." *Folklore* 120: 291–306.

monstrous humans. Monstrous races appear in the popular ancient fictional romances about the adventures of the Macedonian conqueror Alexander the Great in the East.

Legends of monstrous races survived the coming of Christianity and persisted into the Middle Ages. There was a Christian debate as to whether or not these monsters possessed souls and could be saved, which was usually resolved in the monster's favor as they were also creations of God. The most extreme example of this phenomenon is seen in the common medieval legend that St. Christopher was a monster from the race of people having a dog's head. Islam faced a similar question as to whether monsters could be saved on dealing with the jinn, and it generally viewed the jinn as potentially savable. Christians also assimilated monsters through the idea that evil monsters were the children of the first murderer, Cain, or inheritors of his curse. This idea appears in the writings of Augustine of Hippo (354–430) and the Anglo-Saxon epic *Beowulf.* Monstrous races appeared in the writings of medieval travelers such as the fourteenth-century (and possibly fictitious) Englishman Sir John Mandeville, whose work was frequently translated and printed into the sixteenth century.

The European discovery of America led to the relocation of many of the monstrous races to the remote and unexplored (by Europeans) regions of the American continents. (Columbus himself remarked on the absence of the traditional monsters in the American islands he encountered, with the exception of "monstrous" cannibals.) Failure to encounter these monsters anywhere led to a growing skepticism about their existence by the late sixteenth century, although there was an occasional exception, such as the Gonzales family of Canary Islanders whose faces were completely covered with hair and who were described as monsters. European ideas about literally outlandish monsters also contributed to the development of ideas about race. The bodies of racial "others," whether internal such as the Jew or external such as the Black African, were believed to be essentially monstrous.

The association of monsters with the remote and unknown can also be seen in the belief in sea monsters in many cultures. Associated both with danger and the unknown, the sea was the natural haunt of monsters stretching from the Leviathan of the Book of Job to the Sirens of the Odyssey to the Kraken of Scandinavian epic. The *shen* of China was a shape-shifting, illusion-creating sea monster variously pictured as a giant clam or aquatic dragon. Scottish legends of sea monsters eventually led to the legend of the Loch Ness monster, believed by many people to be a reality today. Many of the largest and most terrifying sea monsters were based on whales.

Monsters were not always positioned in remote areas but often in the heart of human society. Many phenomena that would now be characterized as the result of a birth defect or mutation, such as conjoined twins, were characterized in premodern society as monstrous. The traditional Aristotelian theory of the generation of monsters was that they were caused by an excess or defect in the seed or a failure in the developmental process. (This led to the idea of a monster as an "error of nature.") In some sense, every human conception that did not lead to the ultimate goal—an anatomically complete, healthy human male—was monstrous, and some Aristotelians took this so far as to describe women as monsters.

The Aristotelian theory of individual monsters being the result of a failed natural process, however, coexisted with another ancient theory—that monsters were prodigies, divine warnings. Monsters were included in the lists of prodigies in the works of ancient Roman historians, and their interpretation was among the responsibilities of the Roman College of Augurs. (Monsters were also potential evidence of a dynasty's loss of the mandate of heaven in traditional China.) Christians adopted the classical tradition, situating monsters along with other prodigies as providential warnings from God. The appearance of monsters was also a possible sign of the fast-approaching end of the world. Monstrous beings appear in apocalyptic literature, including the Book of Revelation with its seven-headed beast arising from the depths of the sea.

Christian providential interpretation of monsters reached its postclassical height in early modern Europe, where monsters were pressed into service in the debates over the Reformation and the wars of religion. These disputes were not always marginal but involved leading figures of the age. Two of the most prominent leaders of the German Reformation, Martin Luther (1483–1546) and Phillip Melanchthon (1497–1560), wrote a pamphlet about two recent monsters, the "Papal Ass" and the "Monk-Calf." The Papal Ass had been found in the Tiber River in Rome in 1495 and combined the body of a woman, the head of an ass, and several monstrous limbs, while the Monk-Calf found in Germany in 1522 was hairless but had a mark on its head resembling a monk's tonsure. Luther and Melancthon offer a close reading of these monster's bodies (or texts describing them) in light of the issues of the Reformation. By making Popes and monks into monsters, these phenomena discredited Catholicism. (The case of the Monk-Calf was particularly urgent, as Luther's Catholic opponents identified that the warning God was sending as one of the monstrosities was of Luther, an Augustinian friar, himself.) The English Civil war of the mid-seventeenth century was accompanied by a plethora of accounts of headless births or births with deformed heads, signifying a crisis of authority as the head was the ruler of the body.

Monstrous births, however, could be interpreted not just in the light of collective sin, but also of individual sin or weakness, particularly that of the mother. Monstrous births could be the product of bestiality, whether committed by a man or woman. This went back to the ancient Greek myth of the minotaur, begotten by a bull on Queen Pasiphae. Monsters could also be the product of an unregulated maternal imagination, as a mother thinking of or viewing a thing other than her husband at the moment of conception or during pregnancy could generate a monster in the womb. A woman who craved a lobster could give birth to a monster resembling one, for example. The belief in the power of the maternal imagination to shape the child into a monster was not restricted to Europe; in South India, it was believed that women in their first pregnancies should not see a temple car with the statue of a god lest they give birth to a monster. Monsters had multiple meanings; the same monster could be viewed as both a collective warning from God to a sinful society and a punishment to a sinful individual, sometimes by the same person. One particularly historically significant case was the monsters born to Anne Hutchinson (1591–1643) and Mary Dyer (1611–1660), the women leaders of the "Antinomian" movement that challenged New England Puritanism. In the eyes of Puritan ministers, the stillborn monsters that the women gave birth to showed not only God's condemnation of their movement but also their role as sinful, heretical women.

Monsters were not merely punishments for sin, however; they were also examples of the power and fecundity of nature. Stuffed and otherwise preserved monsters were collected and appeared in cabinets of curiosities, and living monsters, such as the omnipresent dwarfs, entertained nobles at courts or commoners at fairs. Monsters were pictured and discussed in learned treatises and popular broadsides, and writers such as Giambattista Della Porta (1535–1615) gave instructions on how to create artificial monsters through natural magic. These varied between techniques to produce actual monsters, such as the breeding of two-legged dogs by repeated mutilation, and illusionary techniques, such as the distorting mirrors that could replace the head of a viewer with that of an animal's.

Monsters were also the subject of science. Sixteenth-century medical men, such as the French surgeon Ambroise Pare (1510–1590), wrote treatises discussing human and divine causes of monstrosity. The English philosopher of science Francis Bacon (1561–1626) suggested that the study of monsters could reveal truths about the order of nature. Reports of monstrous births and monstrous individuals, both human and animal, appeared frequently in seventeenth-century scientific publications such as *Philosophical Transactions of the Royal Society of London*. Scientists performed and published dissections of monsters, tending by the eighteenth century to explain the purpose of their research as casting light on normal development rather than exploring monsters for their own sake. The desire to examine or dissect a monster could involve scientists in conflict with the monster's family or owners, and scientists grew increasingly suspicious of fake or fraudulent monsters. The connection between science and popular monster culture went both ways, and by the late seventeenth century, promoters of monster exhibits were using scientific language and presenting their monsters as educational.

The belief that monsters were prodigies bearing providential or allegorical messages, as a two-headed baby presaged division in a kingdom, waned in the Enlightenment. Such beliefs were increasingly identified with intellectually marginalized populations, including the lower classes, religious dissidents, and women. However, monsters and their causes did remain the object of scientific curiosity. This tradition eventually developed into "teratology," literally the science of monsters, the study of developmental abnormalities in living things (including plants). In a broad sense, this can be seen as the revival of the Aristotelian approach.

Despite the naturalization of monsters and the end of belief in the traditional monstrous races, human culture has not lost its ability to create new monsters. From the vampire of the eighteenth century to the monstrous alien "Grays" of twentieth-century UFO mythology, monsters continue to shape and be shaped by human society.

See also: Dragons; Intersex Conditions; Jinn; Lycanthropy; Menstruation; Pregnancy and Childbirth; Prodigies; Vampires; Yokai

Further Reading

Hanafi, Zakiya. 2000. *The Monster in the Machine: Magic, Medicine and the Marvelous in the Time of the Scientific Revolution.* Durham, NC: Duke University Press; Mittman, Asa Simon and Peter J. Dendle, eds. 2012. *The Ashgate Research Companion to Monsters and the Monstrous.* Burlington, VT: Ashgate; Platt, Peter G., ed. 1999. *Wonders, Marvels and Monsters in Early Modern Culture.* Newark: University of Delaware Press; White, David Gordon. 1991. *Myths of the Dog-Man.* Chicago: University of Chicago Press.

Moon

The Moon has many associations that attract superstitions and supernatural beliefs. It complements the Sun as the luminary of the night, it moves swiftly through the heavens, and it has a very visible cycle of phases connected with both the tides and the female menstrual cycle. It also has prominent markings.

Moon deities are usually female. The Greek Selene, the Roman Luna, the Zulu iNyanga, and the Korean Myeongwol are among the many Moon goddesses. Moon gods are usually found in pantheons with female solar deities, as the Japanese Moon god Tsukiyomi is the estranged lover of the Sun goddess Amaterasu, head of the pantheon. One exception to this rule is the Egyptian pantheon, which includes male deities for both Sun and Moon, Ra and Khonshu. Moon deities do not generally occupy the leading position in a pantheon, one exception being the Moon god Sin in early Mesopotamia.

The Moon is generally considered feminine in Western astrology. It governs the fourth sign of the zodiac, Cancer. In addition to signifying women generally, particularly those in authority, the Moon signifies travelers. Constantly changing, the Moon signifies changeability. As the Sun corresponded to the heart, so the Moon did to the liver. The Moon was also magically and alchemically associated with silver as the Sun with gold.

The ancient Greeks and Romans considered the Moon goddess, known as Hecate or Selene, to be the ruler of magic, and many spells preserved from the ancient world include invocations to her. Magical rites were referred to as "calling down the Moon." The association of the full Moon with lycanthropy, so that werewolves were believed to assume the shape of wolves on the full Moon, however, emerged relatively late. Full Moons generally harbored danger, in that those who stared at the full Moon too long or were exposed to moonlight excessively could be moonstruck, literally "lunatics." Exposure to moonlight was also believed to cause fish to rot. The full Moon was generally believed to be disturbing, and there is a common belief even now that the world is more disorderly and crime is higher during a full Moon. In China, however, the full Moon was usually believed to be auspicious, its roundness a mark of perfection.

The Moon plays a prominent role in the apocalyptic beliefs of the Abrahamic faiths. The book of Joel in the Hebrew Bible and the book of Acts in the New Testament both describe the Moon turning to blood before the terrible day of the Lord. This is probably a reference to the "blood moon," a phenomenon sometimes accompanying lunar eclipses when the Moon appears red. The Quran's Sura of the Moon describes the Moon splitting into two as a sign of the apocalypse and a warning to believers to adhere to the truth. The Moon was generally considered auspicious in Islamic culture, particularly in the Ottoman Empire, which eventually adopted the crescent Moon as a symbol.

In the West, moonlight was associated with curing warts, an idea that goes back to the ancients and could originate from the idea that the waning Moon would cause warts to wane. The English natural philosopher Sir Kenelm Digby (1603–1665) recommended that warts be cured by being washed in moonlight reflected in a silver bowl.

There were many superstitions about the proper phase of the Moon to carry out various activities. Generally, the time when the visible portion of the Moon was increasing, or "waxing," was considered better for new undertakings than the time when it was shrinking, or "waning." The Roman natural historian Pliny the Elder (d. 79 CE) described the waxing Moon as filling the Earth with life, a phenomenon exemplified by the growth of shellfish. The lunar cycle had particularly pronounced impact in the area of sexuality and procreation. One reason for the connection of the Moon with the feminine is the linkage between the Moon's cycle and the female menstrual cycle. There are folk-beliefs that menstruation is more likely to start at the new or full Moon. The Moon also affected pregnancy—Pliny claimed that the only children delivered in seven months were conceived either the day before or the day after the full Moon or in the dark of the Moon.

It was fortunate to be conceived or born in the Moon's increase, unfortunate to be born while it was waning or dark. Weddings should take place only during the Moon's increase. There were exceptions to the favorability of the waxing Moon. Such was the pull of the waxing Moon on the tides that it was dangerous to be bled during the increase.

Lunar cycle superstitions also played a major role in agriculture. It was usually fortunate to sow or undertake other activities such as curing bacon in the waxing Moon. Pennsylvania Germans believed that the day before the new Moon—the "dark of the Moon"—was the best time to plant potatoes. The lunar cycle was often thought to be particularly influential on the course of a disease. Galen, one of the most influential of ancient physicians, endorsed the idea of the Moon's influence on health. His theory of "critical days," the days of most activity during an illness that determine whether or not the patient would recover, depended on the lunar phases. (Although some physicians who believed in critical days denied their connection to the Moon.)

The marks on the surface of the Moon have contributed to the myth of the Moon being inhabited. The most common Western version of this legend is the man on the Moon, according to one version of the tale exiled to the Moon for picking up sticks on Sunday. The Chinese had several inhabited Moon myths, including a goddess, a hare, and a toad on the Moon. The Aztecs claimed that an angry god had flung a rabbit at the Moon as an insult. The Maya and the Khoikhoi of southern Africa also identified the figure on the Moon as a hare or rabbit.

See also: Astrology; Diana; Eclipses; Hares and Rabbits; Menstruation; Pigs; Pregnancy and Childbirth; Silver; Sun; Toads

Further Reading

Gruber, Christiane. 2019. *The Moon: A Voyage through Time.* Toronto: Aga Khan Museum; Montgomery, Scott L. 1999. *The Moon and the Western Imagination.* Tucson: University of Arizona Press.

Moxibustion. See Acupuncture

Mummies

Mummies, long-enduring objects, have had striking different cultural meanings and uses over the millennia. In ancient Egypt, the preparation of a mummy was an elaborate process oriented to the afterlife. Mummification was carried out by priests, as it required both technical and spiritual expertise. Rituals drove malevolent supernatural beings, such as the god Set, away from the embalming space as the body was mummified. By removing all moisture from the body and wrapping it in resin and protective linen cloth, mummification created a body free from decomposition, a fitting home for the soul as it entered the afterlife. The wrapping also incorporated protective amulets. Another ritual "opened" the senses so that the mummified person would be able to enjoy the things provided in the tomb. Mummification was an expensive process usually restricted to pharaohs and society's elite, although for reasons that remain unclear animals including baboons, crocodiles, and cats were sometimes mummified. The practice of mummification ended in the Roman period.

In medieval and early modern Europe, mummies were thought to be medically useful. This idea seems to have been originally based on the idea that bitumen, a substance used to wrap mummies, was also used as medicine by Arab physicians under the name of "mumia." This idea began with naturally occurring bitumen, but quickly extended to the bitumen and other substances used in mummy wrapping. From there, the idea spread that mummified bodies themselves were of medicinal value. The Muslim authorities in Egypt frowned upon the trade, but there was an active but illicit business of exporting mummies and powdered mummies to Europe from the fifteenth to the eighteenth century. Such was the demand for mummy that fake mummy was produced from the dried bodies of travelers who had died in the desert and other recent sources. Medical experts warned against the use of fake mummy, some of which they suspected was even made from the bodies of executed criminals. Mummy was considered particularly valuable in treating falls, bleeding, and bruises. It fell out of fashion in the eighteenth century, although there are occasional references to mummy medicine into the twentieth.

The idea of the mummy as an undead creature, whether a romantic hero or heroine or a threatening monster, emerged in the nineteenth century in parallel with increasing scholarly and popular knowledge of ancient Egypt. This period also saw the emergence of the idea of the "mummy's curse," which threatened archeologists and tomb robbers with death.

See also: Cats; Curses; Death

Further Reading

Elliott, Chris. 2017. "Bandages, Bitumen, Bodies and Business: Egyptian Mummies as Raw Materials." *Aegyptiaca: Journal of the History of Reception of Ancient Egypt* 1: 26–46.

N

Natural Magic

"Natural magic" was a concept devised in the European Renaissance to denote magic that involved manipulating the hidden or "occult" properties of various substances and objects, as opposed to magical operations such as necromancy or scrying that involved angels or devils. Natural magicians argued that their practice was religiously permissible, in that it did not involve dealing with devils. "Natural Magic" became a catch-all term, not only overlapping with alchemy and astrology but also including subjects such as magnetism, salves and ointments, and the magical or healing properties of gems. It drew on a wide range of ancient, medieval, and Arabic sources. Natural magicians were also interested in testing popular superstitions or identifying their basis. They distinguished themselves from natural philosophers by taking an activist position, seeking knowledge not for knowledge's sake but for use. For example, the skilled natural magician was expected to be knowledgeable of how to use lenses and mirrors to play tricks with light optics, but not necessarily of academic optical theory.

Natural magic frequently aimed to startle and amaze. A classical piece of natural magic was the sixteenth-century English philosopher John Dee's invention of an artificial insect to fly across the set of a play. Natural magic's entertainment value made it popular in courtly settings, but it also reached out to a wider population in books of secrets, widely distributed collections of recipes and formulae.

Phenomena thought demonic could be explained by natural magic, as the eminent Neapolitan Giambattista Della Porta (1535–1615), author of *Magia Naturalis* (1558), claimed that the ointment that allegedly gave witches the ability to fly was really a hallucinogen working through the natural properties of the ingredients. Despite its difference from witchcraft, however, natural magic attracted the interest and suspicion of the Catholic Church in the late sixteenth century, as well as that of many Protestants. The Church wished to monopolize supernatural power and associated natural magic with witchcraft and demonic magic. The Inquisition suppressed Della Porta's writings and forced a group of natural magicians he founded in Naples to disband. But in the long run, a more effective strategy for combating natural magic was promoting the natural philosophy of the Scientific Revolution, whose material and mechanist approach to the universe left no occult powers to serve as the basis for natural magic. The tradition of natural magic continued at a popular level, and the late seventeenth-century public performer of entertaining experiments for salon, lecture hall, or courtly audiences was the heir of the natural magician.

See also: Alchemy; Amulets and Talismans; Mirrors; Precious Stones; Royal Touch; Weapon-Salve

Further Reading

Eamon, William. 1994. *Science and the Secrets of Nature: Books of Secrets in Medieval and Early Modern Culture.* Princeton, NJ: Princeton University Press.

Naturphilosophie

Naturphilosophie, the "philosophy of nature" was founded by the German professor Friedrich W.J. Schelling (1775–1854), who published *Ideas for a Philosophy of Nature* in 1797. Although strongest in Germany, the movement also greatly influenced Scandinavia, which looked to Germany for intellectual and cultural leadership. *Naturphilosophie* was intellectual resistance to the mechanism and materialism that dominated Enlightenment science. *Naturphilosophes* saw the investigation of nature as a spiritual quest with an ultimately spiritual goal. Nature did not exist outside humans as a material phenomenon but was ultimately a product of the human spirit. *Naturphilosophie*, unlike much Enlightenment science, was not expected or desired to lead to technological progress. The mission of *Naturphilosophie* was ultimately to restore, on a higher level, the original unity of man and nature. This unity existed in the Golden Age or, for more Christian *Naturphilosophs*, before the fall of man, when the products of the human spirit became separated from the human spirit itself.

Nature itself was moved by transcendent, nonmaterial forces. For example, a nonmechanical drive to organization manifested itself both in the crystallization of minerals and the growth of living things. *Naturphilosophs* studied these transcendent forces and the systems that they shaped. Their approach to these forces was dualist—*Naturphilosophs* saw the world as governed by pairs of opposed forces and placed great emphasis on symmetry. (The *Naturphilosoph* Johann Wilhelm Ritter (1776–1810) discovered ultraviolet light after hearing about the discovery of infrared light by reasoning that there must be something at the other end of the scale.) Phenomena taken in isolation were not the proper object of study, as they could be for "reductionist" scientists. *Naturphilosophs* were holists, emphasizing entities as wholes that worked in a certain way rather than collections of parts. Mechanical science was not wrong so much as radically incomplete. Only the *Naturphilosoph*, possessing spiritual awareness, could truly understand nature—traditional scientists merely explored the outward seeming.

Naturphilosophie had the most impact on biology. *Naturphilosophs* treated the life sciences as the model for the others, applying biological concepts of development to the inanimate universe. Seeing a universe structured by cosmic dualities, *Naturphilosophs* identified these dualities as male and female and conceived of the animate universe as a pregnant female. In biology itself, *Naturphilosophs* emphasized parallels between the anatomy of different creatures, seeing a range of geometrical variations on a few basic forms. They were more concerned with abstract variation than what the organs and parts of the body actually did. The biologist and professor Lorenz Oken (1779–1851), for example, held that all bones were modifications of the vertebra, suggesting that even the skull was created by the fusing of four vertebra. Polarity became an organizing concept for physiologists, reaching an extreme in the works of the poet physician Joseph Gorres (1776–1848), whose *Exposition of Physiology* (1803) and other works set forth an elaborate system of parallels between physiological and cosmic forces.

After the 1820s, *Naturphilosophie* lost most of its vitality. New German scientific leaders such as the organic chemist Justus von Liebig (1803–1873), who famously referred to *Naturphilosophie* as the "Black Death," asserted that mechanical and reductionist science, including mechanical and reductionist biology, was the only proper science, ridiculing *Naturphilosophie*'s transcendent orientation.

See also: Brunonianism; Dowsing

Further Reading

Cunningham, Andrew and Nicholas Jardine, eds. 1990. *Romanticism and the Sciences*. Cambridge, UK: Cambridge University Press; Jardine, Nicholas. 1996. "*Naturphilosophie* and the Kingdoms of Nature." In N. Jardine, J. A. Secord, and E. C. Spary, eds. *Cultures of Natural History*. Cambridge, UK: Cambridge University Press. Pp. 230–245.

Nazar. See Amulets and Talismans

Necromancy

Necromancy emerged in the ancient Mediterranean as the summoning of the dead. Originally, it was a Latin portmanteau of "Nekroi" meaning dead and "mantia" referring to divination. Although the word was not used in biblical Hebrew, the witch of Endor who summoned the spirit of the deceased prophet Samuel for King Saul would have been identified as a necromancer. Classical writers such as the Roman poet Lucan also showed witches as calling up the spirits of the dead. The usual purpose of this exercise was to gain knowledge that the dead were believed to possess. That made necromancy a form of divination.

Christianity refuted the legitimacy of classical necromancy by denying that magicians had the power to summon the dead from the afterlife. If a magician called up an entity claiming to represent a dead person, it was actually a demon impersonating that person. Necromancy continued into the Christian era, but it was now viewed primarily as the summoning of demons. The European Middle Ages from the twelfth to

A fanciful eighteenth-century representation of a necromantic rite, by Giovanni Battista Tiepolo. (The Metropolitan Museum of Art, New York, Bequest of Phyllis Massar, 2011)

the fifteenth century was the golden age of necromancy, viewed as the practice of summoning and commanding spirits, most frequently demons but also angels and planetary "astral spirits." While female necromancers frequently appeared in the ancient literature, the medieval image of the necromancer was male. The identification of necromancy as a learned art meant it was principally practiced by the clergy, including at the highest level. At least four popes during the Middle Ages were accused of necromancy by their political enemies: Sylvester II (Pope, 999–1003), who was accused of having learned necromancy in Islamic Spain, Gregory VII (Pope 1073–1085), Boniface VIII (Pope, 1294–1303), and Benedict XIII (Pope, Avignon Succession, 1394–1403). These accusations were products of the fierce political conflicts emerging around these popes, and not necessarily actual papal necromancy.

Necromancy was embodied in Latin books called grimoires that were clandestinely circulated. These grimoires were sometimes ascribed to the biblical King Solomon, considered an authority on necromancy. The ancient Roman poet Virgil was also considered to have been a great necromancer. Necromancers were frequently charged with associating with Muslims and Jews and learning magic from them.

Medieval necromancers employed spirits for many purposes, including the gaining of knowledge. This knowledge could be specific, as in finding the identity of a thief who had stolen from the necromancer, or general, as in learning arts and science. Another necromantic power was coercing people to do the necromancer's will, including forcing women to have sex with him. Necromancers could also work powerful and convincing illusions. Necromantic procedures included invocations and the drawing of magic circles in which demons and other spirits could be summoned. Necromancers drew complex patterns on parchment, often using the blood of a human or animal as the ink. They also wrote out verbal formulae, including names of God or invocations honoring God or the saints. Necromancers also employed scrying as a divination technique, looking into such surfaces as mirrors or fingernails to acquire hidden knowledge. Necromancy was connected to other magical sciences such as astrology—some necromantic rituals had to take place at astrologically determined times, and astral images, many derived from the magical traditions of the Islamic world, were frequently used by necromancers.

Necromancers defended their art by claiming it was permissible to invoke and command demons if one did not honor or worship them. Theologians said that any kind of dealing with demons was impermissible, even the use of natural knowledge gained from demons. In the eyes of the Church, while many forms of magic relied implicitly on the power of demons, the necromancer differed from other magic users in that he explicitly and knowingly invoked the power of demons, making necromancy a much more serious charge than other forms of magic. Some books of necromancy did include formulas for praising and honoring demons or even engaging in ascetic practices such as fasting in their honor. Demons were also honored with the sacrifice of birds or animals. Food could also be offered to the demons. However, the necromancer was not only the suppliant of demons but also their master, and many necromantic formulae included commands given to demons, frequently in the name of God or the saints. (The tradition of Christian exorcism, in which the exorcist commands demons to leave a person or place in the name of God, influenced the necromantic attitude to demons.) The necromancer did not put himself beyond God's mercy, but necromancy could only be atoned for with a life of rigorous penance, and even the repentant necromancer could still be troubled with demons. The Inquisition was the foe of necromancy, torturing and burning necromancers and destroying their books (although many survive to the present day). Clerical authorities increasingly interpreted necromancy in terms of the Satanic Pact, as necromancers got their power from Satan in exchange for their souls. This idea of the satanic necromancer would contribute to the image of the witch during the early modern witch hunt.

Although necromancy lost much of its vitality and popularity after the Middle Ages, it continued into the early modern period, and eventually contributed to the emergence of the modern Western tradition of ritual magic.

See also: Astrology; Blood; Death; Divination; Love Magic; Satanic Pact; Scrying; Witchcraft

Further Reading

Fanger, Claire, ed. 1998. *Conjuring Spirits: Texts and Traditions of Medieval Ritual Magic*. Phoenix Mill: Sutton; Kieckhefer, Richard. 1989. *Magic in the Middle Ages*. Cambridge, UK: Cambridge University Press.

Neidan. See Alchemy

New World Inferiority

From the beginnings of the European invasion of the Americas, there was a tension between the belief that the Americas was the earthly paradise with abundant resources and the belief that the people, and eventually the nature, of the New World were inferior. This belief grew stronger in the eighteenth century, but also faced vigorous rebuttal from Americans.

At first, much of this debate took place in astrological terms. The earliest areas of the Americas to be colonized were considerably to the south of Europe, so Europeans were observing an unfamiliar sky. The first natural historian of Spanish America, Gonzalo Fernandez de Oviedo (1487–1557), argued in his *General History of the Indies* (1535) that the southern stars induced cowardice and weakness among those whom they ruled. The Franciscan friar Bernardino de Sahagun (c. 1499–1590), author of the *General History of the Things of New Spain*, was more sympathetic to the Natives, but argued that the corrupting influence of the stars made Natives Americans unfit for the priesthood and weakened Creoles, the descendants of European settlers in the Americas, as well. Sahagun and others asserted that the only way of overcoming this astrologically based weakness was through strict discipline, a regime that Sahagun claimed had characterized the preconquest Aztec Empire. This argument, common among Creoles, justified harsh exploitation. The Dominican Gregorio Garcia (d. 1627) argued that the American stars and climate explained how the Carthaginians, who he believed were the original settlers of the Americas, had degenerated from their brave ancestors into "slothful," "cowardly" Natives. The decline of learned astrology in the seventeenth century marginalized this argument, but the idea that Native Americans, despite their physical hardihood and endurance, were somehow inferior persisted, partially based on Native susceptibility to imported disease.

In the eighteenth century, European natural historians maintained that the entire biology of the New World was inferior to that of the Old World. The most famous zoologist of the century, the Comte du Buffon (1707–1788) pointed to cases where New World animals were smaller than those of the Old World, as pumas were smaller than lions. He argued that the Americas, for reasons of climate and humidity, produced generally inferior animals. Those that should be large, such as mammals, were smaller, while animals that should be small, such as insects and frogs, were bigger. Moreover, Eurasian animals such as horses degenerated in the Americas. Buffon's argument was expanded and amplified by a Dutch scholar, Cornelius de Pauw (1739–1799), author of *Philosophical Researches on the Americans or Memoirs for the History of the Human Species* (1768). De Pauw extended Buffon's claims about the inferiority of American animals to claims of the inferiority of American people, both Native and European descended. De Pauw argued that nature in the Americas was not

The Mystery of the Opossum

When European settlers arrived in the Americas, they were completely unfamiliar with marsupial animals, found only in Australia and the Americas. The only American marsupial is the opossum whose unfamiliarity attracted a host of legends. William Byrd, an eighteenth-century Virginia landowner, wrote a book on Virginia's natural history mostly with an eye to practical considerations. But the opossum appears as an utterly bizarre creature:

"This animal is found nowhere else in the whole world except in America; it is a phenomenon among all animals living on land. Its shape and color are similar to a badger. The male's genital member sticks out in the back, because of which they turn their backs to each other at the time of copulation. The female bears her young in the teats, or breasts, where they grow. At first, they are no larger than a pea and cling fast to them before they seem even to be alive. This animal has another stomach, in addition to the natural one, in which it carries its young—after they fall away from the breasts—until they can help themselves. Meanwhile, they run in and out of it as they desire until they become large. If a cat has nine lives, as one commonly says, then this animal has certainly nineteen, for if all the bones in its body are broken, so that it lies there as if dead, still it recovers again in a short time and very soon gets well again."

Source: Byrd, William. 1940. *William Byrd's Natural History of Virginia or the Newly Discovered Eden*. Richmond, VA: Dietz. Pp. 85–86.

immature, as Buffon described it, but senile. Native Americans were smaller, weaker, and less intelligent than Europeans. Native American women were so ugly as to be indistinguishable from Native men. In fact, De Pauw suggested that Native Americans were not really human at all. Although his principal focus was on people, De Pauw drew heavily on Buffon to suggest that the inferior environment of the Americas affected animals as well, both those indigenous to America and those imported there—dogs lost their bark, camels their reproductive ability.

The argument on the inferiority of American nature expanded into all kinds of detailed claims, such as that American birds could not sing beautifully like the nightingales of Europe. Some made great play with the beardlessness of Native Americans or Native men's purportedly lower sex drive in comparison with European men. The most controversial claim gave a new scientific gloss to the old anti-Creole prejudices by asserting that Europeans themselves inevitably degenerated in the Americas. This aroused the pride of scientists in both North and South America at a time when in both the British and Spanish Empires Creole intellectuals were feeling an increasing sense of their own identity that would eventually be expressed in independence struggles. Creoles praised the abundance, health, and wealth of their lands and the strength of their peoples, making exactly the opposite claims to those of Buffon, De Pauw, and other European naturalists. Some works by Creoles included discussions of natural history meant to disprove the assertions of Buffon and De Pauw. One well-known example is Thomas Jefferson's *Notes on the State of Virginia* (1785). Jefferson's discussion of the bones of fossil mammoths found in America is aimed at Buffon's assertion that nature was less active in the Americas. Jefferson claimed that the mammoths, which he believed might still be found in the American interior, were far larger than the elephants of Asia and Africa. Jefferson's fellow scientist and revolutionary Benjamin Franklin (1706–1790) was also a defender of American nature.

Spanish America, particularly Creole Jesuits, made a particularly strong contribution to the case against New World inferiority. The Mexican Jesuit Francesco Saverio Clavigero's four-volume *Ancient History of Mexico* (1780–1781) praised the sweet song of Mexican birds, rebutting European claims that American birds could not sing, and the abundance and hardihood of American crops such as corn and chocolate. Nor were America's Natives physical weaklings. Clavigero had seen them carry burdens heavy enough to amaze a European philosopher such as De Pauw. Giovanni Ignazio Molina (1740–1829), a Creole Jesuit from Chile, published *Natural History of Chile* (1782), which pointed out that Buffon never visited the Americas himself and instead drew on unreliable travelers' accounts. He charged that European naturalists also made mistakes by not taking American nature on its own terms, instead applying European names without regard for whether the American plants and animals they were describing were really the same as the European creatures with that name. Despite these efforts, however, the idea of the inferiority of American nature persisted into the nineteenth century.

See also: Astrology; Decay of Nature; Race

Further Reading

Canizares-Esguerra, Jorge. 2006. "New World, New Stars: Patriotic Astrology and the Invention of Amerindian and Creole Bodies in Colonial Spanish America, 1600–1650." In *Nature, Empire and Nation: Explorations of the History of Science in the Iberian World.* Stanford, CA: Stanford University Press. Pp. 64–95; Gerbi, Antonello. 1973. *The Dispute of the New World: The History of a Polemic, 1750–1900.* Translated by Jeremy Moyle. Revised Edition. Pittsburgh: University of Pittsburgh Press.

Nudan. See Alchemy

Obeah

Obeah originated as a term for magic practiced by slaves and other African-descended persons in the British Caribbean, particularly Jamaica. It persisted after the abolition of slavery and till the present day. The origin of Obeah can be difficult to understand since most of our surviving sources for its early history are slaveowners or other whites, who often identified rebellious slaves or people respected in the African-descended community as obeah practitioners. The earliest written uses of the term date from the early eighteenth century, when African slavery in the Caribbean had been established for over a century. Some scholars argue that "obeah" is a label imposed by whites on diverse Afro-Caribbean practices rather than a unified system.

Obeah originates in the practices and beliefs of West Africans, although scholars of the subject disagree about which West African group is the main origin. Unlike the voudun of Haiti or the santeria of the Spanish Caribbean, obeah does not incorporate the worship of gods. Obeah practitioners carried out some of the same functions as cunning people in Europe, such as divination, healing, love spells, or the locating stolen goods. Obeah workers also claimed to have the ability to protect against malevolent supernatural forces. They were also associated with more malevolent practices such as curses and poisonings. (The connection of obeah with alleged poisonings was particularly important to slaveowners, who depended on their slaves for food preparation.) One connection between obeah and slave revolts was the belief that obeah practitioners had the ability to protect rebels against bullets. Queen Nanny, the leader of the early eighteenth-century Maroons, a community of runaway slaves and their descendants, was credited with obeah powers. Her success against the British in the First Maroon War (1728–1740) was attributed to her ability to use obeah both to provide defenses for her people and to attack British soldiers. Obeah was banned in Jamaica after the suppression of the slave rebellion known as "Tacky's War" in 1760, although this prohibition was ineffective. In the early nineteenth century, Matthew Lewis (1775–1818), the owner of a Jamaican slave plantation, described an old man as carrying a bag filled with cat's ears, animal feet, fish bones, and alligator teeth for working obeah.

The use of obeah for gain or extortion remains illegal in Jamaica, although it has been legalized or decriminalized throughout much of the rest of the English-speaking Caribbean, and even in Jamaica, the law is seldom enforced. Although the connection of obeah to West African culture remains strong, there are also a few cases dating to the nineteenth century of non-African-descended people working obeah.

See also: Cats; Curses; Defensive Magic; Divination

Further Reading

Lewis, Matthew. 1999. *Journal of a West-India Proprietor Kept during a Residence in the Island of Jamaica.* Oxford: Oxford University Press; Paton, Diana. 2015. *The Cultural Politics of Obeah: Religion, Colonialism and Modernity in the Caribbean World.* Cambridge, UK: Cambridge University Press.

Omamori. See Amulets and Talismans

Oneiromancy. See Dreams

Oni. See Yokai

Onions

Fiery, pungent, and widely distributed, onions have acquired many superstitious and magical associations. The ancient Egyptians revered onions, and Egyptians, mummified or not, were often buried with parts of the onion plant. (The Roman satirist Juvenal claimed that Egyptians so revered onions that it was reckoned sinful to eat them, but this is undoubtedly humorous exaggeration.) The onion was frequently believed to have the power to heal or ward off disease. Roman natural historian Pliny the Elder (d. 79 CE) believed onions, made into a poultice with vinegar or with wine and butter, were effective in treating wounds, particularly dog bites. Pliny, along with other ancient writers on medicine, also recommended onions for diseases of the eyes, recommending that the sufferer smell the onion until tears came to their eyes, although it was even better if their eyes were rubbed with onion juice. Onions were also good for toothaches and the bites of snakes and scorpions.

Perhaps due to their pungent smell, onions were used to protect from plague, smallpox, cholera, and other contagious diseases. Protective onions could be carried on the body or in the pocket. Cut up, they could be placed in a room to absorb the disease or its causes. Onions are also used in traditional Chinese medicine. Eaten, onions are believed to regulate the body's qi, or life energy. The onion was also recommended by Indian physicians for heart and joint illnesses as well as indigestion.

Onions could be viewed both as favorable and unfavorable plants. Onions were favorably contrasted with garlic in the Christian and Islamic legend that when Satan was expelled from the garden of Eden onions grew from his right footprint and garlic from his left. The nineteenth-century American grimoire, *The Golden Wheel Dream-book and Fortune-teller*, describes dreaming of eating onions as a good omen and of throwing them away as a bad one. However, there was also a tradition that having cut onions in the house was bad luck. Pennsylvania Germans believed that if a planter missed a row when planting onions, someone would die.

Onions were used in divination. Early modern European women had numerous techniques for using onions to divine the identity of a future sweetheart or spouse. One technique was to take a group of onions, name them after different men, and observe which one sprouted first. The one that onion was named after would be the future partner. Another common method was to put an onion under the pillow to dream of a future husband. This was frequently associated with the eve of St. Thomas's Day, the winter Solstice December 21. Other days for onion-related love divination included Christmas and St. Valentine's Day.

Other magical uses for the onion included an ancient Roman charm employing onions, hair, and pilchards to summon thunder and lightning. Attributed to the legendary second king of Rome, Numa Pompilius, this charm was still in use in the period of the Roman Empire according to the Greek scholar Plutarch (b. 46 CE–d. after 119 CE). In the Mediterranean area and much of Europe, the onion was believed to be a defense against the evil eye or malevolent witchcraft.

See also: Defensive Magic; Divination; Evil Eye; Garlic; Herbs; Luck; Snakes; Witchcraft

Further Reading

Wilson, Eddie W. 1953. "The Onion in Folk Belief." *Western Folklore* 12: 94–104.

Oracle Bones

One of the earliest known divination techniques is the use of oracle bones in the semi-legendary Shang Dynasty of China (c. 1600–1046 BC). Shang court officials divined the future by heating the shoulder blades of cattle and tortoise shells and reading the cracks and the sounds of the cracks the heat produced. (Some scholars have argued that turtle shells were used because their rounded shape meant they represented the cosmos, although Shang diviners used the plastron, the bottom flat shell, in preference to the carapace, the domed top shell.) Turtle shells were included in the tributes dependent rulers sent to the Shang monarch. Before the bones were heated, small holes or "pits" were drilled in them to encourage cracking, and the questions to be asked were inscribed on the bones in "oracle bone script," the earliest form of Chinese writing to survive in significant quantity. Sometimes the result of the divination was inscribed on the bone after use. The bones themselves were preserved for a short period of time, but then buried in large pits dug into the Earth rather than permanently preserved. Tortoise shells and bones were buried in separate pits. The process took place in the temples of ancestral spirits.

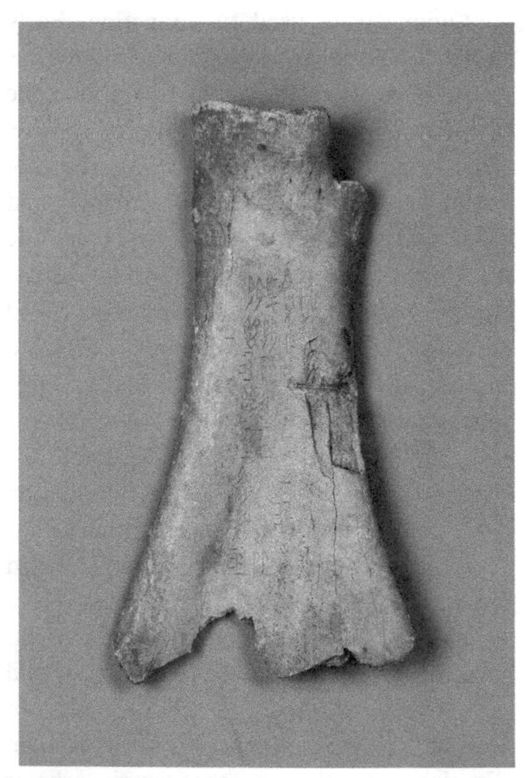

A Shang Dynasty oracle bone. (The Metropolitan Museum of Art, New York, Gift of Paul E. Manheim, 1967)

Oracle bone divination was carried out by diviners working for the Shang monarch and was concerned with public questions. Private citizens were not allowed access to oracle bones. Public questions included those having to do with agriculture, such as rain. Surviving accounts of oracles sometimes claim that rain will come regardless of human action, but also sometimes claim that nature spirits or ancestral spirits will send rain if honored with sacrifice. War, hunting, and the health of the king and royal family were also topics of oracles. As the divination tradition developed in the Shang Dynasty, the range of questions asked grew narrower and the divinations themselves more consistently positive. The Shang monarch himself did more and more of the divining, eclipsing the role of professional diviners.

Although it never entirely died out, after the fall of the Shang Dynasty oracle bone divination was supplanted by other methods such as the use of the *Yijing* or "Book of Changes" as divination spread beyond the court more broadly in Chinese society. When farmers happened upon oracle bones millennia later on abandoned Shang Dynasty sites, they treated them as "dragon bones" and used them medically by grinding them and consuming the powder.

See also: Divination; Dragons; *Yijing*

Further Reading

Smith, Richard J. 1991. *Fortune-Tellers and Philosophers: Divination in Traditional Chinese Society*. Boulder, CO: Westview Press.

Otoliths, Fish

Otoliths or "ear stones" are the hard stones found in the ears. The otoliths of fish in particular have a variety of magical and medical uses. Otoliths were worn as amulets, sometimes in costly and elaborate settings, and ground into powder for human consumption. The Roman natural historian Pliny the Elder (d. 79 CE) distinguishes between the medical uses of various otoliths, all of which are associated with food fishes, whose otoliths were the kinds humans were most likely to encounter. The otoliths of a fish called Bacchus treated kidney stones; another fish called Asellus provided stones that, when bound up with linen and applied to the patient, were an excellent treatment for fevers, although the otoliths had to be gathered during a full Moon. The otolith of the fish called Cinaedus was used to divine weather conditions at sea, although Pliny does not say how it was so employed.

Isidore of Seville, an early medieval encyclopedist, does not mention the medical uses of otoliths but does give a brief description of the use of otoliths to divine weather conditions at sea. In the later medieval literature, the marine use of otoliths was not to predict weather changes, but to protect sailors from inclement weather. Medical uses of otoliths also appeared in Renaissance writing. Various otoliths, either worn in contact with the skin or ground into a powder and taken with a liquid, could treat fevers, sudden attacks of pain, or colic. Like Pliny, Renaissance medical writers also asserted that otoliths were effective against kidney stones. Interest in otoliths among learned medical writers diminished in the Enlightenment, but they remained popular in folk medicine, particularly among fishing communities; otolith amulets are still worn in Spain to protect from fever. Fishing communities in Brazil both wear otoliths and drink ground otoliths to protect from or treat asthma, urinary problems, back pain, and kidney stones. The otoliths of the Whitemouth croaker (*Micropogonias furnieri*) are boiled to produce a tea Brazilian fishermen claim protects from its sting, although the Whitemouth croaker does not have a sting.

Otoliths are used medically in Asian traditions as well. The otolith of the yellow croaker, a fish found off the waters of Taiwan, is known as yunaoshi and used to treat inflammations of the nasal membranes in traditional Chinese medicine, although it is difficult to distinguish between otoliths, and not all otoliths sold as coming from yellow croakers actually do so. In Turkey, otoliths ground into a powder are used to treat urinary tract infections.

See also: Amulets and Talismans; Divination

Further Reading

Duffin, Christopher John. 2007. "Fish Otoliths and Folklore: A Survey." *Folklore* 118: 78–90.

P

Palmistry

Palmistry, also known as chiromancy, is the practice of reading the lines of the hand to divine an individual's future. It is part of a group of divinatory practices treating the human body as a source of divinatory knowledge. Like many other forms of divination, palmistry first appeared in the ancient Near East. Passages in the Bible such as Proverbs 3:16, "Length of Days is in her right hand, and in her left hand riches and honor," were interpreted as endorsing divination through the hand. The philosopher Aristotle (384–322 BCE) stated a basic chiromantic principle in his *History of Animals*, that the lines of the palm are correlated with length of life. Short-lived persons have palm lines that do not go all the way across the hand, while the palm lines of long-lived people do go all the way across. Chiromancy also appeared relatively early in India, where it was the subject of a poem credited to the legendary sage Valmiki, better known for his authorship of the epic poem *The Ramayana.* The poem's title translates as *The Teachings of Valmiki on Male Palmistry*, indicating that like some later chiromancers the early Indians distinguished between the techniques appropriate to read the hands of men and women.

As it developed, palmistry became connected with astrology as features of the hand were identified with the seven traditional astrological planets. It was incorporated into Kabbalah through the mention of palmistry and metoposcopy (divination through the lines of the forehead) in the *Zohar*, the founding text of Kabbalah. Eminent kabbalists were believed also to be experts at reading the palm. Chiromancy, like other divination forms, was also widely practiced in the Arabic Islamic world, despite the fact that many religious authorities condemned it along with other forms of divination. Chiromancy is also practiced in China, from which it spread to other areas of East Asia including Korea, Japan, Vietnam, and Thailand.

Like other Islamic knowledges, palmistry made it into Christian European culture with the great translation projects of the Middle Ages. For Renaissance magicians such as the German Heinrich Cornelius Agrippa (1486–1535), palmistry fit into the macrocosm/microcosm picture of the relation of the universe and the human body. The correlation of the features of the hand with the astrological planets was part of this. Although many later writers tried to separate palmistry from other forms of magic as its own distinct science, astrologers like the Englishmen George Wharton (1617–1681) and Richard Saunders (1613–1675) continued to write works linking astrology and chiromancy. The Flemish Jesuit Martin del Rio (1551–1608), one of Europe's most influential writers on magic and demonology, treated magical palmistry as illicit and was one of the first writers to identify it specifically with Romani women. The treatises on chiromancy written in the seventeenth century and later by gentlemen took care to distinguish themselves from the fortune-telling of "gypsy" women in the marketplace. Chiromancy continued to be

An eighteenth-century palm reading session, by Joseph Haynes after William Hogarth. (The Metropolitan Museum of Art, New York, Harris Brisbane Dick Fund, 1932)

the subject of elaborate treatises into the eighteenth century, and despite condemnation by the leaders of the Enlightenment, it continued to be widely practiced into the nineteenth century and to the present day.

Western chiromancers usually preferred to read the right hand, reflecting the traditional preference for the right. (However, Saunders preferred the left, due to its connection with the heart.) In other palmistry systems, the best hand to examine could depend on the sex and age of the person being read. Various combinations of lines and markings enabled palmists to predict the length and circumstances of life and death. Major lines included the Line of Life, the Line of Fortune, and the Line of the Liver. Other lines might not appear on every hand. Besides the palm, other features of the hand to be read included the fingers and fingernails. Distinguishing between the hand's "natural" features and those acquired by accident or long labor was important, as the latter lacked divinatory meaning. Like astrology, palmistry could reveal both a person's character and the events of their life. Careful interpretation of the hand (there were many systems) could reveal not just the length of life but the cause of death as well. Chinese palm readers also used chiromancy as a form of medical diagnosis, correlating the fingers with the major systems of the body, such as the liver with the index finger.

See also: Astrology; Death; Divination; Kabbalah; Left-Handedness; Macrocosm/Microcosm; Physiognomy

Further Reading

Lynn, Michael R. 2018. "The Curious Science: Chiromancy in Early Modern France." *Magic, Ritual and Religion* 13: 447–480; Rutkowski, Pawel. 2019. "Through the Body: Chiromancy in 17th-Century England." *Świat i Słowo* 1(32): 33–44; Scholem, Gershom. 1972. "Chiromancy (Palmistry)." In *Encyclopedia Judaica volume 5.* Jerusalem: Keter. Cols. 477–479.

Physiognomy

Physiognomy is divination by a person's physical appearance. It is most often used to divine information about a person's character rather than their fate or life course. In addition to the physiognomy of individuals, physiognomy could be applied to races and ethnic groups. Although physiognomy can apply to any part of the body—palmistry was often considered a branch of physiognomy—it was most often applied to the face and head. Metoposcopy, the art of reading character from the lines of the forehead, was another branch of physiognomy. Physiognomy is a very popular and enduring divination method, as it appeals to the widespread prejudice in favor of attractive people. Physiognomists usually identify beautiful bodies and body parts with moral virtue. Gender also affects physiognomic evaluation; the same features may not be considered appropriate in a man or a woman.

Early evidence of physiognomy comes independently from the ancient Mediterranean, India, and China. The philosopher Aristotle described physiognomy as based on the idea that the passions that shape the soul also shape the body. The earliest known physiognomic text in the West, *Physiognomica,* is ascribed, probably falsely, to Aristotle and deals with the connection between facial features and specific character traits. It also discusses how a person's resemblance to an animal signifies an affinity with the character ascribed to that animal—thus, a man who resembles a fox was expected to share the fox's cunningness. The philosopher Socrates was the subject of a famous anecdote, in which a physiognomist described him as a corrupt and wicked man, to the indignation of Socrates's disciples. Socrates himself, however, agreed that the physiognomist had actually described his nature, but that he had overcome these base desires. A similar anecdote would later be told of the "Father of Medicine" Hippocrates.

As it developed, physiognomy became bound up with humoral theory. A person's dominant humour affected both their character and their appearance. The person whose face was normally red was sanguine by nature and had the character traits of a sanguine person, such as quickness to anger. Other physiognomic beliefs depended on the relation of particular organs to character traits, such as the belief that a man with a small head, and presumably a small brain, must be stupid. The linkage of a person's resemblance to an animal to their sharing the characteristics of an animal also continued to appear in physiognomic texts. Physiognomy was closely connected to medicine, and discussions of physiognomy would appear in medical treatises. Knowledge of physiognomy was recommended to monarchs choosing advisers, slaveowners buying slaves, and prospective husbands choosing wives.

There are some references to physiognomy in the Hebrew Bible, such as Proverbs 17:24, which in a translation found mostly in

On the Origin of Physiognomy

"The various thoughts which arise in the mind, the different passions which agitate the soul of man, are respectively connected with his features and the external parts of his frame; and so intimate is their correspondence, that the expression of the countenance, more rapid than speech, betrays his sentiments and emotions, and gives to his utterance energy and animation. The one was designed as a mirror in which we might behold the other reflected; but the vicious study dissimulation; they endeavour to lock their passions and vices within their own breasts, and, by a virtuous exterior, to conceal the characteristic expression of villainy. In vain, however, does hypocrisy tender them her aid: the outward figure and form of the man are forced to a resemblance of the internal model, and the dispositions of the heart are almost invariably depicted on the countenance. These facts were observed and verified, and such was the origin of physiognomy."

Source: *The Pocket Lavater.* 1817. New York: Van Winkle and Wiley. Pp. 13–14.

physiognomic texts reads "The countenance of the wise sheweth wisdom." Although the Bible does not set forth a physiognomic system, these references helped legitimize physiognomy for Christians and Jews. Early Christians largely accepted physiognomic theory, portraying themselves with positive qualities of both soul and body and their persecutors as physically and spiritually inferior. The ancient legacy of physiognomy was taken up in both the medieval Christian and Islamic worlds. The pseudo-Aristotelian *Secret of Secrets*, an Arabic compilation, attributed the defective character of blonde and blue-eyed peoples, both stupid and quick to anger, to having been "undercooked" in the womb. Arab men buying female slaves could refer to physiognomic tracts explaining both how to evaluate women's genitals based on the shapes of their mouths and how to judge a prospective slave's degree of sexual desire. A red mouth, for example, indicated a woman who took great pleasure in sex.

Physiognomy was widely studied in the Latin Christian world, even extending to university teaching. The famous mathematician and magician Michael Scot (1175–1232) wrote an influential book of physiognomy, mostly derived from Arabic sources, which offered physiognomy as a way to quickly distinguish between good and evil persons. However, Scot warns, the physiognomist must be careful not to quickly draw a conclusion based on one body part only, for that conclusion may be modified by observations of other body parts.

Physiognomy continued into the Renaissance. The Neapolitan natural magician Giambattista Della Porta (1535–1615) wrote an encyclopedic treatment of physiognomy that was treated as authoritative into the nineteenth century. Della Porta synthesizes a great deal of material from medicine, humoral theory, astrology, and the study of herbs and natural magic to produce a system for judging not just humans but a variety of natural phenomena from their appearances.

Della Porta in effect extended physiognomy as a science describing a universe in which every phenomenon appearing to the senses could be interpreted as a sign.

Unlike many other forms of divination, physiognomy survived the European Enlightenment. One of the most influential physiognomic writers was the Swiss clergyman Johann Kaspar Lavater (1741–1801) whose physiognomic writings were popular both in their original German and in French and English translation in Britain and the United States. Although Lavater's work was vigorously contested, physiognomy continued to be widely supported into the nineteenth century, when it influenced literature and the development of criminology and scientific racism. Phrenology, the popular nineteenth-century science of identifying personality traits by feeling the bumps on the skull, was also an offshoot of physiognomy.

Physiognomy in China was more oriented toward telling an individual's fortune than the character-based systems of the West, although it also analyzed character. Chinese physiognomists also relied on a macrocosm/microcosm system, as an individual's body was shaped by the same forces as shaped the cosmos. Like Western physiognomists, Chinese physiognomists saw the head as the most important area to analyze. The head was a microcosm of the microcosm as its features paralleled those of the body. One Chinese system divided the head into a hundred parts, each representing a potential year of a person's life. Physiognomy was not only recommended to rulers as a way of recruiting officials but also practiced by Buddhist monks and other social outsiders as a way of telling individual fortunes.

See also: Divination; Foxes; Gender and Sexual Difference; Horses and Donkeys; Humours, Theory of; Macrocosm/Microcosm; Palmistry

Further Reading

Johnson, J. Cale and Alessandro Stavru, eds. 2019. *Visualizing the Invisible with the Human Body: Physiognomy and Ekphrasis in the Ancient World.* Berlin/Boston: De Gruyter; Swain, Simon, ed. 2007. *Seeing the Face, Seeing the Soul: Polemon's Physiognomy from Classical Antiquity to Medieval Islam.* Oxford and New York: Oxford University Press; Wang, Xing. 2020. *Physiognomy in Ming China; Fortune and the Body.* Leiden and Boston: Brill.

Pigs

The pig, widely cultivated across Eurasia, has acquired a rich set of associations both positive and negative. The negative associations are primarily religious, while the positive ones are related to wealth and fertility. The pig has a particularly disfavored status in Abrahamic religions, its consumption being forbidden to observant Jews and Muslims, as well as to some Christian denominations such as the Ethiopian Orthodox or Seventh-Day Adventists. (The rabbis did allow an exemption if a person was starving and pork was the only food available.) Jewish avoidance of pork was an early distinguishing feature of their community in the eyes of Greeks and Romans. (The Romans were particularly fond of pork.) The Roman historian Tacitus (c. 56 CE–c. 120 CE) stated that the reason for the dietary law was that Jews had earlier suffered a plague resembling a skin disease that they suffered due to pigs and had avoided them ever since. (The association of Jews with disease is a common anti-Semitic trope.) Pigs also feature in an unflattering context in the gospels, in which Jesus drives demons possessing a man into a herd of pigs. Although most Christians were allowed to eat pork, the idea that the pig was disfavored continued. In many, although not all, Christian societies, it was considered bad luck to see a pig when setting out on a journey. Scots were particularly known for disfavoring pork, to the extent that English people sometimes asserted that their northern neighbors were descended from Jews. Even the word "pig" was taboo in Scottish and Northern English settings, and hearing it required touching iron or saying "cauld iron" to take away the evil power. This even applied in Church, if the minister was reading a passage from the Bible that mentioned pigs. In Islamic dream interpretation, a pig represents a worthless person.

However, pigs also have positive associations due to their connection with wealth. A fat pig signified prosperity. In some parts of Europe, such as Slovenia, encountering a pig was considered a sign of good fortune. A money safe in the form of a pig, or "piggy bank," was believed to attract wealth. The nineteenth-century American grimoire *The Golden Wheel Dream-book and Fortune-teller* claimed that dreaming of pigs prophesied success both economic and romantic. Pigs could also contribute to good health. There is an Irish superstition that childhood diseases such as mumps could be cured by transferring them to a pig. The afflicted child would rub the pig's back.

Like other domestic animals, pigs required care and were subject to bewitchment. One way of preventing this was to slit their ears. Pigs were to be slaughtered, and bacon cured, only in the waxing phase of the Moon. The meat of pigs slaughtered in the Moon's waning phase would shrink rather than expand when cooked.

The pig is viewed favorably in Chinese superstition, associated with wealth and good fortune. The pig is the last animal in the twelve-year Chinese animal cycle, and persons born in the year of the pig are believed to be generous, noble, and trusting. In China and Korea, to dream of pigs is a good omen. In traditional Chinese medicine, pork is considered a "warm" food. In nineteenth-century South India, anointing a person with pig's fat was considered a cure for cholera.

See also: Iron; Moon

Further Reading

Opie, Iona and Moira Tatem, eds. 1989. *A Dictionary of Superstitions.* Oxford: Oxford University Press.

Possession and Exorcism

Possession occurs when a supernatural entity takes over or "possesses" a human body. The supernatural entity can be a god, a divine servant such as an angel, a demon, or a ghost. The idea of possession is found in most cultures and is frequently linked to exorcism, the process by which a human expert, a cleric or a magical specialist, drives the supernatural entity out of the body. Although possession can be desired, as in the case of shamanistic religions in which the religious professional serves as a conduit between the gods and the human community, it is more often conceived of as a problem or even an assault. Speaking in strange and unknown languages, using a voice different from a person's normal one, such as a woman speaking with a deep, masculine voice, and physical symptoms such as seizures and headaches can all be evidence of possession. Vomiting of unusual objects, such as pins, needles, and stones, was sometimes a symptom. A person knowing things that they had no reason to know could be drawing on the knowledge of the possessing spirit. Possession could be used as an explanation for phenomena that would now most commonly be described as symptoms of mental illness. In addition to freeing possessed persons, exorcists could also drive malevolent spirits from buildings or other places they haunted.

A 1790 representation of a Christian exorcism, by Daniel Nikolaus Chodowiecki. ("Exorcisme," by Daniel Nikolaus Chodowiecki, 1790. Rijksmuseum.)

Traditions of benevolent possession are found in ancient Greece, where priestesses "inspired" by a god could deliver prophecies like those of the oracle of Delphi. Another tradition of divine possession is that of the African-based religions of the Americas such as Haitian voudun, where one of the main functions of the priest or priestess is to serve as a "horse" for the god who possesses them, the "divine horseman." For voudun and other practitioners of African-derived religions, divine possession, which always takes place for a limited period of time, is a desirable state.

The concept of malevolent possession goes back at least as far as the ancient Mesopotamians, who greatly feared possession by demons and ghosts and developed techniques such as the use of defensive magic to protect themselves. Amulets, herbs, and incantations have all been used to prevent or remedy possession. Possession and exorcism appear in the Jewish Bible in the episode where David drives a possessing spirit out of Saul with the power of music. Jesus is portrayed in the New Testament as an exorcist of demons, and exorcism became an intrinsic part of Christianity. Skepticism about possession could easily be portrayed as skepticism about the Bible itself. The Christian tradition focuses almost exclusively on demons as the possessors and on the possessed as an unwilling victim.

Early tradition focuses on exorcism as a miraculous gift of Jesus and the saints, but like other religious functions, it became increasingly bureaucratized and routinized by the Church, treated as a power that could be wielded by ordained priests. The first formal ritual of exorcism in the Catholic tradition was published in 1614, at a time of heightened interest in possession and exorcism. The majority of the possessed in the era and a large majority of the widely publicized possession cases were women, usually young women and girls. Possession offered young women, usually powerless in early modern society, a way to be heard and even to be the center of attention. The most dramatic and well-publicized possession cases took place in French and southern Netherlands convents and involved large numbers of young women claiming to be possessed. Convent life, with its strict routine and repression of sexuality and other forms of emotional expression, imposed a heavy psychological burden on young women, many of whom were there for reasons of family economics and lacked religious vocation. Possession could excuse norm violation, as in the case of the Italian heretic abbess Benedetta Carlini (1590–1661), who claimed to be possessed by a male angel called Splenditello when she made love to her fellow nun Bartolomea Crivelli.

One reason for the popularity of demonic possession in the Reformation era was confessional conflict. Both Catholics and Protestants believed in demonic possession, but they dealt with it in different ways. Here Catholics had the advantage, since exorcism was and is a ritual of the church, and usually believed by Catholics to be invariably effective if done properly. Protestants believed that exorcistic rituals were idolatrous and Catholic belief in their power arrogant, but could offer nothing better than prayers and Bible reading. Those advocating a particular religious position used the possessed to vindicate it. The Frenchwoman Marthe Brossier owed much of her fame as a possessed person to the support of Catholic religious orders. Her possessing demon Beezelbub expressed his fear and respect for Catholic priests and his contempt for Protestant ministers, who he claimed were allies of the demons. In England, the use of exorcisms by puritans advocating for a more Protestant Church of England eventually caused religious and political leaders to take a more skeptical view of possession and exorcism in general.

There were several problems with using possessions in religious controversy. The unrestrained and sometimes unpredictable drama and the power claimed by the usually young and female possessed troubled religious authorities generally, and even in Catholic areas, possession cases were declining after the mid-seventeenth century. Possessions were also notoriously easy to fake, and many well-known "victims" of possession, including Brossier, were eventually revealed as frauds. Physicians were particularly likely to be skeptical of demonic possession, preferring to blame symptoms on conditions on which they could claim expertise. Even if a possession was believed genuine, the question remained as to whether the demons could be believed. Wringing truth from a demon was one among the abilities claimed by exorcists.

Although most episodes of possession involved the possessor, the possessed, and possibly an exorcist, during the early modern European witch hunt the concepts of witchcraft and possession became intertwined and witches were frequently blamed for possessions. Both Protestants and Catholics accepted the linkage of possession to witchcraft, viewing it as one of the most serious among the many evil deeds witches performed. There was a gap between the position of the theologians and demonologists, that witchcraft was one possible cause of possession and the popular view that possession was always the work of a witch. Possessed people were often used to identify witches, one of the most famous being the young girls whose accusations launched the Salem witch trials, although there were many others. One advantage with blaming possession on witches is that a stubborn case of possession, resisting exorcism and prayer, could be cured by the execution of the witch. But the mere claim of a possessed person that she saw a given individual tormenting her or inflicting demons on her was also suspect, as the devil was a master of deceit and not above framing the innocent. Growing skepticism about the authenticity of accusation brought an end to the Salem episode, and the reaction against it ended witch hunting in New England. Similar doubts led to the end of high-profile possession cases in France after the Provencal case of Marie Catherine Cadire from 1729 to 1731.

Belief in possession and exorcism, however, did not end with the end of the witch hunt or the rise of the Enlightenment, particularly among Catholics. In the middle of the eighteenth century, the German Catholic priest Johan Josef Gassner (1727–1779) became a celebrity, performing thousands of exorcisms. Gassner associated demonic possession not just with mental conditions but also with physical disease, and his exorcisms, performed on Protestants and Catholics alike, were presented as a form of healing. Gassner did distinguish between demonically caused illnesses he could heal and "natural" ones against which he was powerless. As was true previously, the majority of possessed people he healed were women, about 60 percent. Gassner's exorcisms were eventually halted by papal command, but exorcism continues to flourish in this day in the Catholic church, with a revival in popularity in the West after the popularity of the movie *The Exorcist.*

In the ancient Mediterranean, Jews were often credited with the power of exorcism even by pagans. Judaism developed a tradition of exorcism carried out by a pious man, or sage, and directed either at demons or at malevolent ghosts called "dybbuks" in the Ashekenazic community. As opposed to Christian exorcists, for whom exorcism was often a struggle of brute force, Jewish exorcists placed great emphasis on learning the reason the victim was possessed and the "true name" of the possessing entity. Knowing the true name enabled the exorcist to cast the possessing spirt from the body of the possessed. Exorcism also involved the repetition of prayers and passages from the Hebrew Bible.

In Islamic tradition, possession can be by demons or by jinn, a non-demonic race of spiritual beings. Jinn may possess individuals for revenge, if the individual has somehow harmed the jinn or its relatives, or for love, if a jinn has fallen in love with the person it is possessing. Islamic exorcists can use religious rituals, including the recitation of passages of the Quran or the names of God. Islamic jurists draw a distinction between these permissible practices and illegitimate "sorcerous" practices such as the use of amulets. In religiously mixed societies, a powerful exorcist can appeal across religious lines, and there are cases of Christian priests whose reputations for expelling demons and jinn have attracted possessed Muslims.

Possession by malevolent ghosts is also common in East Asia. East Asian exorcists frequently enquire about the purpose of the ghost in the hopes that seeing to the ghost's needs would end the possession. For example, the ghost of a dead person buried without the proper rituals or honors could possess or haunt an individual, and an exorcist would learn of the spirits need and see to its fulfillment, thus ending the possession. In the Han Dynasty, China developed a tradition of professional exorcists known as fangxiang, who specialized in exorcising places and protecting dead bodies from corpse-eating demons by exorcising funerals. The fangxiang was adopted into Japan as the Shinto religious hososhi. Both cultures have other types of exorcist as well. In Japan, the tradition of Buddhist "mountain ascetics" known as shugendo has produced exorcists and offered the unusual example of an exorcistic tradition that allows women as well as men to be exorcists. Shugendo ascetics are both exorcists and possessed, deriving their power to cast out malevolent spirits from being possessed by gods and other divine spirits. The repetition of sacred texts, in this case Buddhist sutras, is again part of the exorcism process.

Beliefs in possession by malevolent spirits are also common in many African societies, Christian, Islamic, and polytheist. Some societies draw gender-based distinctions, with different types of spirits possessing men and women, with different effects. As elsewhere, religious professionals can free the possessed through the performance of exorcist rituals.

A much more benevolent form of possession was characteristic of nineteenth-century spiritualism, when a "medium," usually female, became temporarily possessed by a "spirit guide," often the spirit of a person who had died. The purpose of the possessing spirit was usually to enlighten the medium's hearers. This type of benevolent possession evolved into "channeling," as practiced in modern New Age circles.

See also: Amulets and Talismans; Defensive Magic; Jinn; Madness; Witchcraft

Further Reading

Brown, Judith C. 1986. *Immodest Acts: The Life of a Lesbian Nun in Renaissance Italy.* New York: Oxford University Press; Maarouf, Mohammed. 2007. *Jinn Eviction as a Discourse of Power: A Multidisciplinary Approach to Moroccan Magical Beliefs and Practices.* Leiden: Brill; Midelfort, H. C. Erik. 2005. *Exorcism and Enlightenment: Johann Joseph Gassner and the Demons of Eighteenth-Century Germany.* New Haven, CT: Yale University Press; Walker, D. P. 1981. *Unclean Spirits: Possession and Exorcism in France and England in the late Sixteenth and early Seventeenth Centuries.* Philadelphia: University of Pennsylvania Press.

Powder of Sympathy. See Weapon-Salve

Pre-Adamites

Although the idea that Adam was not the first human had been occasionally put forth and speculated about in Judaism and Christianity for centuries, it was first systematically developed by the French Protestant Isaac La Peyrere (1596–1676) in his book *Prae-Adamitae* (1655), translated into English as *Men before Adam* (1656). "Pre-Adamatism" held that the Bible, including Genesis, was a history of the Jews, not of the entire human race. Adam was the ancestor of the Jews, not all of humanity, and was created into a world that was already inhabited by people. This meant that different branches of the human race could have been separately created with separate ancestry, a doctrine that would later come to be called "Polygenism." (La Peyrere also claimed that the Flood of Noah had not covered the entire world, so different branches of humanity other than Noah's descendants, the Jews, survived it.) La Peyrere's theory solved certain problems both within the Biblical narrative and in its relationship with the evidence of the world. The classic biblical mystery of the origin of Cain's wife was solved by suggesting that she was a member of a pre-Adamite people. The theory also offered as a way to reconcile the "short" six-thousand-year biblical narrative with the growing evidence from European contacts with China, Mesoamerica, Egypt, and other foreign civilizations that humans had been on Earth for much longer. The pre-Adamite theory was, however, incompatible with a literal reading of the biblical narrative and called into question the idea of humans inheriting original sin from Adam, an idea central to Christianity. It was vigorously condemned by Protestant and Catholic religious authorities and attracted numerous scholarly refutations.

Despite religious opposition, pre-Adamitism was the object of growing interest from some late seventeenth-century scientists, who applied it to racial questions that the original theory was not designed to address. La Peyrere's work was not primarily directed at the European encounter with Africans and Native Americans—for him, they, like the Europeans, were equally the descendants of the non-Jewish, pre-Adamite creation and thus outside the biblical narrative. However, as the idea developed after La Peyrere, his original position that the Jews were the sole legitimate descendants of Adam was quietly abandoned in favor of one that Europeans occupied that "favored" status. This change was driven by material interests. Some slaveowners in the Americas supported a non-Adamite origin for their slaves, on the grounds that would justify keeping them as slaves. The English chemist and scientific leader Robert Boyle (1627–1691), who had a strong Christian religious commitment, opposed the growth of pre-Adamitism on religious grounds, both as a challenge to the veracity and completeness of the biblical account and as a barrier to efforts to convert slaves to Christianity, a project close to his heart. In the Enlightenment, pre-Adamitism was secularized into the doctrine of "polygenism," the belief that rather than springing from a single ancestor different branches of humanity had different origins. Various pre-Adamite theories, mixing racism and religion, were put forward in the nineteenth century, often by slaveowners. Pre-Adamitism was eventually marginalized by the triumph of Darwinian evolution, which while denying the biblical account of creation supported the idea of a common ancestry for humanity.

Ideas about pre-Adamites also existed in medieval Islam, some arguing for separate human creations, some referring to the jinn as pre-Adamite.

See also: Jinn; Race

Further Reading

Popkin, Richard H. 1997. *Isaac La Peyrere (1596–): His Life, Work and Influence.* Leiden and New York: Brill.

Precious Stones

Visually striking jewels, gems, and precious stones were widely attributed with supernatural power. The magical uses of jewels ranged from defensive magic to healing to scrying. Precious stones were also attributed with medicinal virtues.

The mysterious origins of gems attracted a variety of beliefs. Some ancient and medieval writers including Roman natural historian Pliny the Elder (d. 79 CE) believed transparent crystal to be ice that had hardened into rock. Pliny stated that only "pure" water, rainfall and melted snow, could be transformed into crystal. Gems were also treated as solidified products of the blood, tears, or other bodily fluids of gods, dragons, and other supernatural beings.

The powers of jewels varied, often based on their appearance. Red stones such as rubies or "bloodstones," green heliotropic rocks with spots of red resembling blood, were believed to have the power to stop or slow the flow of blood. Green stones such as emeralds were believed to be good for the eyes and yellow stones for jaundice.

The more valuable jewels—the diamond, emerald, sapphire, and ruby—were often associated with good fortune, perhaps because they were associated with wealthy people. (The gem most associated with misfortune was the opal. It was believed that if an opal was set in an engagement ring, the bride would swiftly become a widow.) Pliny describes gems that could be worn to protect against drunkenness or harmful magic, but clearly scorns these beliefs. He is more credulous about a stone called coralloachates found on the island of Crete, which protected against the sting of scorpions. (Like many of the gems to which Pliny refers, the exact meaning of coralloachates has been lost.) The healing and other beneficent powers of stones could be accessed by wearing them as jewelry or simply pressing them to the affected flesh. Jewels were also gendered female; the second-century CE Greek dream writer Artemidorus states that to dream of jewels indicates good fortune and prosperity for a woman but is a bad omen for a man.

In Western, Islamic, and Indian magic, jewels were correlated with other magical phenomena, such as astrological signs and planets and alchemical metals. The correlation of jewels and astrological signs originated in the medieval Islamic world. Wearing a jewel would channel the influence of its planet. In India, where the sapphire was associated with Saturn, an unfavorable planet, wearing a sapphire would channel Saturn's influence in a benevolent way. Engraving a figure representing a planet on a stone under the influence of that planet would create a powerful talisman.

The "lapidary," a list of stones and their properties, or "virtues," was a genre with roots in the classical world. One of the most influential writings on jewels, gems, and their properties in the Western tradition was the *Book of Minerals* of the cleric and philosopher Albertus Magnus (1200–1280). These works did not make a strong distinction between mineral gems and jewels and organic products such as pearls or toadstones. All were "stones."

Renaissance physicians continued to use jewels as treatments or preventative, but increasingly distinguished between superstitious belief in the magical powers of jewels and rational exploitation of their "natural" characteristics. Belief in the powers of jewels was incorporated into the tradition of natural magic. This went along with a tendency to apply gems medically by grinding them into a powder and drinking them mixed with fluid, a technique that except in rare cases was not applied to more valuable jewels.

The diamond, the most precious of jewels, is usually associated with good fortune. However, particularly large, well-known, or valuable diamonds such as the Hope Diamond are believed to be associated with curses that would bring misfortune to their owners. The diamond was believed to be both a remedy against poison and disease and a powerful poison itself. The English natural philosopher Sir Thomas Browne (1605–1682) recounts but does not endorse the superstition that a diamond laid under the pillow would reveal an unchaste wife.

In addition to promoting love and good fortune, emeralds were believed to be powerful against poison as well as diseases such as dysentery and epilepsy. The emerald was also thought to protect against evil spirits. In Islamic dream interpretation, the emerald is a very positive sign, signifying martyrdom or an alternative way to enter paradise.

Like diamonds and emeralds, sapphires were believed to encourage love and good fortune. There is a Buddhist belief associating sapphires with meditation. According to some legends, the ring of King Solomon bearing his seal was made from sapphire. (The English writer Gervase of Tilbury (1150–1220) claimed that Solomon had been the first to discover the magical powers of gems.) In Islamic dream interpretation, a sapphire represented a noble and prosperous woman. The association of sapphire with the preservation of chastity made it a popular choice in the Middle Ages for the rings of wealthy clerics. Sapphire was also believed to be helpful in treating diseases of the eye, a theory endorsed by Albertus Magnus, who claimed the jewel should be dipped in cold water before being applied to the eye. The Flemish alchemist Jan Baptist Van Helmont (1580–1644) recommended that a good-quality sapphire be used to stroke the skin around a plague boil. The jewel would draw out the infectious matter, assuming the case was not too far advanced.

Jade, a popular medium for Chinese art, was also treated by Chinese as medicinally valuable, with a variety of formulae for combining powdered jade with other substances for cures for various diseases. Properly prepared, jade strengthened the body, purified the blood, and contributed to overall health. In the West, jade was more narrowly defined as good for persons afflicted with kidney stones, who wore jade bracelets or jewelry in the hope of a cure. Kidney stone sufferers wearing jade included the skeptical French Enlightenment philosopher Voltaire (1694–1778).

See also: Alectorius; Amber; Amulets and Talismans; Astrology; Bezoars; Fossils; Luck; Natural Magic; Otoliths, Fish; Solomon; Swallow Stones

Further Reading

Harris, Nichola Erin. 2009. "The Idea of Lapidary Medicine: Its Circulation and Practical Applications in Medieval and Early Modern England: 1000–1750." Ph.D. Dissertation, Rutgers University; Kunz, George Frederick. 1913. *The Curious Lore of Precious Stones.* London and Philadelphia: Lippincott.

Pregnancy and Childbirth

Pregnancy and pregnant women, vital for the survival of any society, have been subject to a range of superstitious and magical beliefs. Given the risk of death accompanying childbirth in many premodern societies, pregnant women had a unique status between life and death.

Legends associated heroes or godlings with their mother's prolonged or otherwise unusual pregnancies. Laozi, the founder of Daoism, was supposed to have spent many years, one common figure being eighty-one, in his mother's womb before being born as an old man. Mary the mother of Jesus, founder of Christianity, had a pregnancy that was unusual not in length but in being unconnected to sexual intercourse. The Greek goddess Athena was born from the head of her father, the chief god Zeus.

There were a variety of superstitions about how a woman could become pregnant or prevent pregnancy. Even something as mundane as sitting in a chair that a pregnant woman had previously sat in or putting on a coat that belonged to a pregnant woman could be regarded as leading to pregnancy. Fertility in women generally could be connected with agricultural fertility; in some nut-growing regions, a good harvest of nuts was held to indicate the birth of many babies. This could work both ways; a pregnant woman shaking a fruit tree would cause it to bear heavily. Amulets, talismans, and herbs, including mandrake and mistletoe, could all promote fertility. Particular springs or baths could promote fertility in women who drank their waters or bathed in them. Religious rituals and pilgrimages or contact with certain relics were also believed to promote healthy pregnancy. Herbs were also used as contraceptives or abortifacients.

Like other specifically female bodily conditions, such as menstruation, pregnancy was considered unclean or ritually defiling in many cultures. This is reflected in the purification ceremony carried out for

Mary Toft

In 1726, a poor Englishwoman from the village of Godalming, Mary Toft, claimed to have given birth to stillborn rabbits. Toft claimed her strange births had been caused by having been startled by rabbits when pregnant. Although several physicians found her story credible, an investigation by a local magistrate found out that Toft's husband had been buying young rabbits before her first delivery, and the arrest of a porter who was smuggling another rabbit to her revealed Toft as a fraud. Sir Richard Manningham (1690–1759), a London physician and Toft skeptic, extorted a confession from her by threatening to perform a painful operation to examine her for malformed genitals. Toft, who claimed that other people put her up to the fraud, was imprisoned for four months and then released without charges.

Prompted by the case, the physician James Blondel (d. 1734) wrote *The Strength of the Imagination in Pregnant Women Examined* (1727), the first book-length attack on the "power of the imagination" theory. Using a basically mechanical theory of life, Blondel excluded maternal influences on fetal development. The Toft case was remembered for decades as an example of fraud and credulity.

a woman forty days after childbirth. It is found in Jewish law and carried over into Christianity. In the Christian denominations that continue the practice, it is referred to as the "churching" of women. A similar idea is found in Hinduism, where the period of impurity is eleven days. In early modern England, a pregnant woman was considered disqualified either from being a godmother or taking an oath in court. In some versions of this belief, a pregnant woman standing as godmother would cause her godchild to die soon after. Pennsylvania Germans had the belief that a child would die if its hair was cut by a pregnant woman. They also believed that pregnant women, like menstruating women, would foul vinegar in a cask by contact.

Many techniques existed for determining or predicting the sex of an unborn infant, the single characteristic that would have the most impact on the child's future. The position of the child in the womb, whether carried high or low or mostly to the left or right, was held to reveal its sex. A pregnant woman suffering intense headaches was carrying a boy. Intense morning sickness could indicate a boy or a girl, depending on the culture and region. One common divination technique in Western societies was to hold a small weight (possibly a wedding ring) on a string over a pregnant woman's stomach. If the bob swung back and forth, the child was a boy—if in a circle, a girl.

There were various astronomical systems relating the time of conception to the baby's sex, as in babies conceived at a specific time during the lunar cycle were more likely to be boys than girls. The age of the mother or the father was also held to influence the child's sex.

Very commonly across civilizations and cultures, people believed that the environment in which pregnant women lived would shape the outcome of their pregnancies. One common version of this belief was the "power of the imagination"—the idea that the pregnant woman's imagination, working on the evidence of her senses, could shape the child in her womb. Numerous tales cautioned the woman and her family to avoid visual stimuli that could result in an ugly, deformed, or monstrous baby. The birthing chamber was often kept free of any form of representational imagery for this reason. Substances a woman brought into her body could also affect her baby by their appearance. The French letter-writer Mme. de Sevigne (1626—1696) claimed that a noblewoman who had drunk too much chocolate while pregnant had given birth to a Black baby. (Europeans had a particular horror at the idea that a child would be born Black and recommended that Black people and their images be kept away from pregnant women.) A woman who gazed upon a corpse while pregnant could give birth to a child that was excessively pale. In South India, a pregnant woman gazing at a temple car carrying the statue of a god risked giving birth to a monster.

A pregnant woman's cravings were frequently associated with the body of her child. If a craving was not satisfied, the thing craved, such as a piece of fruit, would appear as a birthmark on the child. In Mediterranean societies, denying a pregnant woman's craving was considered very bad luck. The *Distaff Gospel*, a late medieval French compilation of folk-beliefs, goes so far as to suggest that no kind of food that was not readily available be mentioned in front of a pregnant woman, lest it induce an unsatisfiable craving and thus mar the child. A rat, mouse, or weasel jumping over the belly of a pregnant woman could also induce a birthmark. A far more dangerous condition, in which the child was born strangled by its own umbilical cord, could be brought on if an expectant mother stepped over a rope. Samoans believe that a woman underwent the same risk if she wore a floral arrangement around her neck. Misfortunes for the pregnant woman were not always misfortunes for the child; in South India, there was a belief that if a pregnant woman was stung by a scorpion and survived, her child would be immune to scorpion stings their whole life. The dreams of a pregnant woman could also be considered prophetic of the destiny of her child.

In India, pregnant women were warned against working during an eclipse, as their child would be born with a mark or deformity associated with their occupation. Thus if her work involved cutting, the child would be born with a cleft lip, bearing the mark of the cut. In Europe, where cleft lips were referred to as "harelips" for their resemblance to the mouth of a hare, they were viewed as having been caused by a pregnant woman's being startled by a hare.

A deformed or monstrous birth could also be a punishment for the sins of the mother or, less commonly, the father. During the English Civil War of the mid-seventeenth century, a pregnant woman who cursed the Parliamentarians, or "Roundheads," by stating that she would rather have a child with no head than a roundhead was punished by giving birth to a headless child, at least according to one Parliamentarian journalist.

There were various techniques for ensuring an easy delivery. In ancient Greece and other societies, it was believed vital that the birthing chamber be free of knots. Locks were also unlocked at the time of delivery. Mesoamericans believed the tail of an opossum speeded delivery. Having a second pregnant woman in the same house or room as the woman giving birth was advised against in early modern Europe. Techniques to accelerate a protracted labor included making noise, to the extent of firing a gun. Various forms of defensive magic could protect mother and child against malevolent supernatural entities. (Pregnant women and newborn children were often believed particularly vulnerable to the evil eye.) In South India, smearing the wall of the room where the woman was giving birth with cow dung and cotton seed protected the room from demons. Scots used urine for similar purposes, sprinkling urine over the bed and doorposts of the room in which the woman giving birth would be lying, to protect the mother and child from the malevolent magic of the fairies. (Fairies were particularly dangerous at this time, due to their propensity for kidnapping human children and replacing them with changelings.)

Women who died when pregnant or in childbirth were particularly likely to become malignant ghosts. In Islamic dream interpretation, to dream of pregnancy signified material gain.

See also: Amulets and Talismans; Bathing; Caul; Defensive Magic; Dreams; Eclipses; Evil Eye; Excrement; Fairies; Ghosts; Hand of Glory; Herbs: Mandrakes; Menstruation; Mistletoe; Monsters

Further Reading

Fogel, Edwin Miller. 1915. *Beliefs and Superstitions of the Pennsylvania Germans.* Philadelphia: American Germanica Press; Opie, Iona and Moira Tatem, eds. 1989. *A Dictionary of Superstitions.* Oxford: Oxford University Press; Thurston, Edgar. 1912. *Omens and Superstitions of Southern India.* London: T. Fisher Unwin.

Prodigies

A prodigy is a strange and aberrant event outside the normal order of nature held to have a meaning for human history. This meaning can be negative or positive. Negatively, prodigies show a change is coming, a war or political disaster. Prodigies were common in Chinese histories, where they were evidence of a dynasty's loss of the mandate of heaven. Thus, an increase in prodigies heralded a dynasty's losing power and a period of civil strife or foreign invasion until a new dynasty was established. Prodigies appear in the *Spring and Autumn Annals*, one of the five Confucian classics, supposedly written by Confucius himself. In the *Analects*, however, Confucius warns against concern with prodigies and the supernatural. Chinese historians distinguished between those prodigies whose interest lay solely in their novelty—to be excluded from serious histories—and those that were omens of subsequent events. The Chinese philosopher Wang Chong (25–100 CE) went beyond this, to argue that prodigies and omens were not signs from heaven, as heaven was too distant to be concerned over human affairs. Wang Chong does not exclude the possibility of prodigies and omens being significant, but denies that his contemporaries had the ability to interpret them. Excessive concentration on prodigies and omens meant that people were no longer recognizing and listening to the sages among them.

Prodigies were common in classical Greek and Roman historiography—the work of the Roman historian Livy (59 BCE–17 CE) in particular is full of them. There were so many prodigies in Livy that he was the main source for a subsequent writer, the fourth- or early fifth-century Julius Obsequens, whose *Book of Prodigies* catalogs prodigies in Rome from 190 BCE to 11 BCE. The Romans considered prodigies warnings from the gods, and in many cases, they were followed by an expiatory sacrifice in the hope of averting the divine wrath.

Prodigies could also have a positive meaning. This was particularly true of biographical prodigies, which were frequently held to mark out people with a special destiny. The biographies of Roman Emperors by Suetonius (c. 69 CE–after 122 CE), *Lives of the Twelve Caesars*, include biographical prodigies marking out each man for his career as the ruler of Rome. The death of "great men" as well as their birth was heralded by prodigies, such as the star that heralded the death of Julius Caesar. The change from the Roman Republic to the empire was accompanied by a shift from public prodigies to those associated with particular individuals, often emperors or future emperors, although public prodigies did not disappear and there is some evidence of their persistence into the fourth century.

Prodigies could take many forms. Natural disasters such as floods and earthquakes were sometimes viewed as prodigies. Comets and supernovae were among the most spectacular prodigies and among the ones most likely to be seen as positive. A very common kind of prodigy was the monstrous birth, either of humans or animals (or sometimes a blend of both). Rains of blood, or frogs, were prodigies. Animals talking, spontaneous sex changes, lightning bolts hitting objects of significance—all could be included in the category of the prodigious.

The rise of Christianity connected prodigies more directly to God, particularly with the widespread idea that they were warnings from God to repent. The Star of Bethlehem is one of the most famous of biographical prodigies. An increase in prodigies and wonders was also commonly believed to be a sign of the imminent end of the world. Many of the events recounted in the apocalyptic books of the Bible such as Daniel and Revelation could be viewed as prodigies. Prodigies differed from miracles in that they were generally not held to be violations of the established course of nature requiring a direct application of divine power.

Medieval chroniclers recounted prodigies, but the Renaissance saw even more prodigies with the revival of the classical historical tradition. As hard-headed a political analyst as Niccolo Machiavelli (1469–1527) endorsed the idea that great events were heralded by prodigies in his *Discourses on the first Ten Books of Livy* (1531). The idea that prodigies were signs from God made them useful in the polemics of the Protestant Reformation and subsequent religious conflicts, from the Thirty Years' War in Germany and Central Europe to the Wars of the Three Kingdoms in Britain and Ireland. The belief that prodigies were moral rather than political warnings calling for repentance and moral reform also persisted.

The rise of the sciences in the seventeenth-century West meant that many prodigies were now explained as "natural" events, deprived of the extra-natural role of the prodigy as a sign or warning. (Science also inspired some efforts to collect and interpret prodigies more systematically.) In some cases, such as England after the Restoration of Charles II in 1660, "natural" explanations were promoted by governments fearing that prodigies would otherwise encourage rebellion. The Enlightenment meant the end of prodigies in the learned writing of history and theology. Belief in the political or moral significance of prodigies was stigmatized as "superstition," relegated to the politically marginalized and to the poor and female.

See also: Apparitions; Comets; Death; Decay of Nature; Divination; Intersex Conditions; Lightning; Monsters, Sex Change

Further Reading

Burns, William E. 2002. *An Age of Wonders: Prodigies, Politics and Providence in England, 1657–1727.* Manchester:

Manchester University Press; Daston, Lorraine and Katherine Park. 1998. *Wonders and the Order of Nature 1150–1750.* New York: Zone; Niccoli, Ottavia. 1990. *Prophecy and People in Renaissance Italy.* Translated by Lydia G. Cochrane. Princeton, NJ: Princeton University Press; Puett, Michael. 2005. "Listening to Sages: Divination, Omens and the Rhetoric of Antiquity in Wang Chong's Lunheng." *Oriens Extremus* 45: 271–281; Santangelo, Federico. 2019. "Prodigies in the Early Principate?" In Lindsay G. Driediger-Murphy and Esther Eidinow, eds. *Ancient Divination and Experience.* Oxford: Oxford University Press. Pp. 154–177.

Qilin. See Unicorn

R

Rabdomancy. See Dowsing

Rabies. See Dogs

Race

Although the idea of differences between peoples, physical and cultural, has a long history, the idea of race emerged slowly and competed with other systems of classification, including according to political allegiance, culture, religion, and language. Ancient Greeks, including Aristotle, distinguished between "Greeks" and "Barbarians" according to certain aspects of Greek culture that people of non-Greek heritage could, at least in theory, acquire. Physical and social characteristics were often linked with climate—the dark skin of Africans, for example, was linked with the hot Sun of the tropics. Ancient Jews defined their community as possessing a particular relationship with God, a "covenant" others could join only through a process of religious conversion and the renunciation of their earlier religion. Although this concept is distinct from that of race, the Jewish Bible also offered a scheme for classifying peoples according to descent from the three sons of Noah, Shem, Japeth, and Ham. The idea that the descendants of Ham's son Canaan was cursed—the "Curse of Ham"—would be greatly influential in the shaping of anti-Black racism in the Abrahamic traditions. Ancient and Medieval Mediterranean and Chinese peoples also believed in "monstrous races"—peoples whose intelligence and culture were in many ways human but were also physically distinct, possessing such unique features as dog's heads or only one leg. In addition to sources in ancient Greco-Roman writers such as Herodotus and Pliny the Elder, the monstrous races also drew on the biblical and postbiblical accounts of Gog and Magog.

Although Jews continued to define themselves as a religious community throughout the Middle Ages, they were increasingly viewed by their Christian neighbors as a racial community defined by a common descent. Such ideas as the belief that Jewish men menstruated were characteristic of biological rather than religious distinctiveness. Such beliefs put those Jews who had converted to Christianity and their descendants under a cloud of suspicion that could never be dispelled. Black Africans were also frequently the subject of racial characterization in the Middle Ages as intellectually backward and brutish. Much of this came from Muslim Arabs and was associated with the Indian Ocean slave trade in which Africans were sold by predominantly Muslim slave traders. Arabs also justified what they were doing in terms of the biblical "curse of Ham" although it does not appear in Islam's sacred book, the Quran. Muslim geographers and ethnographers also adopted ancient ideas about the influence of climate on the development of different peoples.

The influx of new information about the peoples of the world in sixteenth-century Europe posed questions concerning differences of peoples and cultures in a particularly urgent form. Native Americans posed problems for European classifiers who traced the descent of foreign peoples from Noah or argued that the natives of America were descendants of the ten lost tribes of Israel. The question of whether Native Americans had souls and were therefore both human and potentially Christian was settled in the affirmative by a papal bull, *Sublimi Deus*, in 1537. Vigorous debate on whether Americans were barbarians in the Aristotelian sense, and therefore naturally slaves, was carried on in sixteenth-century Spanish universities. Both sides produced voluminous treatises, one side stigmatizing the differences between Native American and European society and the other arguing that Natives were not barbarous, merely idolaters in need of the Christian revelation. Opponents of the natural slavery theory won the debate in the Spanish intellectual world although their influence on Spanish practice was slight. The lack of evidence for the traditional monstrous races as the world became increasingly known caused them to lose credibility, although some continued to believe in their existence.

In the Renaissance, the category "race" did not have the centrality that it attained later. Religion remained the most important way of categorizing the world's peoples to most early moderns. A roughly fourfold division of the world into Christians, Jews, Muslims, and "Idolaters" was common. The climactic theory of human difference continued to have supporters. The dominant planet and astrological sign of a given area could also determine or influence the nature of its people. Most Europeans believed in the superiority of their cultures over those they encountered elsewhere, although a few used the experience of other cultures to argue against assumptions of European or Christian superiority. African slavery was often explained and legitimated by Africans' non-Christian beliefs or by the curse of Ham. The idea of a "pre-Adamite" creation of humans outside the biblical narrative of Adam and Eve and their descendants although originally put forward to distinguish between the Jews and everybody else was adapted to provide an ideological justification for European superiority.

Schemes for classification of human races, along the lines of classifications of plants and animals, emerged in the late seventeenth century as part of the "Scientific Revolution." The French physician traveler Francois Bernier (1620–1688) drew on his experience of India to write *New Division of the World among the Different Species or Races of Men that Inhabit it* (1684), one of the earliest works to use "race" in the modern sense. Bernier divided humanity into a small number of groups based on skin color, physiognomy, and areas of habitation. "Whites" included Native Americans, Middle Easterners, and Indians as well as Europeans. The greater darkness of the former groups was caused by exposure to the Sun, an environmental factor that did not affect their essential racial nature. Africans by contrast were "Blacks" as their darkness was innate, not a product of their environment. Chinese and Japanese were yet another race, while Laplanders fell into their own categories.

The eighteenth century saw the further development of the scientific concept of human races and of its application to explain and justify social inequalities. The greatest of the classifiers of plants, the Swede Carl Linnaeus (1707–1778), also set forth a system for the classification of human beings based on skin color and habitat—White Europeans, Red Americans, Black Africans, and Yellow Asians. Linnaeus's fourfold classification was influenced by archaic theories of the four humours and the four elements. The German professor Johann Friedrich Blumenbach (1752–1840) produced the most detailed and influential system of racial classification, drawing on extensive evidence to divide humans into five categories—Caucasian, Mongolian, Ethiopian, American, and Malay. Blumenbach, who rejected ideas of racial hierarchy, was the first to identify white people as "Caucasian." He believed that the original humans were Caucasians and that they had been affected by the different environments in which they had settled. However, Blumenbach rejected ideas that non-Caucasians were mentally or spiritually inferiors. Believers in the theory that the life forms of the Americas were inherently inferior, such as the French naturalist the Comte de Buffon (1707–1788), used the idea of degeneration to argue for the inferiority of Native Americans and to claim that European colonists in the Americas were heading the same way.

Although climactic considerations were not forgotten, European thinkers increasingly ascribed differences between human groups to inherited differences, independent of environment. Ideas about race were connected to ideas about gender. The relative profusion of facial hair in the European male, particularly as compared to the African, the Native American, and all females, brought forth profuse tributes to the dignity and majesty of the "philosopher's beard." The genitals of African women were believed to be formed differently than those of European women.

The conformation of the skull began to rival skin color as a criterion of race. A pioneer in skull measurement was the Dutch anatomist and painter Petrus Camper (1722–1789). His concern with accurate artistic representation led him to devise the concept of the facial angle formed by the horizontal line joining the ear to the base of the nose and the vertical line joining the upper jaw to the most protruding point of the forehead. It provided a way to quantify differences in the shape of skulls. Researchers in the nineteenth century ranged human skulls by the facial angle, finding it most acute in Africans and least acute in Europeans. Whereas Camper had considered the least acute facial angles to be simply the most beautiful, subsequent craniometricians elevated the facial angle to a general measurement of human worth. Eventually, scientists began to measure the volume of the skull, asserting that smaller skulls indicated smaller brains and therefore a lower level of intelligence and character. Collections of skulls became an important part of the equipment of racial classifiers, such as Blumenbach or the French comparative anatomist Georges Cuvier (1769–1832). Both Blumenbach, a supporter of the fundamental equality of human beings, and Cuvier, a believer in the inferiority of non-European peoples, appealed to the evidence of the skulls.

Emerging systems of racial hierarchy invariably placed Europeans on top and usually placed Black Africans at the bottom. There was a growing emphasis in European and Euro-American anti-Black literature on the alleged physical, intellectual, and moral deficiencies of African and African-descended people, frequently linking them to animals. The German anatomist Samuel Thomas von Soemmerring (1755–1830) dissected bodies of Africans to publish *On the Physical differences between the Moor and the European* (1784), asserting that Africans were physically closer to apes than to Europeans. Black inferiority was often used to justify slavery, particularly by people coming from European colonies where Black slavery was the basis of the economy. The Jamaican slaveowner Edward Long's (1734–1813) *The History of Jamaica* (1774), a work written to defend slavery, described Blacks as a separate species similar to apes. However, opponents of slavery did not necessarily reject theories of Black inferiority. Soemmering himself opposed slavery, and another opponent, the Scottish philosopher David Hume (1711–1776), held extreme views on Black intellectual inferiority, comparing Blacks who had acquired proficiency in European intellectual disciplines to trained parrots. Intellectually able Blacks such as the American surveyor and astronomer Benjamin Banneker (1731–1806) were described as persons of mixed race, whose abilities sprung from their white inheritance.

Most eighteenth-century racial theorists did not view races as fixed and entirely separate categories, but emphasized the gradations and intermediate conditions. Since all known varieties of humans were interfertile, racial theorists such as Blumenbach acknowledged the "unity of mankind," scorning those who identified Blacks with apes. This belief fit with the Abrahamic belief in the common descent of humans from the original couple, Adam and Eve.

The most extreme view of race, "polygenism," was associated with radical materialists and religious skeptics such as Hume, who denied the Adam and Eve story and with it the common descent of humanity. They saw racial differences as entirely innate, inherited, and fixed and the different human groups as having completely different origins. The existence of persons of mixed race was particularly annoying for such theorists. Although polygenism was only supported by a minority in Europe, it became highly influential among nineteenth-century defenders of slavery in the southern United States.

See also: Blood Libel; Curses; Elemental Systems; Gender and Sexual Difference; Great Chain of Being; Humours, Theory of; Monsters; New World Inferiority; Physiognomy; Pre-Adamites

Further Reading

Goldenberg, David. M. 2003. *The Curse of Ham: Race and Slavery in Early Judaism, Christianity, and Islam*. Princeton, NJ: Princeton University Press; Hannaford, Ivan. 1996. *Race: The History of an Idea in the West.* Washington, DC: The Woodrow Wilson Center Press; Jordan, Winthrop D. 2012. *White over Black: American Attitudes Towards the Negro, 1550–1812.* Second Edition. Chapel Hill: University of North Carolina Press; Schiebinger, Londa. 1993. *Nature's Body: Gender in the Making of Modern Science.* Boston: Beacon Press.

Rain

The supernatural techniques human beings have developed for predicting, causing, and ending rain are innumerable. Rain could be at different times a dire necessity and a disaster, and as a phenomenon that human individuals and communities were deeply involved in it, it attracted superstitious beliefs. The fact that rain falls from the sky also associated it with divine power.

There were numerous signs of approaching rain, and the barrier between "natural" and "superstitious" rain prediction was not always clear. The idea that pain in certain joints presaged rain has some scientific basis, in that some pains can be sensitive to changes in air pressure before a rain. The behavior of animals was also frequently correlated with the imminent arrival of rain. In Europe, a cat licking itself or a spider seeking shelter were heralds of rain. In South India, a dog barking from the roof of a house presaged heavy rains. . In areas marked by a distinct "rainy season," it could be delineated in relation to astronomical cycles, leading to astrological prediction.

In dry spells, there were many ways to encourage rain. In polytheistic systems, rain was often the responsibility of a particular god, and there were proscribed rituals and sacrifices to end a dry spell. Among the best known are the "rain dances" of Native Americans. Zuni rain dancers wore turquoise or other blue items to symbolize rain. However, rain dances are also practiced in Africa, Asia, and parts of Europe. Some scientists have suggested that by stirring up small bits of matter that then seed the clouds, vigorous rain dances are actually effective in causing rain in borderline situations. Dances were not the only method of bringing on rain. Frogs, lovers of wet places, were commonly associated with rain. In South India, there were procedures for carrying a frog around the community when it needed rain.

There were also numerous charms and practices to end rain. In the Anglophone world, the best known are the numerous variations on the children's song, "Rain, rain go away/Come again some other day." There was a superstition in North India that dealers in grain, who wished to stop rain so as to drive prices up, did so by burying water in an earthen pot.

Rain was not only divined for, it also divined other things. Medieval Europe, whose calendar was full of saint's days and religious festivals, developed a rich body of traditions about what rain on particular days meant. Among the best known of these traditions is the English belief that if it rains on St. Swithin's day (July 15) it would rain for the next forty days. Many of the rainy day superstitions were agricultural and varied regionally according to which crops were raised. Apple farmers viewed rain on St. Peter's day (June 29) as evidence that the saint was sending water, ensuring a good apple crop. Rain on a wedding day is sometimes believed to be good luck due to the association of rain with fertility and sometimes bad. *The Golden Wheel Dream-book and Fortune-teller*, a nineteenth-century American book of dream interpretation, identifies dreaming of rain as prophesying success, easy success if the dreamer experiences a gentle rain and success after struggle if the dreamer experiences a storm. Rains of things other than water, such as frogs or blood, were prodigies, signs of great events in the near future.

See also: Divination; Foxes; Frogs; Lightning; Prodigies; Rainbows

Further Reading

Fontane, Felix. n.d. *The Golden Wheel Dream-book and Fortune-teller.* New York: Dick & Fitzgerald; Opie, Iona and Moira Tatern, eds. 1989. *A Dictionary of Superstitions.* Oxford: Oxford University Press.

Rainbows

The beauty and other striking qualities of the rainbow have made it the subject of many legends and superstitions. Judaism and Christianity give positive associations to the rainbow as it appeared after Noah's flood, as a sign of God's promise not to flood the Earth again. (The rainbow does not appear in the Quran, but a hadith, or tradition, endorses the idea that it represents God's promise not to drown the Earth.) There is a Jewish legend that during the life of the revered Rabbi Simon bar Yochai in the second century CE there were no rainbows, the rabbi himself being a sufficient sign of God's forbearance. Joseph Smith (1805–1844), the founder of the Church of the Latter-Day Saints, proclaimed that before the end of the world, God would withdraw rainbows. However, the rainbow can also be viewed as a bad omen of approaching death or other calamities or as a sign of a change in the weather.

In Britain, where the rainbow was often viewed as the sign of an impending death, there was the practice of "breaking the rainbow" by forming a cross out of two sticks (in some versions laying a pebble on each of the four ends). There were also charms that could be repeated to counteract the dreadful power of the rainbow. Among the Indigenous peoples of South America, the rainbow was considered to be a disease-bringer. The Amuesha people of the Peruvian Amazon use the phrase "the rainbow hurt my skin" to describe the onset of disease, and many South American people will close their mouth at the sight of a rainbow to prevent death or disease from entering. The people of the Malabar region of India viewed a rainbow as a good omen if located behind a person or on either side, but a bad omen if directly in front.

Superstitions about the rainbow and weather are numerous, and depending on the position and time of day, the rainbow can be seen as a harbinger either of rain or of dry weather. The Roman natural historian Pliny the Elder (d. 79 CE) denied it was either or that the rainbow was an omen at all, preferring naturalistic explanations. *The Golden Wheel Dream-book and Fortune-teller*, a nineteenth-century American grimoire, identifies the dream of a rainbow as an excellent omen for the dreamer.

The illusionistic nature of the rainbow—the fact that despite appearances it is impossible to find its end or to pass under it—has led to many semiserious legends whose point is that they can never be tested. The best known in the West is the legend of the pot of gold at the end of the rainbow. This legend has become associated with the Irish fairy folk known as leprechauns. Buddhist legend locates a cintimanti or "wish fulfilling jewel" at the end of the rainbow. Some southeastern European cultures have a legend that passing under a rainbow will change a man into a woman or a woman into a man.

See also: Fairies; Sex Change

Further Reading

Lee, Raymond L., Jr. and Alistair B. Fraser. 2001. *The Rainbow Bridge: Rainbows in Art, Myth and Science.* University Park: Penn State Press.

Rats and Mice

Small rodents are some of humanity's oldest and least loved domestic companions and have attracted a range of superstitious beliefs. Among the ancient Greeks and Romans, the gnawing of mice was considered a bad omen—particularly for the person whose clothes, shoes, or other property they were gnawing. These omens could also have a public meaning—the Roman natural historian Pliny the Elder (d. 79 CE) claimed that Rome's Marsian War (91–87 BCE) had been presaged by mice gnawing the sacred silver shields at Lanuvium. This belief persisted into modern times in Europe.

Another way that mice and rats were ominous was through swarming, the sudden appearance of large numbers. The most famous example is the widespread Western superstition that rats presage the sinking of a ship by leaving it en masse. (This sometimes extended to the idea that it was bad luck even to mention rats on shipboard.) Inversely, rats seen entering a ship could be a good omen, while rats leaving a building foretold its collapse. In medieval and early modern Europe, swarms of rats or mice were also believed to forecast the coming of war.

In medieval Europe, St. Gertrude was invoked as a protector against rats and mice. It was claimed that her power kept rodents away from the cheese and bacon of friars and that water from her abbey's well protected against rats and mice. In the twentieth century, the association of the saint and protection against rodents developed into the belief that Gertrude was the patron saint of cats. Early modern Germany featured superstitions around the "rat king," a group of rats whose tails were bound or knotted together. A rat king was a very bad omen, particularly associated with the coming of plague. Since the plague was carried by fleas that infested rats, the association between rats and the plague was based on reality. There were various magical or superstitious techniques for getting rid of rodents in the house or granary, including writing them a letter or charming them with music—the origin of the legend of the "Pied Piper" with his near-magical rat-charming skills. Rats and mice were among the small animals that served early modern English witches as "imps."

Mice, like other small, "verminous" animals, were frequently believed to be produced by spontaneous generation rather than sexual reproduction. In seventeenth-century Europe, there was a "recipe" for creating mice by enclosing sweaty clothing with grains of wheat in a jar for about three weeks.

In Islam, rats and mice were viewed as evil and among the animals that Muhammad said were permissible to kill at all times. It was believed that a rat had dragged a burning wick from a lamp in Muhammad's dwelling in a diabolically inspired attempt to murder the prophet of Islam. In Islamic dream interpretation, a rat or mouse can represent not only evil acts or an enemy in one's own household but also prosperity as rats and mice are less likely to be found in the homes of poor people who do not have excess food for them to eat. However, in the medieval Middle East, mice were also thought to have medical uses—their ashes and body parts treated hemorrhoids, skin diseases, and wounds.

Rats have more positive associations in Hinduism. The goddess Karni Mata is associated with rats and has a famous temple inhabited by thousands of rats believed to be her reincarnated descendants. Hindus believe that these rats, unlike ordinary rats, are not aggressive or disease spreaders. The rat is also one of the twelve animals of the Chinese zodiac.

See also: Amulets and Talismans; Cats; Spontaneous Generation

Further Reading

Lev, Efraim. 2006. "Healing with Animals in the Levant from the 10th to the 18th Century." *Journal of Ethnobiology and Ethnomedicine* 2; Opie, Iona and Moira Tatem, eds. 1989. *A Dictionary of Superstitions.* Oxford: Oxford University Press.

Rattlesnakes. See Snakes

Ravens and Crows

Ravens and crows had a strong association with death in many cultures, but carried different meanings as well.

The idea that ravens have a particular connection with the supernatural goes as far back as the ancient Greeks, for whom the raven as a bird of divination was identified with Apollo. For the Norse, the ravens Hugin and Munin were the messengers and spies of their chief god, Odin. In the Bible, the raven is one of the birds sent from Noah's ark to find land and also the bird that carried food to the prophet Elijah in the wilderness. (The Mesopotamian *Epic of Gilgamesh*, whose flood story precedes the biblical story by centuries, also features a raven let out of the ark that never returns.) A Jewish legend states that one of the ravens that fed Elijah was the raven from the ark. The Talmud states that the raven was one of only three animals that engaged in sex on the ark and was punished for it In the Quran, Cain learns how to dispose of the body of his murdered brother Abel by observing a raven or crow burying its dead mate. A hadith, or tradition of the Prophet Muhammad, states that crows are one of the five animals that no one can ever be blamed for killing—the others being kites, mice and rats, scorpions, and mad dogs.

The idea of these black birds who feed on carrion as harbingers of death goes at least as far back to the ancient Mediterranean. The Roman natural historian Pliny the Elder (d. 79 CE) stated that the raven was particularly ominous when it swallowed its voice as if it were being choked and also believed the crow to be a bird of ill omen. The biographer Plutarch (46–119? CE), in his life of the great Roman orator Marcus Tullius Cicero (106–43 BCE), describes Cicero's murder as having been preceded by literally flocks of ravens, first landing on a ship that he was a passenger on and then later cawing outside the window of his villa. One raven is described as pulling a garment over a sleeping Cicero, showing him kindness when humans were afraid to. Ancient writers also associated crows and ravens with changes in the weather, particularly rain and storms.

Ravens had a long association with divination. Among the titles given to diviners in early medieval Scandinavia and Russia was "raven's priest" and "priest of raven sacrifice." The idea that the raven, particularly the cawing raven, was a bird of ill omen persisted into the medieval and early modern periods. Some argued that the raven's acute sense of smell enabled them to sense the dying, so they could be omens without anything supernatural about it. The medieval church fought this superstition; one medieval penitential suggested seven days of penance for anyone who believed the cawing of a young crow or raven either a good or a bad omen. In the early modern West, some believed that a crow could be a good or bad omen, depending on whether it appeared on the right or the left. In Islamic dream interpretation, the raven signified a powerful person.

As carrion feeders, ravens and crows were associated with battle. The Italian humanist Jovianus Pontanus (1426–1503) claimed to have seen struggles between flocks of ravens and kites in areas where there would be great battles. The Byzantine historian Nicetas Choniates (1155–1217) made a similar claim that a battle between ravens and jackdaws preceded the invasion of the Peloponnese by the nomadic Cumans in 1206. The fact that crows attack other crows led to the development of a medieval legend that they try and execute criminals among their kind. This is one theory for the development of the collective noun "a murder of crows" that appeared in the Middle Ages, disappeared in the fifteenth century, and then was revived in the nineteenth. Crows and ravens were also associated with the demonic. Crows were among the animals that early modern English witches identified as familiars, or "imps." The late sixteenth-century Habsburg magistrate Jakob Bithner reported that after intercourse with Satan, a woman gave birth to a demon in the form of a raven. The demonic bird had actually flown out of her side, leaving a wound.

The raven was believed to eat the eyes of the dead, and for that reason, it was believed to have particularly keen eyesight. In Wales, it was believed that a raven could reward a blind person who was kind to it by restoring their eyesight.

There is a legend that the British monarchy will last only as long as there are ravens roosting in the Tower of London. After WWII, when the ravens had been driven from the Tower by the bombing, they were quickly replaced. Ravens were also associated with the legendary British monarch King Arthur. The idea that Arthur was reborn as a raven first appears in the work of a sixteenth-century Spanish writer and is repeated by another Spaniard, Miguel de Cervantes (1547–1616), in his novel *Don Quixote* whose title character states that for this reason, no Englishman has ever killed a raven. The earliest evidence for this belief from England itself does not appear until the nineteenth century. It comes from Cornwall, a region with many Arthurian associations where it was believed bad luck to kill a raven as one might be Arthur reborn. Another version of the legend identified the bird Arthur had become as a Cornish chough, a type of crow.

Although the idea of the raven as a bird of ill omen appears in East Asia, it is less common than another tradition, which is that the bird is recognized for its filial piety. The Chinese believed that ravens fed their parents. A three-legged crow or raven was believed to inhabit the Sun.

Ravens play a central role in the religion of many Indigenous peoples of the Pacific Northwest. The creator of the world is viewed as a raven god, but the raven is also viewed as a trickster. As a culture hero, Raven is credited with such accomplishments as bringing the Sun and the Moon to the world. Similar beliefs are found among the Indigenous peoples of Eastern Siberia.

See also: Death; Divination

Further Reading

Gippius, Aleksei A. 2018. "An Eleventh-Century Poetic Divination from St. Sophia's, Novgorod." *The Russian Review* 77: 183–199; Green, Caitlin R. 2009. "'But Arthur's Grave Is Nowhere Seen': Twelfth-Century and Later Solutions to Arthur's Current Whereabouts." *Arthuriana*. http://www.arthuriana.co.uk/n&q/return.htm; Schwartz, Donald Ray. 2000. *Noah's Ark: An Annotated Encyclopedia of Every Animal Species in the Hebrew Bible*. Northvale, NJ: Jason Aronson; Wright, Andrew. 2001. "The Death of Cicero. Forming a Tradition: The Contamination of History." *Historia: Zeitschrifte fur Alte Geschichte* 50: 436–452.

Relics

Relics are physical objects associated with religious figures and events. They can be body parts, excreta such as tears or blood, clothes, weapons, or even instruments of torture or execution (one of the most famous relics of Christianity is the true cross, the cross on which Jesus was crucified, which exists only in fragments). Cults of relics existed in many religious traditions, among them Catholic and Orthodox Christianity, Islam, and Buddhism. (Judaism and classical paganism, by contrast, had little interest in relics.) Many relics are physically unprepossessing or even grotesque items, incongruously displayed in ornate altars or reliquaries. Particularly impressive relics or relic collections are the objects of pilgrimage or the center of religious festivals. Relics of religious founders, such as the Buddha, Jesus, or Muhammad are usually the most venerated by their followers.

The origins of relic veneration in Buddhism are debated, but there is evidence that a cult of relics emerged relatively early in the time of the Buddha or shortly after. The most valued relics were those of the Buddha's body. Since the Buddha's body was burned on a funeral pyre, the most common relics of the Buddha were teeth and bones that would have been hard enough to survive the fire. Footprints allegedly of the Buddha's are also viewed as relics. Legends were attached to the travels and circulation of relics. Possession of a particularly venerated relic was associated with political power in much of the Buddhist world, one notable example being the tooth of the Buddha in the Sri Lankan city of Kandy. The water in which the tooth was washed is believed to have healing powers, but Buddhists generally place less emphasis on the miraculous power of relics than do Christians.

In addition to strongly emphasizing the miraculous power of relics, Christian relic veneration identifies a broad range of objects, associated with a broad range of personalities, as relics. The Christian practice of preserving and venerating relics emerged in late antiquity. Relics were associated with particular individuals as part of a cultural transformation during late antiquity (not restricted to Christianity) of associating supernatural power with particular people rather than with places or monuments. Relics associated with Christ were particularly likely to be clothing items or excreta, as the whole body of Christ was believed by Christians to have ascended to heaven, leaving nothing behind. (The one exception was the Holy Foreskin, separated from the young Jesus at his circumcision. Several Holy Foreskins circulated in medieval Europe.) The legend of Helena, the mother of the first Christian Emperor Constantine the Great, discovering the True Cross in Jerusalem was important in establishing a Christian identity for the Roman Empire. The Second Council of Nicea (767) required the enclosure of a relic in the altar when a new church was consecrated, further routinizing relic veneration.

Accounts of miracles worked by relics became a standard feature of saint's biographies, in some cases even eclipsing the events of the saint's life. According to Church authorities, the relics or the saint did not work these miracles themselves. The actual cause was God's power working through the relic at the behest of the saint. However, ordinary Christians hoping for a miracle—or fearing divine wrath—may not have made this distinction.

Relics were sometimes believed to have wills of their own and to choose the place in which they resided. This could even legitimate relic theft, a common activity in the Middle Ages. What appeared to be a theft was merely an action of carrying out the relic's will to reside in a different place. One of the most famous relic thefts was the theft of the remains of St. Mark from their resting place in Alexandria to the city of Venice, which became known as the Republic of St. Mark. The circulation of fraudulent relics was common, and characters such as Geoffrey Chaucer's Pardoner, an unscrupulous seller of relics of dubious provenance, were found throughout medieval Europe. The proliferation of relics of the True Cross was particularly notorious, although some claimed that God could miraculously multiply the wood of the Cross to create additional relics.

With the development of the doctrine of purgatory in the Middle Ages, relics, such as the intervention of the saints they represented, were often associated with relief from suffering in purgatory. Judicial

Holy Lance

The Holy Lance is the legendary spear that a Roman soldier used to pierce the side of Jesus Christ while Jesus hung on the cross. The story appears in the Gospel of John. The Lance is also known as the "Spear of Longinus" after the Roman who wielded it, who according to some legends became a Christian after being healed by the miraculous mixture of blood and water that flowed from the wound. The Lance, if it still existed, would have been bathed in the blood of Christ himself. There are several alleged Lances of doubtful authenticity, including one at the Vatican given as a present to the pope by an Ottoman Sultan, one in Armenia (which may be the Lance allegedly found during the First Crusade), and one in Vienna included as part of the regalia of the Holy Roman Empire.

proceedings incorporated oaths sworn on relics, as the saint, personally offended, would wreak havoc on the oath-breaker. Enormous collections of relics marked the prestige of rulers. The greatest collection of all was at Rome, and the papal capital's immense spiritual power was an important part of papal primacy.

The power of relics became an important subject of intra-Christian dispute with the Protestant Reformation of the sixteenth century. Protestants shunned relic veneration, which they viewed as similar to pagan idolatry. (Ironically, Frederick the Wise of Saxony, who protected the founder of Protestantism, Martin Luther, was also one of the foremost relic collectors in Europe.) Catholic polemicists responded to Protestant attacks by reaffirming the power of relics and recounting the miracles wrought by them in anti-Protestant pamphlets and tracts.

Relics are also venerated in the Orthodox Christian tradition. As was the Catholic practice, the altar of an Orthodox Church incorporated relics. Fragments of the True Cross play an important role in Orthodox religion, but the physical remains of the saints are also widely venerated. For Orthodox believers, sanctity transforms the body as well as the soul, so the physical remains of saints are holy.

Both the major branches of Islam, Sunni and Shia, have traditions of relic veneration. Sunni veneration tends to focus on items associated with Muhammad. The treasury of the Ottoman Empire included items alleged to be the sword, beard, battle-standard, and mantle of Muhammad. Reformist Sunni movements such as Wahhabism, a movement that emerged in eighteenth-century Arabia, rejected relic veneration as giving reverence to something other than God. Relic veneration was more accepted and institutionalized in Shia Islam, which had a wide variety of saints and Imams from whom to derive relics. Shia relics were sometimes held in elaborate reliquaries and shrines. Muslims did not generally associate relics with the kind of miraculous power that Christians did.

Belief in relics was important for the emergence of the modern concept of "superstition." The sharp division between Protestants and the "Enlightened" on one side and Catholic and Orthodox believers on the other over the power of relics led relic veneration to become almost the paradigm case of religious "superstition." This in turn affected the study of other religions with the emergence of "religious studies" in the nineteenth century. Western scholars shaped by Protestantism and the Enlightenment tended to stigmatize relic veneration as "superstitious" in comparison to reading religious texts and following ethical norms. This had a particularly strong impact on the study of Buddhism, leading scholars to separate "vulgar" relic veneration from the learned tradition of Buddhist textual practice in a way that did not reflect Buddhist practice. The same cultural changes affected Catholicism, as relic veneration, never formally repudiated, became less central to the religious experience of most Catholics. The Church formally dropped the requirement for relics to be included in new churches in 1969, a change that attracted little interest or opposition from even the most traditionalist of Catholics.

See also: Amulets and Talismans; Blood; Silver; Weapon-Salve

Further Reading

Freeman, Charles. 2011. *Holy Bones, Holy Dust: How Relics Shaped the History of Medieval Europe.* New Haven, CT: Yale University Press; Geary, Patrick J. 1990. *Furta Sacra: Thefts of Relics in the Central Middle Ages.* Princeton, NJ: Princeton University Press; Germano, David and Kevin Trainor, eds. *Embodying the Dharma: Buddhist Relic Veneration in Asia.* Albany: SUNY Press; Hahn, Cynthia and Holger Klein, eds. 2015. *Saints and Sacred Matter: The Cult of Relics in Byzantium and Beyond.* Washington, DC: Dumbarton Oaks Press; Harris, A. Katie. 2014. "Gift, Sale, and Theft: Juan de Ribera and the Sacred Economy of Relics in the Early Modern Mediterranean." *Journal of Early Modern History* 18: 193–226; Nickell, Joe. 2007. *Relics of the Christ.* Lexington: The University of Kentucky Press; Perry, David M. 2015. *Sacred Plunder: Venice and the Aftermath of the Fourth Crusade.* University Park: Penn State Press; Strong, John S. 2004. *Relics of the Buddha.* Princeton, NJ: Princeton University Press.

Roses

Known for not only its beauty and sweet smell but also for its sharp thorns, the rose has attracted many superstitious and magical beliefs. Many cultures have ascribed medical virtues to the rose. In the ancient Mediterranean world, roses were an ingredient in aphrodisiacs and had many medical uses, from whole roses in wine for stomach complaints to rose ointments for cervical inflammation. In Galenic medicine, roses had a cooling quality effective against hot diseases such as fevers. In ayurvedic medicine, the rose is beneficial both spiritually and physically and can also serve as an aphrodisiac. In traditional Chinese medicine, roses help regulate the flow of qi, the body's energy. The rose is considered to be a yang plant, suitable to treat those suffering from an excess of yin. Like other sweet-smelling plants, the rose was believed to be a defense against plague, and during epidemics in medieval and early modern Europe, people wore sachets of roses.

Like the rose flower, the rose gall, a lump appearing on the stem of a rose caused by an insect, was believed to have medical and magical virtues. Worn around the neck, it would cure toothache and whooping cough, and it was placed under the pillow for insomnia.

Roses also had religious meanings. The Romans planted roses around graves to protect the dead from evil. In the Latin novel by Apuleius (c. 124–c. 170 CE), *The Golden Ass*, the hero, magically transformed into a donkey, is restored by eating roses under the guidance of the goddess Isis. The red rose, according to Christian legend, got its color from the blood of Christ shed at the crucifixion. The legendary Rose of Jericho was supposed to bloom on Christmas Eve to mark the birth of Christ. The Rosicrucians brought together the rose and cross to form a magical and alchemical symbol of their identity. Pennsylvania Germans believed that a child would have red cheeks if their baptismal water was poured over a rosebush.

In eighteenth- and nineteenth-century England, there was a widespread divination technique in which an unmarried woman picked a rose on Midsummer Day, wrapped it in paper, and then wore it on Christmas Day; the man who would take it from her would be her future husband. Petals falling from a rose signify a death, and roses that bloom out of season signify misfortune. In Islamic dream interpretation, roses are generally favorable—a bouquet of roses signifies kisses, one after the other—but not always—to dream of a rosebud means a miscarriage. *The Golden Wheel Dreambook and Fortune-teller*, a nineteenth-century American dream guide, emphasized the "thorn" aspect, suggesting that dreams of roses brought trouble, as dreaming of white roses signified a marriage that would bring vexation.

The English natural philosopher Sir Kenelm Digby (1603–1665) tells the story of a woman with such a violent antipathy to roses that one lying upon her cheek raised a blister.

See also: Aphrodisiacs; Divination; Herbs; Rosicrucianism

Further Reading

Opie, Iona and Moira Tatern, eds. 1989. *A Dictionary of Superstitions.* Oxford: Oxford University Press.

Rosicrucianism

The Rosicrucian movement, aimed at uniting spiritual alchemy and religious and social reform, emerged in early seventeenth-century Germany. It was associated with two anonymous manifestoes, one in German, *The Discovery of the Fraternity of the Most Noble Order of the Rosy Cross* (*Fama Fraternitatis*), first printed along with related writings in 1614, but circulating in manuscript earlier, and *The Confession of the Laudable Fraternity of the Most Honorable Order of the Rosy Cross, Written to all the Learned of Europe* (*Confessio Fraternitatis*) (1615). The manifestoes told the story of a mysterious brotherhood of healers and spiritual adepts founded by the fictional fifteenth-century German Knight Christian Rosenkreutz (Christian Rosycross). The third in the series, *The Chemical Wedding of Christian Rosenkreutz* (1616), an alchemical romance also published anonymously, is known to be the work of the Lutheran theologian Johann Valentin Andreae (1586–1654), who was probably also the author of *The Discovery.* Rosenkreutz is described as learning mystical wisdom in a journey to Egypt, traditionally identified with magic since ancient times. Andreae emphasized the elements of Rosicrucianism based on religious renewal and later turned against magical Rosicrucianism. This early Rosicrucianism was associated with Protestantism in the early stages of the Thirty Years' War (1618–1648), Europe's last great war of religion.

Rosicrucian discourse was incredibly eclectic, drawing from alchemy, astrology, Paracelsianism, Hermeticism, Christian mysticism and millenarianism, the lore of chivalric orders, and the Kabbalah. It emphasized macrocosm/microcosm analogies and the raising of matter to a divine

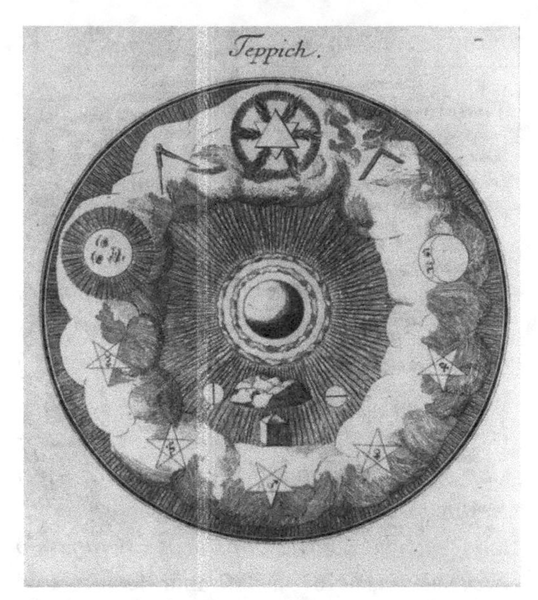

An eighteenth-century Rosicrucian allegorical image. (Library of Congress, Prints and Photographs Collection, LC-USZ62-79702)

status through magical and religious action. Astrologically, the Rosicrucian manifestoes asserted that the world was entering the era of Mercury, identified with Hermes Trismegistus as the Greek Hermes was the equivalent of the Latin Mercury. The full wisdom of the Rosicrucians, however, was not included in the manifestoes, but was available only to the initiates of the Rosicrucian order. The true knowledge of the Rosicrucians, which included the language of Adam, was not to be shared with the common people. Application of Rosicrucian wisdom would bring about an earthly paradise. This would be connected to the coming of a perfectly pure man, who would not be Christ but would complete His redemptive work. In a Paracelsian fashion, the Rosicrucian writings interpreted the philosopher's stone as a universal medicine.

The Rosicrucian order itself was originally a myth, although many readers of the Rosicrucian tracts believed that such an order existed—the English magician Robert Fludd (1574–1637), who used Rosicrucian language in his writings, was even irked that the order had not contacted him. Rosicrucian language and imagery was adopted by many individuals and groups in the seventeenth century, and it was strongly influenced the mythology of Freemasonry. In the eighteenth century, an actual Rosicrucian order emerged, as part of a rejection of Enlightenment materialism and rationalism.

See also: Alchemy; Astrology; Hermes Trismegistus; Kabbalah; Macrocosm/Microcosm

Further Reading

Akerman, Susanna. 1998. *Rose Cross over the Baltic: The Spread of Rosicrucianism in Northern Europe.* Leiden: Brill; Yates, Frances. 1972. *The Rosicrucian Enlightenment.* London: Routledge and Kegan Paul.

Royal Touch

In the Middle Ages and early modern periods, English and French monarchs both claimed to receive on their coronation a God-given power to cure scrofula by touching the victim and making the sign of the cross, the "Royal Touch." Scrofula, a disease of malnutrition that caused unsightly lumps, was referred to as the "King's Evil." Sometimes hundreds of sufferers would line up to be touched by the ruling monarch. These occasions often took place on or near important dates on the religious calendar such as Easter and came to include the gift of a small coin to the sufferer. Over the decades, the coins increased in value. (These were not the only objects with a power to heal. Irish peasants kept pieces of paper believed to be soaked with a king's blood to cure scrofula.) There was never a requirement that the healed person could be a subject of the king; kings could heal foreigners as well as their own people.

An Alternative Scrofula Cure

Touch was not the only way to cure scrofula; long before the first reference to the Royal Touch, Pliny the Elder dealt with how to cure the disease using fish parts:

"For the cure of scrofula, it is a good plan to prick the sores with the small bone that is found in the tail of the fish known as the sea frog, care being taken to avoid making a wound and to repeat the operation daily until a perfect cure is effected. The same property, too, belongs to the sting of the pastinaca, and to the sea hare, applied topically to the sores, but in both cases, due care must be taken to remove them in an instant. Shells of sea urchins are bruised, also, and applied with vinegar—shells also of sea-scolopendræ,3 applied with honey, and river crabs pounded or calcined and applied with honey. Bones, too, of the sæpia, triturated and applied with stale axle-grease, are marvelously useful for this purpose."

Source: Pliny the Elder. *Natural History* XXXII.28 from *The Natural History of Pliny*. Translated, with copious notes and illustrations, by the late John Bostock and H. T. Riley. Six Volumes. London: H. G. Bohn.

The earliest evidence of the French kings' claims to heal scrofula comes from the eleventh century. The origins of the English kings' claims are murkier, but they seem to have emerged in the twelfth century under French influence. Belief in the Royal Touch was widely held, although some churchmen were suspicious of it. They believed that it falsely ascribed to non-saintly temporal monarchs a miraculous power that should rightfully be the preserve of the Church. Despite their complaints, the Royal Touch was widely invoked in the praise of monarchs and was regarded as evidence of royal legitimacy. In 1340, near the beginning of the Hundred Years War, when the king of England Edward III (1312–1377) claimed to be the rightful king of France, an ambassador of King Edward suggested that the French king, Philip VI (1293–1350) demonstrate his rightful claim by curing with the Royal Touch, although the test was never made. The power of the true king to heal was also an issue in the great mid-fifteenth-century struggle of rival claimants to the English throne known as the Wars of the Roses.

During the Renaissance, some natural philosophers, mostly in Italy, accepted the miracle, but attempted to explain it by non-supernatural means. They claimed the healing power was a piece of natural magic, a power belonging to the family of the monarch, but no more divine than the ability to heal possessed by certain jewels and herbs. The Italian physician and astrologer Girolamo Cardano (1501–1576) believed that the king of France carried concealed herbs that actually caused the miraculous cures. The Royal Touch was also involved in the controversies of the Protestant Reformation. The founder of Protestantism, Martin Luther (1483–1546), seems to have believed in it, but some Protestants, reflecting Protestant skepticism about miracles, were reluctant to ascribe divine powers to a human being (or to the coins given out by the king to the sufferers at the ceremony, often kept as talismans). Nonetheless, the Protestant rulers of England continued to carry out the Touch, not without resistance from Catholics. The Netherlands Jesuit demonologist Martin del Rio (1551–1608) suggested that when Elizabeth I of England (1533–1603) claimed to heal victims of scrofula with her royal touch, she could be a witch using powers derived from Satan or perhaps simply a fraud. In any event, as a heretic she could not possess truly miraculous powers.

Charles II of Great Britain (1630–1685) was exiled from his native land after the English Revolution overthrew the monarchy and his father, Charles I (1600–1649), was executed by the victors. Charles, who had touched for the King's Evil while in exile, returned to England as its king in 1660—the "Restoration"—and as part of an effort to make England a thoroughly royalist country again, touched many thousands of his subjects, reasserting the divine origin and sacred nature of kingship. Charles's laying on of the Royal Touch was a religious occasion, following a prescribed Church of England service. The scrofula victims knelt before the king while a clergyman read a Bible passage about miraculous healing. The sufferers then withdrew and came back for a second time so that the king could hang around their necks a specially minted gold "touch piece" on a white ribbon. These touch pieces themselves were thought to have healing powers, and some believed that the cure would last only as long as the recipient kept and appropriately honored the touch piece. Charles, who carried out weekly touchings on Fridays, touched nearly one hundred thousand sufferers during his twenty-five year reign. His successor, James II (1633–1701), was also an avid toucher, and the Royal Touch reached its height in England during their reigns.

The Royal Touch was seldom challenged during the Scientific Revolution of the seventeenth century. Charles II was the same monarch who first chartered the Royal Society of London, the first enduring scientific society and a leading light of early modern science. The overthrow of James II in the revolution of 1688 brought in King William III (1650–1702), who did not practice the rite, although his successor, Queen Anne (1665–1714), did. The subsequent accession of the German Hanoverian Dynasty to the British throne in 1714 marked the final end of touching by kings for scrofula in England, although the Stuarts in exile, who still claimed to be the legitimate kings, continued to carry it out until the last claimant in the direct line died in 1807. The practice persisted in France until the French Revolution of 1789, with a perfunctory revival under the counterrevolutionary king Charles X (1757–1836) in 1825. Despite the ambivalent attitude of many early modern scientists and physicians, disbelief in the Royal Touch and other forms of touching for healing became widespread in Europe's educated elite during the eighteenth-century Enlightenment.

See also: Amulets and Talismans; Blood; Natural Magic; Seventh Children

Further Reading

Bloch, Marc. 1973. *The Royal Touch*. Translated by J. E. Anderson. London: Routledge; Brogan, Stephen. 2015. *The Royal Touch in Early Modern England: Politics, Medicine and Sin*. Woodbridge, Suffolk and Rochester, NY: Boydell & Brewer.

S

Sabianism

A kind of magically and hermetically influenced paganism known as Sabianism survived the coming of both Christianity and Islam to the Middle East by centuries. It was centered in the city of Harran in northern Mesopotamia. Harran had a history stretching back to ancient Mesopotamia as a center for the worship of the Moon god Sin. The community in Harran seems to have continued to practice paganism while the Roman Empire was Christianized because it was remote from the centers of Christian power in the empire and disputed between Rome and the more tolerant Sassanian Persia. Harran became a refuge for Neoplatonic pagan philosophers expelled from the Roman Empire and a center of the teaching of astronomy and astrology. Because a pagan people called the "Sabians" were mentioned in the Quran as a religious community to be tolerated, such as Jews and Christians, the Harran community was tolerated after the conquest of the area by Islamic forces. The link between the Sabian community mentioned in the Quran and the Sabians of Harran is tenuous at best, though. They also legitimized themselves in the eyes of their conquerors by claiming Hermes Trismegistus, known to and revered by the Arabs as Idris, as their prophet, although none of the Arabic Hermetic writings can be traced to a Sabian origin.

The Sabians of Harran were identified with occult and magical wisdom. Some Islamic sources refer to them as "star worshippers," indicating that they may have been thought to have a particular expertise on astrology. Worship of the planets was compatible with a philosophical monotheism that saw the ultimate God as too remote to be interested in being worshipped and the planets as his servants or angels. The development of the very influential schema linking astral gods, the seven planets, and seven metals was attributed to the Sabians. Harran, dedicated to the Moon, was believed to be one of seven cities throughout the world dedicated to the seven astrological planets. Eminent Sabians included the astronomer, mathematician, and astrologer Thabit Ibn Qurra (826–901). A work on the construction of astrological talismans surviving only in Latin translation, *On Images*, is attributed to Thabit Ibn Qurra, although not all scholars agree on the accuracy of the attribution. Another important Arabic astronomer/astrologer, Al-Battani (858–929) may have been a Sabian or a descendant of Sabians who converted to Islam. The Sabians also influenced Islamic alchemy and esotericism in the form of the writings of the mysterious group the Ikhwan al-Safa, or Brethren of Purity. (The Brethren claimed, falsely, that Pythagoras, the ancient Greek philosopher they revered, was from Harran.) Subsequent occultists and astrologers looked to a mythologized version of the Sabian tradition for the transmission of astral magic and Hermeticism from the ancient to the medieval world. "Sabian" has served as a brand for such occultist projects as the Sabian symbols, a set of 360 astrological symbols, one for every degree of the zodiac, and the Sabian

Assembly, a still existing occultist group. Both were the work of the American astrologer and esoterist Marc Edmund Jones (1888–1980). The original Sabian community itself, however, appears to have disappeared in the eleventh century, and the city of Harran was destroyed by the Mongols in the thirteenth century.

See also: Alchemy; Amulets and Talismans; Astrology; Hermes Trismegistus; Moon

Further Reading

Green, Tamara M. 1992. *The City of the Moon God: Religious Traditions of Harran.* Leiden: Brill; Pingree, David. 2002. "The Sabians of Harran and the Classical Tradition." *International Journal of the Classical Tradition* 9: 8–35.

Saffron

This golden herb, a product of a crocus plant, was used as both a food ingredient and a dye. It was believed to have healing powers. Frescoes from the Greek island of Santorini show some of the first representations of gathering and using wild saffron, including one showing a woman using it to treat a girl's wounded foot. Bathing in saffron was believed to cure aching muscles and speed the healing of wounds. The first-century CE Greco-Roman botanist Dioscorides identified it as an aphrodisiac, a belief that persisted into the early modern era. Saffron also appears as an aphrodisiac in the *Kama Sutra*, an ancient Indian guide to leading a pleasant life.

Persians, who were the first people to deliberately cultivate saffron, considered it a "hot" food and paired it with "cold" foods such as rice, as classical Persian cuisine was based on a balance of cold and hot. In China, where saffron was used only sparingly in cooking, it was believed to be a purifier and an aider in maintaining qi, the force of life. Saffron could defend the body against demonic possession and serve as a "universal antidote" against poison. The physicians of the Islamic world viewed saffron as a valuable treatment for many conditions, including diseases of the eye and liver. It was one of the ingredients of theriac, the wonder drug of the medieval Christian and Islamic worlds. Christian Europeans believed saffron, like other sweet-smelling substances, was a preventative against the Black Death. Not only saffron itself but also the root of the saffron plant was incorporated into medicinal drinks. Saffron was incorporated into ointments to treat skin conditions and was believed to be good for "melancholy"—in modern terms, an antidepressant. Pennsylvania German farmers gave it mixed with milk to dogs who suffered from poisoning. Giraldus Cambresis (c. 1146–c. 1223), a Welsh bishop, described an underground fairyland whose diminutive inhabitants subsisted on a diet of milk-sop flavored with saffron. In Islamic dream interpretation, if a woman dreams of herself grinding saffron, it means she is a lesbian. *The Golden Wheel Dream-Book and Fortune-Teller*, a nineteenth-century American grimoire, describes dreaming of saffron as auguring health and wealth.

See also: Aphrodisiacs; Bathing; Fairies; Herbs; Theriac

Further Reading

Ganeshram, Ramin. 2020. *Saffron: A Global History.* London: Reaktion Books.

Sage

Sage is a common term for several plants of the genus *Salvia*. Common sage (*Salvia Officinalis*), a herb indigenous to the

Mediterranean, has spread throughout the world and acquired an elaborate mythology. Roman naturalist Pliny the Elder (d. 79 CE) describes it as effective against a variety of diseases as well as the poison of snakes and stingrays. He also asserts that applied to the body, it enables the womb to expel a dead fetus. However, as is common with Pliny, it is not always clear which plant he is talking about, and some scholars argue that ancient medicinal references are really to three-lobed sage (*Salvia Fructicosa*), another species of Mediterranean sage.

Sage continued to be used as a medicinal herb in the Middle Ages, where it was common in monastic gardens. It was also associated with long life generally, as in the saying from the medieval didactic poem associated with the pioneering medical school at Salerno, *Regimen Sanitatis Salernitanum* (The Rule of Health of Salerno): "Why should a man die who has sage in his garden?" The German nun and polymath Hildegarde of Bingen (1098–1179) and others considered it useful in maintaining the balance of humours in the body in that it counteracted an excess of phlegm. It was also believed to encourage fertility in women.

In England, where legend says sage grows by the roadside because it originally fell from the pouches of marching Roman soldiers, it has a great reputation. (There is evidence that both *Salvia Officinalis* and *Salvia Fructicosa* were introduced in the late sixteenth century.) In addition to its healing properties, sage was used in divination. There were various formulas for how an unmarried girl could pick a specified number of leaves on a specific day of the year and see her future husband (or, according to some versions, a coffin if she was to die unmarried). The plant continued its association with health. In Sussex, there was a belief that eating sage before breakfast seven or nine days in a row would cure the ague. A common saying was that eating sage in May would lead to a long life. It was also recommended as good for the memory.

Although European sage species were introduced into the Americas along with European settlement, there was a preexisting culture surrounding a related plant, indigenous to Baja California and Southern California, known as white sage (*Salvia Apiana*). Along with medical uses, some Native cultures burned white sage as a way of ritually purifying spaces. (Burning of the indigenous sage species for similar purposes is found in Greece.) Among Native and Metis communities, sage is associated with luck. It is chewed before entering situations where luck is desired, such as bingo games.

See also: Divination; Herbs; Humours, Theory of; Luck

Further Reading

Rivera, Diego, Conchita Obón and Francisco Cano. 1994. "The Botany, History and Traditional Uses of Three-Lobed Sage (Salvia Fruticosa Miller) (Labiatae)." *Economic Botany* 48: 190–195.

Sapphire. See Precious Stones

Satanic Pact

The Satanic Pact is an agreement made between a human and Satan, or in some cases a lesser demon, in which the human exchanges their soul for a benefit. It is a characteristically Christian idea. Its earliest appearance is in the sixth-century legend of Theophilus of Adana, a Greek clerk who made a pact renouncing God and the Virgin

Mary in order to gain a promotion to bishop and was then rescued by Mary. The Theophilus legend first appears in Latin Christendom in the ninth century, and the idea of the Satanic Pact spread with it. Gerbert of Aurillac (c. 940–1003), Pope Sylvester II, whose learning aroused suspicion, was widely believed to have made a pact with a demon in exchange for the papacy.

Although Theophilus's and Gerbert's pacts were made for promotions and it remained possible to trade one's soul for mundane things such as wealth, power, or sex, the Satanic Pact became increasingly identified with wizards and witches who traded their hopes of salvation specifically for magical power. In the Middle Ages, the concept of the Satanic Pact became identified exclusively with ritual magicians, particularly necromancers. Ritual magicians themselves denied that they were contracting with or subordinating themselves to demons, claiming instead that they were commanding them. Theologians replied that any dealings with demons, whether or not the magician explicitly made an exchange with them. involved an "implicit pact," as demons would not grant powers in exchange for nothing. By contracting with the devil, the witch or magician was giving him homage due only to God. This enabled the crime of magic to be defined as heresy or idolatry rather than simply using magic to cause harm. However, in the Middle Ages even theologians and Inquisitors who denied that any human had the power to control demons usually placed the devil and the magician on a footing of some equality, at least temporarily, until the magician's inevitable death and damnation. This ascription of quasi-equality to magician and demon continued to the early modern period. Most versions of the Faust legend, which originated in the sixteenth century, showed the learned male magician Faust commanding devils, although of course he lost everything at the end. By contrast, the witch hunt that originated in the fifteenth century viewed unlearned female witches as Satan's wretched slaves and the agreement itself as degrading to the witch.

The doctrine of the Satanic Pact was legally useful as it shifted the definition of the witch away from evil actions and toward the demonic source of a witch's powers. Any exercise of magical powers, whether or not it caused actual harm, was damnable as demonic. European law in the witch hunt era moved from condemning witches for evil deeds to treating the Satanic Pact and the witch's subordination to the devil as the primary offense, although in practice witches accused of evil deeds in addition to the pact were more likely to be convicted. Demonologists who stressed the Satanic Pact also almost invariably called for harsher persecution of cunning folk and ritual magicians as well as witches. The outward benevolence of a cunning person or the learning of a ritual magician disguised the alliance with the forces of darkness that were the ultimate source of his or her powers.

Descriptions of how the Satanic Pact was made varied greatly. One common element as the idea developed was that it was not merely a prosaic legal contract, but involved paying homage to and worshipping the devil, as well as denying Christ, God, and, in Catholic areas, the Virgin Mary and the saints. The witch's or magician's agreement to abandon baptism and Christian allegiance could be symbolized by trampling on a cross. The new agreement between the witch and her master was often sealed by the granting of the devil's mark at this time. The homage the witch paid to the devil was also often seen as particularly humiliating and degrading. The witch was often seen as submitting sexually to the devil or kissing his buttocks or anus while the magician, although equally damned, was not degraded in this way. The social and gender subordination of the female witch was also reflected in the fact that ordinarily she received very little from the devil in exchange for her soul and service. Satan or other demons gave witches not vast wealth, but often a single coin that turned into a pebble, leaves, or dung. The one thing that the devil did freely grant in return for service was the magical ability to perform evil deeds, often giving the new witch instruments or potions for this purpose and even punishing witches who failed to perform enough wicked actions.

Although homage and adoration could constitute a pact by itself, formal written pacts were often claimed to exist, often described as being written in blood. The Satanic Pact was never ironclad, although Satan would go to great lengths to keep witches on his side. Some witches told of how Satan prevented them from reconciling themselves with the Church, but stories of God overruling the Satanic Pact (in Catholic areas; this role was often played by the Virgin Mary in the tradition begun by the original Theophilus story) were common and reinforced the idea that the devil was powerless in comparison to God.

The theory of the Satanic Pact was found among both Protestant and Catholic learned demonologists. However, it was not simply imposed on popular witchcraft beliefs that focused primarily on the evil deeds of the witch. Belief in an agreement between the witch and Satan could also be found in popular culture. Accused witches themselves sometimes recounted their agreements with Satan in uncoerced testimony. Satan, or a lesser devil, often appeared in these narratives in a moment of depression or despair and offered the witch the ability to gratify her modest desires or take revenge herself on her enemies in exchange for her service.

See also: Blood; Devil's Mark; Necromancy; Witchcraft

Further Reading

Briggs, Robin. 1996. *Witches and Neighbors: The Social and Cultural Context of European Witchcraft.* New York: Penguin; Levack, Brian P. 2016. *The Witch-hunt in Early Modern Europe.* Fourth Edition. London and New York: Routledge; Peters, Edward. 1978. *The Magician, the Witch and the Law.* Philadelphia: University of Pennsylvania Press.

Scrying

Scrying is a collective name for a body of magical techniques for seeing things at a distance, usually by employing an object. Scrying can be passive viewing or two-way interaction. Surfaces used in scrying are usually reflective including polished stones, egg yolks, and pools of water, oil, or ink. (Scrying by gazing into a dish of liquid is also referred to as lecanomancy.) The Hebrew Bible speaks of divination by viewing iron arrowheads that had been polished. Even fingernails could be used for scrying. Despite this variety, scrying is most closely associated with glass or crystalline surfaces, including human-made objects such as mirrors and naturally occurring substances such as crystalline quartz. Probably the best-known scrying tool is the crystal ball, but mirrors are also commonly used, a practice known as catoptromancy. The crystal ball became a common symbol of the magician, whether the learned male magician or the disreputable female Roma fortune-teller.

Evidence for scrying dates back to ancient Babylonia, where bowls of palm oil were used for divination. Ancient scryers viewed their practice as a way of contacting supernatural entities, be they gods, demons, or ghosts. One Jewish divinatory technique recorded in the Babylonian Talmud was known as the "princes of the oil." It involved preparing a thumbnail or the palm of the hand with oil and using it to contact demonic entities, the "princes." The subject, often a child (many ancient scriers used young virgin boys, believed to have "pure" souls), was ritualistically prepared, and then the child gazed into the reflective oil to contact the princes. The princes then revealed the answer to a question posed by the child (under adult direction). The princes of the oil continued to be practiced into the seventeenth century. Another ancient technique known to have been practiced during the time of the Roman Empire was "Lychnomancy" or lamp divination, scrying by gazing at the flames in an oil-fueled lamp. Scrying both in the ancient world and later was frequently used to identify thieves.

Scrying along with other divinatory practices was condemned by Christian religious authorities, but continued among both learned and unlearned practitioners. Among the best-known scryers was the English wizard John Dee (1527–1608), who used a "shew-stone" for conversations with angels, although the actual conversations were carried out by his assistant Edward Kelley (1555–1597). Dee, a pious Christian, placed great emphasis on the angelic nature of his interlocutors, distinguishing himself from those who spoke with demons. Dee and Kelley prepared themselves with prayer and other religious exercises rather than incantations. Among Dee's scrying tools was an obsidian mirror of Aztec origin. Dee may well have been using it in accordance with its original purpose, as mirror scrying is known to have been a practice of ancient Mesoamericans. Dee's older contemporary, the French astrologer and prophet Nostradamus (1503–1556), was also an avid scryer, employing a bowl of water suspended on a brass tripod. He credited scrying with the production of his famous prophecies. On a more mundane level, unlearned practitioners continued to scry to detect thieves or identify future spouses, among other reasons.

See also: Divination; Mirrors

Further Reading

Bilu, Yoram. 1981. "Pondering the 'Princes of the Oil': New Light on an Old Phenomenon." *Journal of Anthropological Research* 37: 269–278; Harkness, Deborah E. 1999. *John Dee's Conversations with Angels: Cabala, Alchemy and the End of Nature.* Cambridge, UK and New York: Cambridge University Press.

Second Sight

Second sight was a magical ability frequently associated with Scottish Highlanders in the early modern period. Persons with second sight had a vivid sight of objects and people who were not physically there. These visions always had a meaning, usually related to events in the life course. To see a shroud, for example, presaged a death. Visions of phantom funerals were also common, although generally the identity of the corpse was not clear. Future spouses could also be seen in the company of the people they would marry. Persons could acquire the second sight in many ways, including physical contact with one already gifted (foot on foot and hand on shoulder was one common way), being born with a caul or on certain holy days, and looking through a knothole.

Although belief in this kind of precognition was common in Northern Europe, it was originally only in the Scottish Highlands where it was referred to as "second sight." The term originated in Christian theology as a way of contrasting spiritual vision with physical vision. It came to be used in Scotland to describe precognitive and clairvoyant ability around the end of the sixteenth century. These early references to second sight were in the context of witch trials, but over the course of the seventeenth century, the phenomena separated. (While Scottish witches were usually female, possessors of the second sight were usually male.) Interest in the second sight grew in Scotland and England in the late seventeenth century. Stories of the second sight are included in *The Secret Commonwealth of Elves, Fauns and Fairies* by the Reverend Robert Kirk (d. 1692), a Highlander, although he claimed not to possess the second sight himself. The Anglo-Irish natural philosopher and chemist Robert Boyle (1627–1691) investigated second sight by collecting case accounts. Boyle and Kirk were interested in the second sight and other "supernatural" phenomena to combat the materialistic and mechanistic "atheism" that they believed on the rise. Early Freemasons in Scotland talked about the second sight as a power the Masonic initiate could attain. One Lowlander attributed the allegedly superior ability of Highland thieves to locate hidden treasures to the second sight. Some claimed that the second sight was limited to not only Highlanders but also the region of the Highlands, claiming that Highlanders who emigrated to America lost the ability. The second sight remained a topic of debate into the Enlightenment.

See also: Caul; Divination; Witchcraft

Further Reading

Kirk, Robert. 2008. *The Secret Commonwealth of Elves, Fauns and Fairies.* Garden City, NY: Dover; Stiubhart, Domhnall Uilleam. 2020. "The Invention of Highland Second Sight." In Julian Goodare and Martha McGill, eds. *The Supernatural in Early Modern Scotland.* Manchester: Manchester University Press. Pp. 178–203.

Seventh Children

The idea that seventh children, and particularly seventh sons, had miraculous powers was common in the premodern West. Seventh sons were more likely to be thought magical than seventh daughters, and many variations of the legend insisted that it only applied to the seventh of seven consecutive sons, with no daughter intervening, or to the seventh sons of seventh sons. In Iberia and Latin America, seventh sons were often associated with shape-changing. A Portuguese legend states that seventh sons turned into asses every Saturday. In parts of Latin America, seventh sons were believed to be werewolves.

Throughout most of Central and Western Europe, the most common legend attached to seventh sons is that they were healers. The earliest evidence for this belief comes from the sixteenth century. (The "Physicians of Myddfai," a semilegendary group of thirteenth-century Welsh herbalists, claimed that warts could be removed by washing them in water in which a seventh son had been baptized, but this may be a nineteenth-century forgery.) There are records of seventh sons who practiced as healers, such as the seventeenth-century son of a farm laborer Richard Gilbert of Somerset. The disease Gilbert claimed he had the power to cure was scrofula, the "King's Evil" that was also cured by the Royal Touch, and the beliefs were related in that

England and France, the countries whose kings were credited with the Royal Touch, were the places where seventh sons cured the same disease. (English kings in the seventeenth century viewed seventh sons as rivals and persecuted them, while French kings mostly ignored them.) German seventh sons cured a variety of diseases, and Biscayan and Catalonian seventh sons specialized in the bites of mad dogs. In Catholic countries, seventh son healers were often associated with the saint invoked for the same disease. Catalonian seventh sons were associated with Saint Quiteria, invoked against rabies, and their laying on of hands to cure mad dog bites was accompanied by a short prayer to the saint. (Irish and Italian seventh sons also had the ability to cure mad dog bites.) French seventh sons were associated with Saint Marcoul, the protector against scrofula, and many performed a pilgrimage to his shrine at Corbeny.

Belief in the seventh sons (and in some cases, seventh daughters who served as healers) persisted into the twentieth century. In nineteenth-century Lancashire, the seventh son of a seventh son frequently received the name "Doctor" in recognition of his healing powers. Seventh sons were also distinguished by the name "Septimus," Latin for "Seven."

See also: Dogs; Lycanthropy; Royal Touch

Further Reading

Bloch, Marc. 1973. *The Royal Touch*. Translated by J. E. Anderson. London: Routledge; Young, Simon. 2019. "What's Up Doc? Seventh Sons in Victorian and Edwardian Lancashire." *Folklore* 130: 395–414.

Sex Change

Many cultures believed that it was possible for people to physically change from one sex to the other, usually from female to male. Stories of such individuals are found in ancient Greek and Roman physicians and natural historians. Sexual transformations were included in lists of prodigies along with other warnings from the gods. The Roman natural historian Pliny the Elder (d. 79 CE) speaks of a girl transformed into a boy who was abandoned on a desert island on the advice of the augurs, professional interpreters of omens. However, Pliny also speaks of transformed women who simply reentered society and even married as men, one of whom he had personally met. (The case of abandonment took place in 171 BCE, which may indicate a change of attitudes in the centuries between the incident and Pliny's own time.) In the West, Aristotelian science explained female-to-male transformations through "heat"—masculine bodies were characterized by heat, female ones by cold, so an influx of heat could transform a female body into a male one by pushing internal organs outward. Since female bodies could become male, but not the reverse, and male bodies were considered more perfect, this was also supported by the idea that nature was always aiming at perfection. In the early modern period, however, Western medical authorities became convinced that these were really cases of hermaphroditism or various intersex conditions such as women with enlarged clitorises rather than real transformations from one sex to another.

China also had a tradition of viewing spontaneous sex changes as omens. Unlike the Western tradition, however, Chinese chroniclers spoke of both male-to-female and female-to-male transformations. During the Han Dynasty, one of the earliest commentaries on the *Spring and Autumn Annuals* linked changes of sex with political imbalance—if too much power was falling into the hands of the common people, men changed into women; if the emperor was ignoring his counselors and becoming a tyrant, women changed into men. Chinese physicians explained sex changes, like hermaphrodites, as products of the imbalance of yin and yang. By the late imperial period, sex change stories were viewed less as omens and more as reflections of the moral universe—the classic example is a virtuous couple with an only daughter being rewarded by her transformation into a son. (The idea of a virtuous man being rewarded by the transformation of his only child, a daughter, into a son also appears in Irish hagiography, where saints perform the literally miraculous transformation through baptizing an infant.)

There is also a body of semiserious beliefs that performing an impossible act will change a person's sex. In southeastern Europe, this was passing under a rainbow or drinking from the water at the rainbow's end. In American folklore, there is the idea that kissing your elbow will change your sex. Some animals, including hares and hyenas, were believed to change sex "naturally."

See also: Gender and Sexual Difference; Hares and Rabbits; Hyenas; Intersex Conditions; Prodigies; Yin/Yang

Further Reading

Beecher, Donald. 2005. "Concerning Sex Changes: The Cultural Significance of a Renaissance Medical Polemic." *The Sixteenth Century Journal* 36: 991–1016; Xie, Wenjuan. 2015. "(Trans)Culturally Transgendered: Reading Transgender Narratives in (Late) Imperial China." Ph.D. Dissertation, University of Alberta.

Sibyls

The sibyls originated as prophetic women in the ancient Mediterranean. The ancient Greeks, our principal sources for the sibyls, identified prophecy with women. (In this, they differed from the Jews, most of whose prophets were male.) Sibyls prophesied by the inspiration of a usually male god, most frequently Apollo, although they usually were not portrayed as overcome by divine madness or ecstasy as were some oracular priestesses. The earliest surviving Greek references to the Sibyl refer to one person; however, the idea of multiple sibyls attached to different locations emerged later.

Like many aspects of Greek religion, the Sibyls were eagerly adopted by the Romans. The Roman antiquary Varro (116–27 BCE) refers to ten sibyls, in Persia, Libya, Delphi, Cimmeria, Erythrae, Samos, Cumaea, Hellespont, Phrygia, and Tibur. The Cumaean Sibyl was the closest to Rome and the most important to the Romans. Sibylline prophecy was closely bound up with the Roman state through the *Sibylline Books*, a collection of ancient prophecies in Greek hexameter verse supposedly purchased from a Sibyl, sometimes identified as the Cumaean Sibyl, by the legendary last king of Rome, Tarquinius Superbus. According to Roman legend, the Sibyl first offered nine books for sale to the king. When refused, she burned three and offered the remaining six at the same price. When refused again, she burned three more and then offered the remaining three at the original price, at which point the king relented. The *Sibylline Books* were under the care of a body of Roman officials charged with that specific task and were consulted during crises of the Roman state, particularly plagues and major military defeats or for the interpretation of prodigies and omens. Roman governments based political and religious decisions on the counsel of the books, although sometimes they went against them, for example when the book was interpreted as advising against the

An engraving of the Phrygian Sibyl by the fifteenth-century Italian artist Francesco Roselli. (The Cleveland Museum of Art, Purchase from the J. H. Wade Fund)

construction of an aqueduct and the Romans went ahead and built it anyway on the grounds of necessity. The original *Sibylline Books* were kept in the Temple of Jupiter Optimus Maximus in Rome and were destroyed in the burning of the Temple in 83 BCE. They were replaced with a collection of Sibylline prophecies carefully collected from the empire, mostly from the Eastern Mediterranean that was viewed as where the Sibyl originated. The custom of consulting the *Sibylline Books* survived the transition from the republic to the empire and the conversion of the empire to Christianity. The last recorded consultation was in 363, when the last pagan Roman Emperor, Julian (331–363), consulted the books to decide whether to invade Persia. According to the Roman historian Ammianus Marcellinus (330–400?), the books advised against the expedition, but Julian went anyway and was killed. In 405, the Vandal Stilicho (359–408), the master of the soldiers and effective ruler of the Western Roman Empire, had the books destroyed because they were being used against him. Only fragments of the *Sibylline Books* survive.

Although the Sibyls were originally pagan, they were adopted by Jews and Christians. The idea that the Sibyls prophesied Christ can be traced to the use of the Cumaean Sibyl by the Roman poet Virgil (70–19 BCE), a pagan but a revered figure in the Christian Middle Ages. Virgil's Fourth Eclogue begins with a prophecy attributed to the Sibyl describing the birth of a male child who will become a divine ruler of the world. Scholars still debate who or what Virgil was actually referring to, but the idea emerged in Christianity that this prophecy referred to Christ. The idea that the Sibyls preached Christianity to the pagans helped foster the image of Christianity as a universal religion by giving it non-Jewish as well as Jewish roots and may have facilitated conversion to Christianity among educated pagans. (This view would be most famously expressed in the ceiling of the Sistine Chapel by Michelangelo Buonarotti, where Sibyls alternate with Jewish prophets.) Church fathers including Lactantius (c. 250–c. 325) and Augustine (354–430) endorsed the idea that Virgil and the Sibyls had prophesied Christ, although others, including Jerome (c. 342–420), rejected it. It is to Lactantius, who in his surviving works made more references to the Sibyls' prophecies than he did to those of the Old Testament, that we owe the preservation of Varro's listing of ten Sibyls, as the original work is lost.

In addition to the Fourth Eclogue, the other key text for the Jewish and Christian Sibyl was the *Sibylline Oracles*, a collection of Greek verses combining Jewish, Christian, and pagan themes. The work was a compilation in twelve books of various texts written over a period of centuries. Scholars believe it was put together in its present form in the sixth century, but there are many references to individual passages that predate this. (There are also references to "Sibylline" works that do not appear in the collection.) Jewish writers, including the historian Flavius Josephus (37–c. 100), pointed to passages from the *Sibylline Oracles* endorsing monotheism and referring to the Greek deities as false gods to vindicate Judaism in the eyes of their Greek and Roman contemporaries. Christian writers pointed to the many sections of the *Sibylline Oracles* that explicitly endorsed Christianity in addition to condemnations of polytheistic paganism.

Although the original text of the *Sibylline Oracles*, like many other Greek writings, were lost in the Latin Middle Ages, the Latin Fourth Eclogue was not, and the idea of the Sibyls as Christian prophets persisted. The Sibyls were frequently represented in Byzantine, medieval, and Renaissance Art. (The number of the Sibyls was increased to twelve with the addition of the Agrippan and European Sibyls, providing a parallel with the Apostles.) In the Western Middle Ages, the term "sibyl" could be applied to any woman claiming to prophesy. Sibyls were also identified with female fairies, who were often given the name "Sibyl."

The Western rediscovery of the Greek text of the *Sibylline Oracles* in the sixteenth century was followed by a debate lasting two centuries over whether they qualified as true prophecy. This interacted with many other debates of the time. Protestants, with their suspicion of any source of supernatural authority outside the Bible, often rejected the Sibyls, but this was far from a hard and fast rule; many Protestant scholars, particularly those sympathetic to the tradition of ancient wisdom, viewed the Sibyls as true prophets. They applied the Sibyl's criticisms of pagan "idolatry" to the Catholic veneration of saints and images. Applying the techniques of humanistic textual criticism developed in the Renaissance, scholars also distinguished between different Sibylline passages, holding some authentic and some later additions or forgeries. Defense of the prophetic status of the Sibyls continued into the eighteenth century.

See also: Divination; Fairies; Hermes Trismegistus; Prodigies

Further Reading

Guillermo, Jorge. 2013. *Sibyls: Power and Prophecy in the Ancient World.* New York: Overlook; Raybould, Robin. 2016. *The Sibyl Series of the Fifteenth Century.* Boston: Brill.

Siddhis. See Yoga

Signatures, Doctrine of. See Herbs

Silver

Silver's luster and value have caused it to be credited with many unique properties. Silver, particularly in the form of coinage, was credited with bringing luck and having the power to ward off the evil eye. In the British Isles in the eighteenth and nineteenth centuries, an engaged woman wearing a silver sixpence in her shoe until her wedding was believed to be protected from the malice of rejected lovers. Silver is also believed to bring good luck to weddings in India, where wedding gifts frequently take the form of silver jewelry. Silver, particularly in the form of bhasma, or ash, is used to treat several conditions in India's system of Ayurvedic medicine, including urinary and respiratory disorders.

In European and Euro-American folklore, a silver bullet had the power to kill certain magical or other beings that were protected against normal bullets. The Scottish Royalist leader John Graham of Claverhouse (1648–1689) was believed to be protected by the devil, and when he was killed in battle, his Covenanting enemies credited it to a silver bullet. The early eighteenth-century Bulgarian rebel Delyo was also credited in folk song with an invulnerability to regular bullets, which eventually required his enemies to kill him with a silver bullet. A witch was killed by a silver bullet in "The Two Brothers," one of the German folk stories collected by the Grimm Brothers in the early nineteenth century. Silver bullets or weapons are also credited with the ability to kill werewolves and vampires; however, this legend does not appear before the nineteenth century, and as late as Bram Stoker's *Dracula* (1897), a vampire is shown handling silver with no difficulty.

Astrologically, silver is associated with the Moon. It is one of the seven alchemical metals, ranking second in perfection only to gold, the other "noble metal." Gold, associated with the Sun and the god Apollo was believed to be masculine, and silver, associated with the Moon goddesses Artemis and Luna, was feminine. Although it did not attract as much interest as gold, alchemists also believed it possible to transform lesser metals into silver, a process known as argyropoeia. In sixteenth-century alchemy, "moonworts," plants with crescent-shaped leaves, were believed to have the ability to transform mercury into silver. In Christianity, the best-known use of silver is the thirty pieces of silver paid to Judas Iscariot to betray Christ. In the Middle Ages, silver coins claimed to be originally part of the thirty were venerated as relics in several places and believed particularly powerful to assist in difficult childbirths. In the late sixteenth century, the knights of Malta, who possessed one of the alleged coins, distributed wax impressions of them covered in silver or gold leaf to pilgrims to carry the blessing.

See also: Alchemy; Amulets and Talismans; Defensive Magic; Evil Eye; Luck; Lycanthropy; Moon; Natural Magic; Relics; Vampires

Further Reading

De Mely, M. F. 1900. "The Silver Pieces of Judas in Medieval Traditions." *American Journal of Numismatics* 34: 69–74; Nummedal, Tara. 2019. *Anna Zieglerin and the Lion's Blood: Alchemy and End Times in Reformation Germany.* Philadelphia: University of Pennsylvania Press.

Singa. See Defensive Magic

Snakes

Snakes, unique among land vertebrates in their lack of limbs, have acquired a vast mythology. A small minority of snake species are poisonous, but the image of poison has shaped the beliefs about all snakes. Snakes were also indissolubly associated with evil for Jews and Christians due to the temptation of Eve in the Garden of Eden by a snake, a snake that Christians, but not most Jews, identified with the devil. (Among Hindus in India and in traditional China, by contrast, snakes were revered and offerings were made to them.) Snakes also shed their skin annually and were thus associated with immortality and the renewal of youth. In as early as the third-millennium BCE Mesopotamian *Epic of Gilgamesh*, it was a snake that stole the herb that renewed youth from the hero. The belief in the snake as a healer extended to the time of the pagan Greeks, whose healing Aesculapius was associated with the snakes that adorned his staff. Snakes were also widely and inaccurately credited with intelligence, hence the biblical admonition to be "wise as serpents." In *Oneirocritica*, the manual of dream interpretation by the second-century CE Greek Artemidorus of Daldis, the dream of a snake has multiple significances, including that of a king because of its power and time because of its length and renewal of youth.

In India, earthquakes were sometimes ascribed to the movements of huge snakes that lived underground and were the supports on which the Earth rested. The snake is one of the twelve animals of the Chinese zodiac and is associated with wisdom.

Hostility to snakes in Western culture is manifested in the common belief that it is good luck to kill a snake or even bad luck to see one and not kill it. The belief that even when killed a snake does not die until sundown was commonly held in the United States. Ireland's freedom from poisonous snakes was ascribed to the power of St. Patrick, who drove them out. Such was Ireland's identification with freedom from poisonous snakes that the medieval British writer Giraldus Cambresis argued in his *Topography of Ireland* that the Isle of Man, in the Irish sea between Britain and Ireland, is shown to be British since there are poisonous snakes there.

Islam like Christianity recommends the killing of snakes, enjoined by a hadith or tradition allegedly originating in the words of Muhammad himself. (The belief that Satan in the Garden of Eden took the form of a snake to tempt Eve is held by some Muslims, but has no Quranic backing.) Another Islamic tradition, however, holds that snakes found in the house should just be asked to leave rather than killed. There is also a belief that jinn sometimes take the form of snakes, and in that case, it is dangerous or bad luck to kill them. White snakes are particularly likely to be benevolent jinn.

The rattlesnake, a group of poisonous snakes restricted to the Americas, acquired a unique mythology. Some Native Americans believed rattlesnakes punished those that broke community customs. Europeans, initially unfamiliar with rattlesnakes, ascribed the power of "fascination" to the rattlesnake, believing that it could hypnotize its victims with its glance or with its rhythmic movements and rattle before killing and eating them. (All snakes were believed to possess the power of fascination, but the rattlesnake was believed to possess it in its highest form.) In colonial Brazil, a tea made from the ground-up rattles of rattlesnakes could cure poisonous snakebites.

The snake's cast-off skin and other parts of its body can have magical or medical properties. In early modern England, it was believed that wearing a snakeskin would enable pregnant women to have easy deliveries. In traditional Chinese medicine, a snakeskin was applied to skin and eye infections. Chinese healers also used the bile and gallbladder of snakes in medicinal preparations, and Brazilians used the fat of the boa constrictor to treat rheumatism. The poison of the adder, a type of venomous snake, was an ingredient in the medieval wonder drug, theriac.

Another powerful item associated with snakes was the "snakestone," also known as the "serpent's egg" or "adder stone." The idea that groups of snakes create stones by their breathing or hissing, particularly in large groups, goes back at least as far as the Roman natural historian Pliny the Elder (d. 79 CE). Pliny claimed that the druids, Celtic enemies to the Romans, carried snakestones the size of apples to win victories in court or favor with rulers. (This sometimes failed; Pliny claims the Roman emperor Claudius executed a man for carrying a snakestone into court.) Renaissance Europeans believed that snakestones formed in the heads of poisonous snakes (like toadstones were formed in the heads of toads) protected against poison, particularly the poison of snakes, and by the mid-seventeenth century, there was a vigorous trade in snakestones imported from India and supposedly formed in the heads of Indian cobras. The snakestones were a topic of great controversy among the scientists of the time. The Jesuit Athanasius Kircher (1602–1680) performed experiments demonstrating the ability of snakestones to suck poison from a wound inflicted on a dog by a poisonous snake. He argued that the snakestone itself was poisonous, and since poisons attracted each other, it drew the poison out of the wound. This led to a prolonged confrontation with another Italian natural philosopher, Francisco Redi (1626–1697), who denied the power of the snakestone entirely and also had experimental evidence to back up his position. Brazilians believed that the snakestone was found in the head of normally ground-dwelling snakes found in trees and that it could repel snakes as well as cure snakebite. The Aguaruna people of the Amazon believe that stones found in snakes can be powerful aphrodisiacs.

See also: Amulets and Talismans; Aphrodisiacs; Dragons; Earthquakes; Jinn; Sage; Spontaneous Generation; Theriac; Toads

Further Reading

Baldwin, Martha. 1995. "The Snakestone Experiments: An Early Modern Medical Debate." *Isis* 86: 394–418; Brown, Michael F. 1986. *Tsewa's Gift: Magic and Meaning in an Amazonian Society.* Washington, DC: Smithsonian Institution; Dobie, J. Frank. 1982. *Rattlesnakes.* Austin: University of Texas Press; Fita, Didac S., Eralda M. Costa Neto and Alexandre Schiavetti. 2010. "'Offensive' Snakes: Cultural Beliefs and Practices Related to Snakebites in a Brazilian Rural Settlement." *Journal of Ethnobiology and Ethnomedicine* 6: 13.

Solomon

The biblical Jewish King Solomon, the son of David, became the prototype of a master magician for Jews, Christians, and Muslims. This begins in the Bible itself, which specifically praises Solomon for his wisdom. He also had a special place in Jewish legend as the builder of the first Temple and as the author of the *Song of Solomon* and *Ecclesiastes*, included in the Jewish canon that became the Christian Old Testament. (He is also claimed as the author of the apocryphal book *The Wisdom of Solomon.*) Early Jewish tradition continued to add to the legend of Solomon's wisdom and power. The Jewish historian Flavius Josephus (37–100? CE), in his *Antiquities of the Jews*, describes Solomon both as a natural philosopher in the Greek tradition, knowing the secrets of plants and animals, and as a powerful exorcist. Josephus speaks of a Jewish contemporary he had seen exorcise demons in the name of Solomon.

In the early centuries of the Common Era, Solomon was claimed in several Jewish texts, including the Babylonian Talmud and the *Testament of Solomon*, to have commanded demons to build the Temple. The *Testament of Solomon*, a Greek language text of unknown origin probably dating from the fourth-century CE, is primarily a report of a series of interrogations performed by Solomon with various demons, compelled to reveal their names, natures, and powers. The *Testament* recounts that Solomon commanded the demons by means of a seal on a ring that he wore, which was a gift from God. Although the seal and ring are not mentioned in the Bible, they and the associated belief in Solomon's power over demons would greatly influence subsequent writers. The name of Solomon frequently appears in amulets and charms, including incantation bowls, to defend against demons. Several patterns, including the six-pointed "Jewish star" now represented on the flag of Israel, have been put forward as the seal or part of it. The five-pointed star or pentagram used in Western ceremonial magic is also credited to Solomon. Another version of the seal is an eight-pointed star formed by superimposing two squares, with the eight corners marked with loops. This version is found in India and Southeast Asia in a Muslim context. Devoid of reference to Solomon, it also appears in Hindu and Buddhist contexts. It is believed to bring good fortune.

Solomon, or "Sulieman bin Daoud," also frequently appears in the Quran as a master of wisdom, a speaker of the language of birds, and a commander of the jinn. The legend of Solomon, embodied in Islam's sacred book, spread as far as Islam did to Southeast Asia and West Africa. (Jews and Christians tend to place more importance on Solomon's father David than on Solomon, while Muslims placc morc importancc on Solomon.) The Muslim version of the legend of Solomon's ring claims that it was inscribed with the profession of faith, "There is no God but God and Muhammad is his prophet." Solomon appears in Islamic

This Iraqi earring dating from the eleventh–thirteenth century CE shows one version of the Seal of Solomon, possibly intended to defend the wearer from harmful magic. (The Metropolitan Museum of Art, New York, Purchase, 1895)

literature as a model of the pious practitioner of magic, who understands God as the ultimate source of his power. He is also viewed in the Islamic tradition as a prophet.

In the Christian Middle Ages and Renaissance, Solomon became associated with ceremonial magic and necromancy with the ascription of the authorship of many grimoires, including *The Greater Key of Solomon* and the *Lesser Key of Solomon.* He was credited with being the founder of the Ars Notoria, a magical system of accelerated learning, and the English writer Gervase of Tilbury (1150–1220) credited him with being the first to discover the magical virtucs of jewels. Solomon, who according to the Bible had an extensive harem, was also credited with devising aphrodisiac spells. Solomon played a particularly important role in Ethiopian Christianity. The last Ethiopian Imperial House, the House of

The Quran on Solomon's Power over the Jinn

"And to Solomon the wind; its morning course was a month's journey, and its evening course was a month's journey. And We made the Fount of Molten Brass to flow for him. And of the jinn, some worked before him by the leave of his Lord; and such of them as swerved away from Our commandment, We would let them taste the chastisement of the Blaze; fashioning for him whatsoever he would—places of worship, statues, porringers like water-troughs, and anchored cooking-pots. 'Labour, O House of David, in thankfulness; for few indeed are those that are thankful among My servants.'

"And when We decreed that he should die, naught indicated to them that he was dead but the Beast of the Earth devouring his staff; and when he fell down, the jinn saw clearly that, had they only known the Unseen, they would not have continued in the humbling chastisement."

Source: A. J. Arberry. 1955. *The Koran Interpreted.* London: Allen and Unwin. *Sura* 34.

Solomon, claimed descent from Menelik, the son of Solomon and the Queen of Sheba. The House of Solomon ruled from 1270 to 1974. Solomon also appears as a master magician and commander of demons in Ethiopian magical writings.

Solomon's relevance in the West receded in the Enlightenment with the decline of necromancy and the rise of biblical scholarship, which called into question his authorship of many of the works attributed to him. However, his role in the building of the First Temple assured him a prominent place in the mythology of Freemasonry, as "Masonry" traces its origins to the stoneworkers who built the Temple.

See also: Amulets and Talismans; Ars Notoria; Coffee; Defensive Magic; Incantation Bowls; Jinn; Mandinga Pouches; Necromancy; Precious Stones

Further Reading

Gallop, Annabel. 2019. "The Ring of Solomon in Southeast Asia." British Library, Asian and African Studies Blog, https://blogs.bl.uk/asian-and-african/2019/11/the-ring-of-solomon-in-southeast-asia.html; Verheyden, Joseph. 2013. *The Figure of Solomon in Jewish, Christian and Islamic Tradition: King, Sage and Architect.* Leiden: Brill.

Solstices and Equinoxes

The solstices are the two days in the year where the night or the day is longest, and the equinoxes are the two days when the lengths of the day and night are equal. The solstices are reversed so that the longest day in the northern hemisphere is the longest night in the southern hemisphere and vice versa. In many calendars, the equinoxes and solstices mark the transitions between seasons. These "cardinal points" of the year were not just markers of time, but had a sacred and magical character in many cultures. The ritual sacrifices performed by the Chinese emperor to maintain the connection of heaven and Earth took place on solstices and equinoxes. They were also particularly important to astrologers. The vernal equinox, referred to as the "first point of Aries," is regarded as the beginning of the astrological year, even though it now occurs in the constellation of Pisces. In East Asia, the dark cold winter solstice was considered the most yin portion of the year, while the summer solstice the most yang.

As turning points of the year, the solstices and equinoxes offered people a chance to shape or at least divine the following period through the actions taken on that day. Ancient Europeans lit bonfires on the summer solstice, to drive away demons and ensure a good harvest. The association of fire with the summer solstice continued after the coming of Christianity, when the solstice was St. John the Baptist's Day eve. According to one medieval legend, the burning of bones on St. John's Day eve drove away dragons. There is a French superstition that for discovering the hair color of one's future bride one could take a brand from the solstice fire, place it under one's pillow, and in the morning, hairs of that color would be wrapped around it. Roots of mugwort gathered on St. John's Day were held to be effective against epilepsy. In Japan, taking a bath with a citrus fruit called yuzu was considered to bring good luck for the coming year. In China and areas influenced by Chinese culture, cold noodles were often consumed so that their yin energy would counteract the strongly yang summer heat.

There were also numerous beliefs associated with the winter solstice on December 21, known in northern Europe as Yule, although it tended to be swallowed up by Christmas a few days later. Some considered the day to be particularly favorable for divination. For believers in the Yin/Yang system, yang foods such as highly seasoned spicy dishes should be consumed during the winter solstice, to counteract its strongly yin nature. Dumplings were also believed to be good foods to eat around the winter solstice.

Solstices and equinoxes frequently played a calendrical role, often related to the agricultural year. In Malabar, the spring equinox was viewed as the beginning of the agricultural year, and it was important that the first thing a person viewed on waking up on the day of the equinox be propitious. The spring equinox was often viewed as a particularly propitious time for planting. In Hungary, there was a belief that onions planted on the spring equinox would have healing powers when harvested.

A common belief in the European and Middle Eastern Jewish communities was that on the solstices and equinoxes for a moment water, and possibly other consumable fluids such as rendered goose fat, turned to blood, which was forbidden for Jewish consumption. There were various explanations for this including the idea that this commemorated Moses striking a rock and drawing blood from it. To be rendered fit for consumption again, the fluids had to be purified—one method being the insertion of a piece of iron. This folk-belief was referred to as early as the tenth century CE. It was not enjoined by any canonical Jewish text, and many Jewish religious authorities opposed it as superstitious.

See also: Astrology; Onions; Sun; Yin/Yang

Further Reading

Carlebach, Elisheva. 2011. *Palaces of Time: Jewish Calendar and Culture in Early Modern Europe.* Cambridge, MA: Belknap.

Spiders

Spiders, with their mysterious power to spin delicate yet strong webs, have attracted many superstitious beliefs. Spiders were even identified with divine beings by many

African and Native American groups. The Roman natural historian Pliny the Elder (d. 79 CE) believed that spiders webs were weather omens. Spiders, he claimed, spun only in cloudy weather, and a great number of spider webs presaged rain. He also claimed that spiders spun their webs higher up when a river was about to flood. Pliny suggested that spiders and spiders webs, when applied to the head, could cure fevers and that applying spider webs to a cut would stop bleeding—an idea that persisted to the twentieth century.

In China, from ancient times the spider, known as the "happy insect," was considered a good omen. In England and Ireland, spiders, worn on the body in a bag or a nut wrapped with silk, could cure the ague or whooping cough. English people also believed that very small spiders, called "money-spinners," would bring wealth if they fell on a person or were even carried in a pocket. *The Golden Wheel Dream-book and Fortune-teller*, a nineteenth-century American dream book, also associated spiders with money, claiming that dreaming of a spider coming toward a person or dropping down in front of them presaged gaining money. Despite these generally positive associations, spiders were also among the small creatures who served early modern English witches as demonic "imps."

There are numerous legends of how spiders, by concealing themselves with webs, saved famous people, including King David, the infant Jesus, and the prophet Muhammad, from their enemies. This also led some Jews, Christians, and Muslims to believe that killing spiders was religiously forbidden or bad luck. The idea that killing a spider or disturbing a cobweb is bad luck also exists outside the religious context. Medieval Italians believed that the bite of a spider caused wild and frenzied dancing—the "tarantelle"—from which the word "tarantula" was eventually coined to describe the large but basically harmless (to humans) spiders of South America.

Like other animals, spiders were believed to have "natural enemies." The Talmud claims that the spider torments the scorpion by entering its ear as an example of the power of the weak over the strong. Sir Thomas Browne (1605–1682), author of *Pseudodoxia Epidemica* (1646), one of the many works written in sixteenth- and seventeenth-century Europe to attack "superstitious" beliefs, used an experiment to attack the traditional belief that spiders and toads would always fight each other. Browne put a toad and some spiders in a jar, noting that the spiders crawled around undisturbed by the toad's presence, while the toad ate them, one by one.

See also: Doll Magic; Luck; Toads

Further Reading

Schwartz, Donald Ray. 2000. *Noah's Ark: An Annotated Encyclopedia of Every Animal Species in the Hebrew Bible.* Northvale, NJ: Jason Aronson.

Spontaneous Generation

Spontaneous generation is the belief that living beings can emerge spontaneously out of nonliving material. The belief that small creatures such as insects or mice—"lower animals"—could emerge from decaying matter was endorsed both by classical natural historians such as Aristotle (384–322 BCE) and Pliny the Elder (d. 79 CE) and by the Bible. Aristotle's *On the Generation of Animals* provided several examples such as flies emerging from dungheaps. Pliny claimed the frogs emerged out of slime. Judges 14:8, in which a hive of bees emerges out of the corpse of a lion, was frequently adduced as an example of spontaneous generation. Spontaneous generation in this sense was generation out of something previously existing, not generation out of nothing, which most authorities believed was possible only for God. (And even God did not always create out of nothing; Genesis refers to Adam as created out of the dust of the ground.)

Spontaneous generation, particularly the spontaneous generation of human beings, could be viewed as a magical process as in the creation of a golem, but as a natural process, it was compatible with a range of natural philosophies that did not draw a sharp division between living and nonliving matter. Aristotelians believed that all matter had the "potential" to be alive. As early modern "mechanical philosophers" such as Rene Descartes (1596–1650) believed that living beings were only matter arranged in a different way from nonliving ones, the fact that they could spontaneously arise from nonliving matter was no more difficult to explain in principle than the sudden transformation of living into nonliving material through death. The popularity of the belief among natural philosophers was reinforced by its common occurrence among ordinary, unlearned people who found it harmonized with their observations of the world.

Early Christian writers such as St. Augustine of Hippo (354–430) who dealt with the subject of spontaneous generation generally endorsed the idea based on both biblical evidence and the support of ancient philosophers. Medieval Scholastic philosopher Thomas Aquinas (1225–1274) argued that life was produced when passive matter was fertilized by an active principle—male sperm in the case of sexual reproduction, the influence of the stars in the case of spontaneous generation. The authority of Aristotle aided the acceptance of ideas about spontaneous generation in the medieval Islamic world. The physician and philosopher Ibn Sina (980–1037) endorsed the idea to the extent that he believed the spontaneous generation of human beings was possible, a claim vehemently rejected by the Spanish philosopher Ibn Rushd (1126–1198). Both philosophers were influential in medieval Latin Europe where they were known as Avicenna and Averroes.

In late seventeenth-century Europe, belief in spontaneous generation was threatened by a growing awareness, thanks to the microscope, of the complexity of very small animals. The multifaceted insect eyes the microscope revealed did not seem like they emerged spontaneously from putrefying matter. Leading microscopists including the Dutch savant Antoni Van Leeuwenhoek (1632–1723) opposed belief in spontaneous generation. The strongest evidence was provided by another microscopist, the Italian physician, poet, courtier, and experimentalist Francesco Redi (1626–1697). Redi sealed away pieces of meat in airtight containers and noted that they did not spontaneously produce maggots or flies. In *Experiments on the Generation of Insects* (1668), Redi claimed that insects must arise from seeds originating in parent insects. Redi's evidence was widely accepted, but did not completely destroy the doctrine of spontaneous generation. Even Redi himself was at a loss to explain insects occurring in oak galls and speculated that they emerged from a perversion of the life force of the tree. The papal physician and pioneering microscopist Marcello Malpighi (1628–1694) later traced the connection between an insect laying its eggs and the later appearance of an oak gall and a mature insect.

The new knowledge on spontaneous generation led to a dispute within the Jewish community on the relationship of science and Jewish law, as interpreted by the rabbis. Rabbinical authorities allowed Jews to kill lice, but not fleas, on the Sabbath, justifying the distinction by claiming that fleas were true animals that reproduced sexually, while lice were spontaneously generated from decaying matter. The Ferrara physician Isaac Lampronti (1679–1756) pointed out that modern scientists had disproved the idea that lice and other small insects were spontaneously generated and that rabbinical law need to change to recognize that. Lampronti's old teacher, the Mantua Rabbi Judah ben Eliezer Briel (1643–1722), argued that the wisdom of Gentile scientists could never be cited against rabbinical authority, which was from God.

The doctrine persisted on the level of even smaller creatures. In the eighteenth century, it was claimed that not insects but microscopic "animalcules" were spontaneously generated. This doctrine was only finally disproved after much debate in the nineteenth century. Even now, the dominant evolutionary theory of the origins of life relies on a one-time spontaneous generation event. Nor has the fact that spontaneous generation has been scientifically disproven necessarily driven the idea out of popular culture; the belief that a horsehair placed in water could spontaneously become a snake persisted into the twentieth century.

See also: Barnacle Goose; Frogs; Golem; Rats and Mice; Snakes

Further Reading

Bertolacci, Amos. 2013. "Averroes against Avicenna on Human Spontaneous Generation: The Starting-Point of a Lasting Debate." In Anna Akasoy and Guido Giglioni, eds. *Renaissance Averroism and Its Aftermath: Arabic Philosophy in Early Modern Europe.* Dordrecht and New York: Springer. Pp. 37–54; Jacob, Francois. 1973. *The Logic of Life: A History of Heredity.* Translated by Betty E. Spillman. New York: Pantheon.

Sun

By far the most conspicuous feature of the sky, the Sun has attracted many magical and superstitious beliefs. In many pantheons, the Sun god or goddess is the chief deity. The Japanese Imperial house traces its descent from the Sun goddess Amaterasu. The image of the Sun being carried around the Earth in a chariot or boat is a common one. Sun deities are usually male, but can be female also. The usual pattern is for Sun and Moon deities to be of different sexes.

The idea that it is better to begin a project or hold a wedding in the time of the Sun's rising from midnight to noon than in the time of its setting from noon to midnight is a common one. Children born during the Sun's rising would do better than children born during its setting. The Sun is the subject of many weather-related superstitions; in Norway, whistling in the direction of the Sun brings rain. In Portugal, if the Sun shines during a rainstorm, a widow may be about to get married. Throughout Eurasia, rainstorms when the Sun is shining are known as fox's weddings. In Nigeria and other African regions, this phenomenon is proverbially associated with lionesses giving birth.

In the British Isles from the Middle Ages to the twentieth century, there was a common belief that the light of the Sun extinguished fires. There were various attempts to come up with a scientific explanation for this phenomenon, from a Renaissance Aristotelian distinction between "external" and "natural" heat to a nineteenth-century classification of the Sun's rays as luminous, heating, and

Parhelia

A parhelion, also known as a sun dog, is a false sun created by the sun's reflection in atmospheric ice crystals. Parhelia usually appear in pairs on either side of the true sun, resulting in references to "three suns in the sky."

Parhelia were often considered to be omens. Given the association between the sun and rulership, as omens they had a public meaning. Multiple suns appearing over Korea in 760 CE were interpreted as an omen of catastrophe. Parhelia that appeared in the city of Stockholm in 1535 were interpreted as prodigies, some claiming that they threatened the King Gustavus Vasa, controversial for his introduction of Lutheranism. The king himself viewed the central sun as a reference to his position as king and the two false suns as representing political opponents conspiring against him.

chemical. In both cases, the Sun's rays were considered antithetical to the nature of the fire's heat and thus extinguished it.

The New Testament Book of Acts, following the Hebrew Book of the prophet Joel, describes how the Sun will be turned to darkness before the day of the Lord. The idea of the darkening of the Sun in the days preceding the apocalypse has been very influential in Christianity; in addition to shaping reactions to solar eclipses, there was the "Dark Day" in New England in 1780, when a mysterious darkness struck the land, and many believed the end of the world was near. Another Christian apocalyptic image involving the Sun is the "woman clothed with the Sun" in the Book of Revelation. Interpretations of this image vary from the Virgin Mary to the Church to the Jewish people, but the Sun usually represents light from God or the gospel. There was a Christian legend that the Sun dances for joy on rising Easter morning, in honor of Christ's resurrection.

The Sun played a central role in Mesoamerican culture. The Aztecs identified themselves as sustaining the Sun in a daily battle with the stars through the sacrifice of hearts and blood. The rising of the Sun and the banishment of the stars from sight every morning was evidence of the Sun's victory. They identified successive creations according to which God occupied the role of Sun. Humanity's current existence was that of the fifth Sun, when Huitzcatlipocli was the Sun god.

The Sun is one of the seven planets in the tradition of Western astrology, governing the Zodiac sign Leo the lion. (The Sun is frequently associated with lions.) The Sun usually exerts a positive astrological influence. The sun sign, the sign in which the Sun appears, is an important part, but only a part, of a traditional horoscope. In the last century, however, it has come to dominate popular astrology.

In macrocosm/microcosm systems, the Sun usually corresponds with the heart. Alchemically, the Sun is identified with gold, at the top of the metallic hierarchy. In Chinese cosmology, the Sun epitomizes the yang force as the Moon epitomizes yin. Given the positive associations of yang, the Sun is viewed in positive terms.

See also: Astrology; Blood; Eclipses; Foxes; Geocentrism; Gold; Macrocosm/Microcosm; Moon; Solstices and Equinoxes; Yin/Yang

Further Reading

Opie, Iona and Moira Tatem, eds. 1989. *A Dictionary of Superstitions.* Oxford: Oxford University Press.

Swallow Stones

The swallow stone or chelidonius is a term for the small, hard stones found in the crops or gizzards of swallows. The idea that the swallow stone has magical or healing powers goes back to the ancient world and was frequently referred to by medieval and early modern writers.

It was generally believed that the stone should be harvested from nestlings rather than adult birds. The first-century CE Greek physician and pharmacologist Dioscorides and others recommended that the stone be gathered at the time of the Moon's increase. It was frequently suggested that the stone would be more powerful if the bird it was taken from had never left the nest. Harvesters distinguished between different types of stone, the most frequent classification being into "black" and "red" stones.

The virtues claimed for swallow stones were many. Roman natural historian Pliny the Elder (d. 79 CE) suggested that tying a stone on to the left arm helped against epilepsy, and the effectiveness of the stone against epilepsy was repeated by many medieval writers. In the Middle Ages, the red stone was thought to heal lunatics as well as headaches and general aches and pains. When placed under the eyelid, they had the power to cure diseases of the eye. There were various procedures for this, some involving grinding the stone to a powder before placing it on the eye. By the late seventeenth century, a "natural" explanation for this alleged phenomenon had evolved—that the stone attracted filth in the eye and that when removed it took the filth with it, leaving the eye clean. A version of the belief found in nineteenth-century Brittany was that swallows had the ability to find stones with the ability to restore sight to the blind, and the way to obtain them was to put out the eyes of fledglings, following which the mother would find the stone and bring it back to the nest for her young.

Medieval authors also recommended the stone for defensive magic against the devil and enchantments. The first-century CE Greco-Egyptian writer on the power of stones Damigeron recommends the swallow stone for general good fortune, including wealth, agreeability, and favor with the great. In this, he was followed by many medieval and early modern writers including the philosopher and natural magician Albertus Magnus (c. 1200–1280). Other than the particular case of eye diseases, the use of the stone was generally to be worn rather than consumed. Many followed Pliny in recommending wearing the stone on the left arm, with many variations regarding the kind and color of cloth it should be wrapped in. One Renaissance medical professor advised his client, the Duke of Milan, that wearing the stone sewn into a linen shirt so as to rest under the left nipple would help with his gout.

See also: Amulets and Talismans; Defensive Magic; Moon; Natural Magic

Further Reading

Duffin, Christopher John. 2013. "Chelidonius: The Swallow Stone." *Folklore* 124: 81–103.

Syphilis

Syphilis appeared in Afro-Eurasia as a new disease beginning at the end of the fifteenth century. The first recorded outbreak occurred among French troops besieging

Naples in 1494–1495. It was widely believed to have originated in the Americas and been brought to Europe by Columbus's sailors, although many modern scholars dispute this. The tendency to identify syphilis with foreign countries continued; the French referred to it as "the Italian disease" and the English as "the French disease." This pattern continued outside Europe; in the Ottoman Empire, syphilis was referred to as "the Christian disease" and in India as "the Portuguese disease." Syphilis, with its sexual transmission, striking disfigurements, which extended to noses falling off, and high rate of fatality, was associated with the wrath of God and punishment for sin. (Over the centuries, syphilis has evolved into a much less disfiguring and fatal disease.) This was set forth in mythological terms in the poem that gave syphilis its name, the Italian poet and physician Girolamo Fracastoro's Latin epic, *Syphilis, or the French Disease* (1530). This is the story of a shepherd named Syphilis who was punished by the god Apollo for blasphemy. Some argued that since syphilis was a punishment for sin, it should not be treated, but it was the consensus of physicians that it could be treated like other diseases.

Since there was a common belief among Europeans that cures originated in the same place as diseases, many argued that the American plant, tobacco, was the cure for the American disease, syphilis. A tea made from the bark of another American plant, guacium, was also believed effective against syphilis. Physicians applied the same techniques they used to deal with other diseases, such as bloodletting and purges. Sweat baths were believed to help the body expel the "poison" of syphilis. Like the Europeans, sixteenth-century Islamic physicians faced the challenge of the "new disease" of syphilis. The first treatise on the subject from a Muslim physician dates from 1569. The author was Imad al-Din Mas'ud Shirazi (1515–1592), a physician at the hospital in the Persian city of Mashhad. Shirazi recommended a plant called "China root" as a treatment for the disease, the same remedy recommended by many European authorities. In China, where the disease was referred to as "Cantonese sores" after a major port for China's trade with Europeans, it was believed to be caused by "sexual toxins."

Until the development of modern drugs, the most common treatment for syphilis was mercury, applied in a variety of ways. The idea of mercury as a treatment for disease went back to the Middle Ages in both the Christian and Islamic traditions, but it became identified particularly with syphilis in the early modern period—hence the saying, "a night with Venus, a lifetime with Mercury." However, mercury is itself a poison, and the treatment often proved more fatal than the disease.

See also: Tobacco

Further Reading

Quetel, Claude. 1992. *History of Syphilis*. Baltimore: Johns Hopkins University Press.

T

Talismans. See Amulets and Talismans

Tarot

The tarot deck, one of the modern world's most popular divination tools, consists of fifty-six cards, the Minor Arcana, divided into four suits with fourteen ranks, and the Major Arcana, twenty-one cards numbered from one to twenty-one and a twenty-second, unnumbered card, the Fool for a total of seventy-eight cards. The Major Arcana is also known as the "Greater Trumps." The four suits are cups, coins, swords, and wands, and the ranks are numbered one through ten, with four "court" cards, the Page, Knight, Queen, and King. Although the tarot deck existed as playing cards for centuries earlier, the use of the deck for divination emerged only in the eighteenth century. Tarot reading is a form of "cartomancy," divination with the use of cards. The French author and former Protestant pastor Antoine Court de Gebelin (1725–1784) asserted in his massive *The Primitive World Analyzed and Compared with the Modern World* (1781) that the cards had originated in ancient Egypt as a version of a text called the *Book of Thoth* and had been brought to Europe during the Roman Empire, from which they had passed into the hands of the papacy. He also linked the cards to Hermeticism. The idea of Egypt as a source of magical and occult wisdom was still powerful in the eighteenth century, particularly among Freemasons such as Court de Gebelin, although there is no evidence of a connection between the tarot deck and Egypt. Court de Gebelin viewed the cards as part of the esoteric tradition stemming from the Hermetic writings and included descriptions and allegorical interpretations of the greater trumps, often linking them to Egyptian culture. The seventeenth trump, the Star, which shows a woman pouring water from two vases, Court de Gebelin claimed, showed the Egyptian goddess Isis pouring out the waters of the Nile. Court de

An eighteenth-century Dutch representation of divination using cards. ("Kaartlegger" by Daniel Nikolaus Chodowiecki. Rijksmuseum,)

Gebelin also linked the twenty-two greater trumps with the twenty-two letters of the Hebrew alphabet, long believed a source of magical power in the kabbalistic tradition. Court de Gebelin included a short essay by Louis Raphaël Lucrèce de Fayolle, the Comte de Mellet (1727–1804), on the use of the cards for divination.

The major champion of tarot divination in the eighteenth century was another Frenchman with Masonic connections, Jean-Baptiste Aliette (also known as "Ettelia," Aliette spelled backward), who published a book on the cards, *Manner of Recreating Oneself with the Card Game called Tarot*. In 1790, Aliette published the first tarot deck designed for divination rather than card-playing. Aliette was also an Egyptophile and Mason, the "Grand Magus" of the "Lodge of the Perfect Initiates of Egypt." He founded a "Secret Society of the Interpreters of Thoth" in 1790.

The divinatory use of the tarot deck was at first largely confined to France, where it became part of the nineteenth-century occult revival.

See also: Divination; Hermes Trismegistus; Kabbalah

Further Reading

Dummett, Michael, Thierry DePaulis and Ronald Decker. 1996. *A Wicked Pack of Cards: The Origins of the Occult Tarot.* New York: Saint Martin's Press; Farley, Helen. 2009. *A Cultural History of Tarot: From Entertainment to Esotericism.* London: I. B. Tauris.

Tasseography. See Coffee; Divination; Tea

Tea

A popular drink in East Asia and eventually in the West as well, tea has become a daily necessity for many and attracted many superstitions and strange beliefs. In China, where tea drinking originated, tea was associated with Shennong, the divine farmer and one of the mythical three sovereigns who preceded the Yellow Emperor. Tea was believed to be suited for drinking by Daoist masters who were seeking clarity. It is generally considered yin, although darker, blacker teas have more yang while greener teas are strongly yin. Tea is considered to have a variety of health benefits.

The idea that tea was connected to spirituality spread from China to Japan, which adopted tea drinking along with other aspects of Chinese culture. The spirituality of tea is embodied in the tea ceremony. However, the belief that tea is physically healthful is also common in Japan. Tea is also widely consumed in India, the world's largest tea producer today, but the history of tea in India before the nineteenth century is not well documented. Tea, particularly when brewed with other substances, was part of Ayurvedic medicine.

The person most responsible for promoting tea in the West was a Dutch physician named Cornelis Bontekoe (1640–1685). His Dutch language *Treatise on the Excellence of the Herb called Tea* (1678) provides a discussion of the medical beliefs of supporters and opponents of tea at the time that it was first being widely drunk in the West as an oversea import from China. Bontekoe himself claimed to have become interested in tea after drinking it regularly for a few weeks and it relieved him of kidney stones. The myths about tea he wanted to dispel included that it was harmful to those constitutions dominated by bile, that it caused epilepsy, and that it made men impotent and women sterile. In addition to the stone, Bontekoe claimed, tea alleviated gout, scurvy, and dysentery among other conditions. It warmed the blood and diluted thick, sluggish blood. Bontekoe followed an alchemically oriented medical tradition that emphasized the importance of maintaining an acid-alkali balance in the body and viewed tea as a helpful alkali for overly acidic bodies. Such was Bontekoe's enthusiasm for tea that under some circumstances, he recommended drinking two hundred cups in a day. The other route by which tea reached Europe was the overland route from China through Central Asia to Russia, where it was consumed by the court and upper-class people; there too tea was originally promoted as a medicine, good for gas and stomach complaints.

As tea consumption spread, it was assimilated to more explicitly magical traditions. There are various systems for divining from the leaves left at the bottom of a cup, known as tea leaf reading or (along with similar practices such as reading coffee grounds) tasseography. (The Western style of tea brewing, which uses whole or partially cut leaves, is more suitable for tasseography than the Chinese style, in which the leaves are ground into a powder before brewing.) Tea leaf reading built on traditions of interpreting patterns left by the sediment in a glass of wine. Interpretations can be built on the positions of the leaves and the shapes they form. Through a process that remains unclear, tea leaf reading in Europe became associated with the Romani people, but it was also practiced by a broad array of middle-class Europeans, especially women, and by the mid-nineteenth century, specially designed cups were being produced for divination. The introduction of tea bags in the late nineteenth century ended tea leaf reading as a common practice and relegated it to occult professionals, many of them Romani.

The preparation and consumption of tea in the West, particularly in nations such as Britain and Russia where it was consumed by all classes, became surrounded by rituals and superstition. In the nineteenth century as tea became widely consumed in Russia, Russian religious conservatives or "Old Believers" considered tea drinking, along with other relatively recently acquired habits, to be damnable and spread the false idea that tea had been condemned by the seventh ecumenical Council of the Church. That Council had met in 787, long before anyone in the West was aware of tea. When the samovar, the vessel used for preparing Russian tea, buzzes or hums, that is a sign of bad luck. In England, bubbles or foam on tea can presage either kisses or coming into money. Putting in milk, a central ingredient of English tea, before sugar leads to bad luck in love. Tea made too weak means a friendship is ending, while too strong means that a new friend is being made. Stirring leaves in a teapot before pouring can be bad luck. There are several English superstitions associating the pouring of tea with fertility. A man and a woman pouring from the same pot will have a child together, or a woman pouring from a pot in another woman's house will have a child. A stem or a piece of tea leaf floating in the cup was considered to herald a stranger visiting. Disposing of the leaves after brewing a cup of tea can be a fraught act. An old English superstition holds that the leaves should be placed in the back of the fireplace rather than be thrown away, as this keeps off poverty. Another English superstition is that throwing spent leaves on the ground in front of a house will protect the dwelling from evil spirits.

See also: Blood; Coffee; Defensive Magic; Divination; Scrying; Yin/Yang

Further Reading

Cook, Harold J. 2007. *Matters of Exchange: Commerce, Medicine and Science in the*

Dutch Golden Age. New Haven, CT and London: Yale University Press; Opie, Iona and Moira Tatern, eds. 1989. *A Dictionary of Superstitions.* Oxford: Oxford University Press; Yoder, Audra Jo. 2016. "Tea Time in Romanov Russia: A Cultural History, 1616–1917." Ph.D. Dissertation, University of North Carolina Chapel Hill.

Tengu. See Yokai

Theriac

Theriac was a compound medicine and poison antidote first formulated in the ancient world and widely adopted in Christian and Islamic medicine. Different compounds have been referred to as theriacs, and some medical writers and physicians distinguished between theriacs. Its earliest surviving mention is in a poem, *Theriaca*, by Nicander of Colophon, a second-century BCE Greek physician. This poem referred to cures for animal poisons, such as the poison of snakes or scorpions. A compound called theriac was devised by Andromachus, the Greek physician to the Roman Emperor Nero in the first century CE, who added the flesh of vipers to a previously existing antidote called mithridatium. (Mithradatium was named after its purported inventor, King Mithradates VI of Pontus (135–63 BCE), who had supposedly made himself immune to poisons by taking small doses of many poisons.) Andromachus innovated by suggesting theriac could cure diseases such as colic, dropsy, and the plague. Theriac benefited from its alleged endorsement by the ancient physician Galen (131–201), the creator of the most influential body of medicine for medieval physicians in western Afro-Eurasia. Although the two writings on theriac appearing in Galen's name are now widely believed to be spurious, they were accepted for many centuries. The author describes different methods of testing theriac (testing poisons on humans was forbidden in the Roman Empire), from giving two roosters a dose of poison but only one of them theriac to giving a human a purgative and a dose of theriac, on the assumption that potent theriac would negate the effects of purgatives as well as poisons. Literally hundreds of ingredients appear in different recipes for theriac over the centuries. Some commonly found ingredients are aloe, myrrh, and opium.

Like ancient Greek medicine in general, theriac passed into the Byzantine and Islamic worlds. The Iranian physician and traveler al-Biruni (973–1048?) suggested theriac against poisons and suggested that one way of testing it was to take a small dose of theriac and then chew on some garlic. True, potent theriac took the smell of garlic off a person's breath. Ibn Sina's *Canon of Medicine*, one of the most influential treatments of Galenic medicine in the medieval Christian and Islamic worlds, suggests theriac as a treatment for many diseases as well as poison. Theriac recipes with a particularly large number of ingredients were referred to as "grand theriacs" and were believed particularly potent.

Knowledge of theriac spread to China through both Islamic and Byzantine contacts. Discussions of theriac and its uses were included in books on the Arab pharmacopeia written in Chinese, and theriac was included in the gifts to the emperor made by Byzantine diplomats. The Chinese seem to have perceived theriac less as an antidote to poisons and more as a treatment for long life, in the tradition of Daoist medical alchemy.

In medieval Europe, knowledge of theriac was introduced through the Arab physicians such as Ibn Rushd, the author of a popular short treatise on theriac. What was widely believed to be the best theriac was manufactured under public authority in the city of Venice, known as "Venice treacle." Other cities such as Venice's great rival Genoa also manufactured their own versions. (Eventually, Jesuits in Brazil would develop a "Brazilian theriac" by substituting native ingredients for hard-to-obtain European ones.) In Renaissance Europe, theriac was popular among physicians influenced by the classical revival in medicine, but Arab and medieval writings continued to be popular—Ibn Rushd's treatise on theriac was printed at least five times between 1497 and 1530. Theriac, and to a lesser degree mithridatium, appeared frequently and were lengthily discussed in pharmacopeias and other medical writings throughout the early modern period. In the course of the sixteenth century, learned writings in Latin were supplemented by vernacular writings aimed at apothecaries and even patients.

Versions of theriac were sold in the streets by mostly Italian unlicensed medical practitioners known as "charlatans." Charlatans and other medical professionals also sold cheaper theriac derivatives such as the popular drug orvietan. Learned physicians greatly resented the charlatans and viewed their theriac as a fake. Theriac and mithridatium were also opposed by a medical movement that favored "simples," remedies derived from a single source, rather than complex compounds such as theriac. Theriac was denounced in a pamphlet, *Antitherica, an Essay on Mithridatium and Theriaca* (1745), by an English physician, William Heberden (1710–1801). It was dropped from English pharmacopeias shortly afterward, but continued to be sold in Continental Europe, including France and Germany, into the nineteenth century.

See also: Alchemy; Ambergris; Saffron; Snakes

Further Reading

Di Gennaro Splendore, Barbara. 2021. "The Triumph of Theriac: Print, Apothecary Publications, and the Commodification of Ancient Antidotes." *Nuncius* 36: 431–470; Dobroruka, Vicente. 2016. "Theriac and Tao: More Aspects on Byzantine Diplomatic Gifts to Tang China." *Journal of Literature and Art Studies* 6: 170–177; Rankin, Alisha. 2021. *The Poison Trials: Wonder Drugs, Experiment and the Battle for Authority in Renaissance Science.* Chicago: University of Chicago Press; Yildirim, Vedat. 2018. "The Comparison of Grand Theriac Formulations in Canon of Avicenna and Two Manuscripts from Early Ottoman Medicine." *Lokman Hekim* 8: 247–260.

Toads

The many species of toads are found throughout every continent save Antarctica and have been endowed with many meanings. The Western tradition has been generally negative about toads, largely due to their ugliness and the poisonous and sometimes hallucinogenic liquids exuded by them as a means of defense. The papal decree *Vox in Rama* (1233) described how people were recruited into "Luciferian" satanic cults by a toad as large as a dog. Toads had great magical power, and many magical practices used by witches, cunning folk, and ordinary people made use of toads, dried toads, powdered dried toads, toad bile, toad feces, and toad blood. The poisonous liquids exuded by a toad's skin associated the toad with poisoning and sickness, to the degree that other parts of a toad, such as its blood or the ashes of its body, or even its urine, were also believed poisonous. The notorious witch and poisoner of seventeenth-century Paris, La Voisin, both milked toads for their poison and then burned them to use the ashes as another poison. The bumpiness of the toad's skin associated it with warts, which were believed to be caused by handling toads or their urine. Paradoxically, the toad was also associated with healing and health—toads were an ingredient in many magical cures of both humans and livestock. In the English county of Cheshire, it was believed that pertussis or "chincough" could be cured by holding a toad momentarily in the mouth of the sufferer, which would result in the disease being transferred from the human to the toad. Powdered toad was a cure for dropsy. Toads were even believed to have the power to dissolve malignant tumors.

Toads also played a prominent role as witch's familiars, particularly in Basque witchcraft, where witches described entire herds of toads at sabbats. For the Basques, familiars nearly always took the form of toads wearing clothes. The toads were fed with a special mixture prepared in the witch's house and were powerful beings that had to be treated well and propitiated. Sometimes a child being taught to be a witch would have the responsibility of herding the toads (shepherding was a common occupation among Basques) while the adult witches were at the sabbat. If the child touched or harmed the toad, they would be punished. In the witch hunts in Basque country during the early seventeenth century, the Spanish Inquisition searched the witch's houses for dressed toads, among other things, but never found one.

In England, the other area where familiars were common, toads frequently, although not exclusively, appeared in that role. Sometimes the mere discovery of a toad in a person's house was treated as evidence of witchcraft. Toads were associated with devils—John Milton's (1608–1674) epic *Paradise Lost* (1667) pictured Satan taking the form of a toad to pour poison into the ear of Eve, and devils were frequently depicted in art as toad-like. Toads did not always appear as the allies of witches; Pennsylvania Germans believed a toad's foot nailed over a stable door protected against witches. Even an outline of a toad's foot drawn in chalk over a door or window helped keep witches out.

Alchemists also employed toads. The monster called the basilisk was born from a rooster's egg hatched by a toad. Toads supposedly generated in their heads a substance called a "toadstone," which could alert one to the presence of poison by heating up or changing color or even serving as a universal antidote if swallowed. Methods for procuring the stone varied, although many including the German philosopher and alchemist Albertus Magnus (1200–1280) believed that in order to be effective it must be taken from a living toad. The sixteenth-century English author Thomas Lupton recommended putting a dead toad in an anthill, where the ants would eat away the flesh, leaving the stone. Alleged toadstones were even set in rings.

Toads were believed to grow in the human body. Some women had allegedly given birth to toads, and if a person swallowed a toad egg, the toad could come to maturity inside them, producing stomach complaints or an abscess until the toad was vomited up or somehow forced to remove itself. When a seventeenth-century Mexican woman, Juana de los Reyes, was exorcised and she vomited a huge toad, it was believed evidence that she was a victim of witchcraft, particularly since after the toad was killed, it kept its shape while the body was burned.

Ancient Mesoamericans had a much more positive attitude to the toad than

Westerners, viewing it as a creative life force associated with the Earth Mother and hallucinogenic mushrooms. Native Americans along the Orinoco river in the early nineteenth century kept toads under pots, rewarding them for good weather and punishing them for bad.

In East Asia, the toad was associated with magic. Great magicians were portrayed as having wise magical toads as companions or friends. The Daoist Immortal Liu Haichan had as a companion the Jin Chan, the three-legged "money toad" associated with prosperity and good luck. Statues or jewelry in the form of a three-legged toad remain common in the Chinese cultural sphere. Not all associations of the toad were positive, however. In the dualistic Chinese system, the toad was associated with yin along with the spider, lizard, centipede, and snake—the "five poisons." The Moon was often believed to be inhabited by a toad, sometimes accompanied by a hare. Chinese healers used toads as medicine, prescribing dried toad venom for inflammation and toothache. There were also stories of a toadstone in China, this one giving the possessor the ability to walk on water.

See also: Defensive Magic; Doll Magic; Frogs; Spiders; Witchcraft

Further Reading

De Graaf, Robert M. 1991. *The Book of the Toad: A Natural and Magical History of Toad-Human Relations*. Rochester, VT: Park Street Press; Henningsen, Gustav. 1980. *The Witches' Advocate: Basque Witchcraft and the Spanish Inquisition (1609–1614)*. Reno: University of Nevada Press.

Tobacco

Tobacco's psychoactive and addictive properties have attracted many superstitious and magical beliefs. Among the Native Americans who first developed and adopted it, tobacco was believed to have sacred or magical power and was incorporated into many rituals and practices. (The high nicotine content of many strains of American tobacco gave it a stronger psychoactive effect than relatively mild modern tobacco.) Gods were identified as smoking tobacco, and tobacco was sacrificed in their honor. Highland Guatemalans believed that the ritual consumption of tobacco enabled them to see the future, and the Chippewa people left a pinch of tobacco on a stone to ward off storms. The Aguaruna people of the Amazon believed that the learning of anen, magical songs, depended on the consumption of tobacco juice mixed with the saliva of the instructor. Native Americans also believed in tobacco's health-giving properties. The Aztecs believed that tobacco, which they consumed in the form of cigars, was a soporific. The Iroquois, pipe-smokers like other Native North Americans, believed tobacco a cure for toothache.

Columbus's entry into the Americas changed tobacco from an American to a global commodity that attracted myths, legends, and superstitions in a broad range of cultures. The newness of tobacco, the fact that tobacco consumption had few precedents in the cultures of Afro-Eurasia, made it the object of intense fascination. Tobacco was greeted in Europe as a health-giving product, although it took some time for European herbalists to distinguish it from henbane. There was a common belief among European physicians that God provided the remedy for a disease in the same place that the disease originated. Therefore, some argued that the American plant, tobacco, was the cure for the American disease, syphilis. The most enthusiastic and influential European supporter of tobacco's medical uses was the Spanish physician Nicolas Monardes (1493–1588), author of the widely translated and reprinted *Historia Medicinal* (three volumes, 1565, 1571, 1574), who believed that tobacco was a painkiller and cured a variety of illnesses, including cancer. Monardes also believed that tobacco was an antidote for the poisons Natives used on their arrows and darts and was also effective against some European poisons. Jean Nicot (1530–1604), the French diplomat and medical writer after whom nicotine is named, also believed that tobacco was good for a variety of ailments. The sixteenth-century Dutch physician Giles Everard took this belief even farther in a 1587 Latin work translated into English as *Panacea or the Universal Medicine Being a Discovery of the Wonderful Virtues of Tobacco* (1659). Such are tobacco's health-giving virtues, Everard suggests, that it may even make physicians obsolete. The first mention of tobacco in English poetry is in Edmund Spenser's *The Faerie Queen* and describes "divine tobacco" as a natural healer, although the image is of the leaves being directly applied to the wound rather than smoked. In the Aristotelian/Galenic terms of European medical philosophy, tobacco was hot and dry and therefore suited to cure illnesses like the "rheum" or cold that were cold and wet. Medical use of tobacco declined in the seventeenth century.

Tobacco had its European Christian opponents as well as supporters, and its opponents described it as literally diabolical. They associated it as an American herb with the idea of Native Americans as devil worshippers. What was the Native practice of taking tobacco in spiritual ceremonies turned into the use of the plant in devil worship for Europeans. As its use became more common, tobacco retained negative associations, though more of unthriftiness and debauchery than satanism. *The Golden Wheel Dream-book and Fortune*-teller, a nineteenth-century American grimoire, took a very dim view of dreams about tobacco, associating them with poverty and loss in real life. Although not directly connected with Native Americans, tobacco was initially viewed as diabolical throughout much of the Islamic world and forbidden by Islamic law, something that did not stop its popularity. Muslim anti-smoking propaganda described how the corpses of tobacco smokers continued to burn in their graves, linking the fires of tobacco smoke to the fires of hell.

Tobacco also had aspects of culture shock in China, where smoking became very popular among the elite in the seventeenth century. Chinese writers attempted to make tobacco more acceptable through expanding on some obscure passages of the Chinese classics to assert that tobacco was originally Chinese. Chinese writers also expounded on tobacco's health-giving effects and even asserted that it enhanced sexual pleasure. Chinese physicians were never as enthusiastic about tobacco as some of their European peers such as Monardes and Everard, but thought that it could be therapeutic when consumed in moderation but dangerous when consumed in excess. The idea that the same substance could be both a medicine and a poison was common in Chinese medicine. Among the diseases that a moderate consumption of tobacco could treat were the intermittent fevers characteristic of southern China. Tobacco was aligned with the yang forces and thus considered appropriate for those diseases and conditions associated with an excess of yin. As a "yang replenisher," tobacco served as a health aid for men whose yang qi had been diminished by sexual overindulgence or overeating. Tobacco was not advised for women whose yin was already being diminished through menstruation. However, women could suffer from yang deficiency, particularly after menopause, and tobacco was sometimes helpful.

The spread of tobacco around the world led to countless superstitions among particular societies. The Paliyans of India viewed tobacco as unlucky because, along with betel nut, it was buried with the dead. Pennsylvania Germans believed that tobacco would keep lice out of a pigsty and a quid of tobacco would soothe a bee sting.

The cigarette emerged in the nineteenth century as an industrial product. One well-known superstition associated with the cigarette is that it is dangerous to light three on a single match.

See also: Chocolate; Syphilis; Yin/Yang

Further Reading

Benedict, Carol. 2011. *Golden-Silk Smoke: A History of Tobacco in China.* Berkeley: University of California Press; Brown, Michael F. 1986. *Tsewa's Gift: Magic and Meaning in an Amazonian Society.* Washington, DC: Smithsonian Institution; Goodman, Jordan. 1993. *Tobacco in History: The Cultures of Dependence.* London and New York: Routledge; Knapp, Jeffrey. 1991. *An Empire Nowhere: England, America and Literature from Utopia to the Tempest.* Berkeley: University of California Press; Norton, Marcy. 2008. *Sacred Gifts, Profane Pleasures: A History of Tobacco and Chocolate in the Atlantic World.* Ithaca, NY and London: Cornell University Press.

Trepanation

Trepanation is a form of surgery that creates a hole in the skull. There were a variety of methods including drilling, cutting, or abrading. It is often survivable, particularly in areas where there is a low risk of infection. From the evidence of preserved skulls, trepanation is one of the oldest surgical procedures, practiced in a variety of civilizations from the Neolithic period onward. There is evidence from Neolithic sites that the discs of skull were removed to make the holes wearable as amulets.

Trepanned skulls of the Paracas, an ancient Peruvian people. (Travelstrategy | Dreamstime.com)

Although there may have been medical reasons for trepanation, involving the removal of skull fragments or foreign matter after a blow to the head, it was also practiced for nonmedical reasons. Trepanning allowed for a release of evil spirits from the head, alleviating madness or seizures. In the Hippocratic literature of ancient Greece, trepanning after a head wound allowed blood to escape before it could decay.

Trepanation was practiced as a cure for epilepsy beginning at least as early as the second century CE by Greek physician Aretaeus the Cappadocian. The justification was that by allowing the escape of "evil air" trepanation would alleviate the conditions that gave rise to seizures. A related practice was to trepan to treat madness or mental illness. The idea was also to release whatever was lodged in the skull causing the problem. This was frequently expressed in terms of the removal of a "stone." Some early modern artists showed trepanations, including Hieronymous Bosch (d. 1516) and Pieter Breugel the Elder (d. 1569). The use of trepanation to treat mental conditions as opposed to injuries to the skull diminished over the early modern period in Europe.

Trepanation was practiced in Africa to treat headaches following a head injury and was also practiced in ancient China. There is a reference to the exposure of the brain through surgery in the *Yellow Emperor's Inner Classic*, considered the foundation of the Chinese medical tradition. Although it long predates the Han Dynasty, trepanation in China is associated with the legendary Han physician and surgeon Hua Tuo, who supposedly practiced it to cure a headache. Later Chinese medical works described trepanation as a way to remove worms and parasites from the skull. However, noninvasive treatments were preferred by Chinese physicians. Among the most active centers of trepanation was the Andes region before the Spanish conquest. However, we have little idea of the reasons for trepanations in the region. Trepanation continues as a traditional therapeutic practice in some contemporary Oceanic and African societies. Trepanation has also been taken up by Western new age and wellness practitioners to expand consciousness and emulate the effects of psychedelic drugs.

See also: Blood; Yellow Emperor

Further Reading

Verano, John W. 2016. *Holes in the Head: The Art and Archeology of Trepanation in Ancient Peru.* Washington DC: Dumbarton Oaks Research Library and Collection.

Twins

Twin births have been seen in different cultures as both auspicious and abominable. Twins can appear in mythology as both hostile (Jacob and Esau, Romulus and Remus) and closely allied, as were the "Great Twin Brethren," Castor and Polynices, among the Greek gods or the "Hero Twins" of Native American legend. Although twins can be male or female, most legendary twins were male pairs.

In Japan, twins are believed shameful, as multiple births are associated with animals. In China, by contrast, twins were often viewed as auspicious, particularly the "dragon and phoenix pair" of twins of opposite sexes.

The Igbo and Yoruba, two neighboring West African peoples, contrast in their view of twins. Twins are of particular importance to the Yoruba, who have the highest rate of twin births of any people in the world. There is some evidence that in early times, twins were sacrificed, but now twins, if treated properly, are believed to bestow great blessings on a Yoruba family. The twin that is born first is believed to be the younger twin, as it is believed that the elder twin sends the younger forth to scout ahead and that they firstborn twin's cry is a signal for the elder twin to emerge. Three days after their birth, an *Ifa* diviner will perform rituals to cleanse them of evil spirits and determine how they are to be raised. Yoruba twins are believed to share a single soul, and if one dies, the survivor's life is threatened. To protect the surviving twin, a special wooden image called an ibeji is created to take the place of the deceased twin. The Igbo, by contrast, viewed twins as a manifestation of the wrath of the gods and left twins, and sometimes their mothers, to die in an area called the "Bad Bush." Like the Japanese, the Igbo viewed multiple births as an animal phenomenon that was highly inappropriate among humans. The practice of twin abandonment was a key element in the conflict between traditional Igbo elders and Christian missionaries, European or African, in the late nineteenth and early twentieth centuries.

In premodern Europe, it was claimed that a twin, or any multiple birth, was evidence of a wife's infidelity, as one man could only beget one child at a time. Therefore, a woman who gave birth to twins or other multiples must have had sex with more than one man. (Twins can be born with different fathers, but this is rare.)

The Native American Mohave saw twins as sent down from heaven, having a special spiritual status not shared by ordinary people. They claimed that a male and female pair of twins had been a married couple in a previous existence.

See also: Ifa Divination; Pregnancy and Childbirth

Further Reading

Bastian, Misty L. 2001. "'The Demon Superstition': Abominable Twins and Mission Culture in Onitsha History." *Ethnology* 40: 13–27; Devereux, George. 1941. "Mohave Beliefs Concerning Twins." *American Anthropologist* 43: 573–592.

U

Unicorn

The idea of a one-horned (the literal meaning of "unicorn") creature otherwise resembling a horse or wild ass can be traced to the ancient world, emerging in the Mediterranean, Persia, India, and China. It might have been influenced by the rhinoceros or by the habit of representing horned antelopes and cattle in profile so that only one horn appeared. The home of the unicorn for Greek and Roman writers was India, and various descriptions of the creature are found in the writings of ancient natural historians and travelers. The unicorn also appears in the earliest Greek and Latin translations of the Bible in place of the Hebrew word "re'em," which refers to a mighty, bull-like animal with one or two horns. This probably originally referred to the aurochs, but the Greek translation (*monokeros*) and the Latin translation *(unicornis)* treated the single horn as its essential quality and thus provided biblical backing both to the unicorn's existence and to its power. The King James English version followed the older translations in referring to a unicorn, although more modern translations call the creature a wild ox.

In the European Middle Ages, an elaborate mythology developed around unicorns, seen as beautiful, shy forest creatures. (The medieval traveler Marco Polo was shocked by the difference between the graceful image of the unicorn and the ugly reality of the rhinoceros.) The idea stemming from late antiquity that the unicorn can only be captured by a virgin woman was originally an allegory of the incarnation, with the virgin playing the role of Mary and the unicorn of Christ. This idea was secularized into an example of the power of woman's love to tame the savage unicorn or of the treachery of love as the unicorn was then captured and killed by the hunters.

The horn of a unicorn was believed to neutralize poison, a belief that may derive from the common belief in India and China that the horn of a rhinoceros protects from poison. (Another version of this legend ascribes the magical power to a jewel growing under the unicorn's horn rather than the horn itself.) The unicorn horns that appear in medieval and Renaissance collections were collected from narwhals, a type of whale with a single tusk resembling a horn or elephant and walrus tusks altered to resemble horns. Narwhal horn was also used to make poison-neutralizing "unicorn" cups and even the royal throne of Denmark built between 1662 and 1671, referred to as the "Unicorn throne." The substance from which the horn of the unicorn was made was called "alicorn," and powdered alicorn was sold as a remedy for a wide variety of diseases. Belief in the medical virtue of alicorn is found in the writings of medieval Arab physicians as well as Western ones.

The lion was believed to be the "natural enemy" of the unicorn. This led to the choice of a unicorn as the heraldic representation of the Kingdom of Scotland after England had chosen the lion. An alternative candidate for the unicorn's natural enemy was the elephant, an idea probably stemming from a confusion between the unicorn

of the Unicorne. 711

OF THE VNICORNE.

Sff 2

An early seventeenth-century unicorn, from Edward Topsell's *The Historie of Four-footed Beasts* (1607) (Library of Congess, Prints and Photographs Division, LC-USZ62-95208)

and the rhinoceros. Classical writers described the unicorn as sharpening its horn on a rock it used as a whetstone before attacking the elephant.

The unicorn in medieval Islamic culture, referred to as the "karkadann," is more closely connected to the rhinoceros than the Western or Chinese unicorns, unsurprising as Muslims, unlike Chinese or Westerners, were in direct contact with rhinoceros-inhabited Africa, India, or Indonesia. The karkadann is described in a more realistic fashion than the Western or Chinese unicorns, but it attracted a few legends. One was that it was particularly fond of the cooing of the ring dove and would stand peaceably under a tree to listen. This is in contrast to the belief that the karkadann was among the fiercest of animals, and Muslims agreed with Westerners that the unicorn was a particular enemy of the elephant. Some Muslim authorities claimed that battles between unicorns and elephants ended in the death of both parties as the unicorn could not dislodge the dead elephant from its horn. The ferocity of the karkadann was also manifested in dreams, where it represented a tyrannical ruler.

Two animals have been put forth as the "Chinese unicorn." One is the qilin. The qilin is described as somewhat resembling an ox or horse, but with scales like a dragon. It can be represented with one horn or with two. The qilin is not believed to be a naturally occurring animal like the Western unicorn, but an "auspicious beast" sent by heaven as a token of good fortune. A qilin was believed to have appeared in the garden of the Yellow Emperor. A qilin attended the birth of Confucius, and the sudden appearance of another qilin dissuaded the Mongol conqueror Chinggis Khan (1158–1227) from an invasion of India. Versions of the qilin have spread to other places influenced by Chinese culture, including Korea, Japan, Vietnam, Thailand, and Tibet. The other candidate is a wild goat-like animal called a zhi, usually represented with a single horn. According to legend, the first zhi was a gift from the gods to the Yellow Emperor. The zhi was most strongly associated with Gao Yao, the legendary just magistrate of the semimythical Emperor Shun. The zhi belonging to Gao Yao possessed the ability to identify the guilty and butt them with its single horn. Representations of the zhi were guarantors of justice associated with courts and judges in Imperial China and were also tomb guardians due to their power against evil spirits. Zhi representations are also found in Korea and Japan.

Belief in the unicorn died slowly. Open skepticism appeared among several writers in the sixteenth century—the Swiss zoologist Konrad Gesner (1516–1565) suggested that the unicorn may have been destroyed in Noah's Flood, an idea revived by the American cartoonist and folk musician Shel Silverstein (1930–1999) in his 1962 song "The Unicorn," a hit for the Irish Rovers in 1968. Gesner also claimed that the unicorn's horn was good against poison, plague, intestinal worms, and epilepsy. The Danish naturalist Ole Wurm's (1588–1654) identification of the "true unicorn's horn" with the narwhal tusk in 1638 dealt a blow to the clearest piece of evidence of the unicorn's existence. The Swedish natural historian Carl Linnaeus (1707–1778) included the unicorn, or "monoceros," in his *Systema Naturae* (1735) as one of the "Paradoxa" animals in which he did not believe, ascribing it to the "fiction of painters." This did not end belief, however. In the late eighteenth century, the German naturalist Eberhard August Wilhelm von Zimmerman (1743–1815) believed the possibility that the unicorn existed could not be ruled out until Central Asia and South America were thoroughly explored, and the raja of Bhutan even claimed to own one. There were rumors of unicorns in Central Africa well into the nineteenth century.

See also: Elephants; Horses and Donkeys; Yellow Emperor

Further Reading

Ettinghausen, Richard. 1950. *The Unicorn.* Washington, DC: Smithsonian Institution; Parker, Jeannie Thomas. *The Mythic Chinese Unicorn.* Victoria, Canada: Friesens Press; Rankin, Alisha. 2021. *The Poison Trials: Wonder Drugs, Experiment and the Battle for Authority in Renaissance Science.* Chicago: University of Chicago Press; Shepard, Odell. 1930. *The Lore of the Unicorn.* Boston and New York: Houghton Mifflin.

Urine. See Excrement

Vampires

The vampire is the European version of the bloodsucking monster found in many cultures. It differs from some other bloodsucking monsters, though, in that it originates as a corpse. The earliest evidence of the term comes from Eastern Europe in the mid-seventeenth century, and Eastern Europe was always the homeland of the vampire. The earliest vampire stories to appear focus on the corpse eating its shroud rather than drinking blood. The monk and Biblical scholar Dom Augustine Calmet (1672–1757), author of *Treatise on Apparitions and on Vampires or Revenants of Hungary and Moravia etc.* (1751), the most exhaustive treatment of the vampire during the eighteenth century, regarded them as a phenomenon that had only appeared during the last sixty years.

The first vampire case to receive extensive publicity outside the place where it originated was that of Peter Poglojowitz in Serbia, which had recently been taken by Austria from the Ottoman Empire. This case occurred in 1725 and was the first of several celebrated vampire cases in the eighteenth-century Habsburg Empire. (There were also more obscure cases, such as a large-scale outbreak in the Serbian garrison city of Kragujevac, known only from a report published in a German journal at the end of the eighteenth century.) These cases involved Catholic Austrian authorities and Orthodox Christian Serbian peasants in a territory Austria had recently conquered from the Ottoman Empire, and Austrians saw vampirism as evidence of the backwardness of the Orthodox population. The villagers among whom vampire cases arose seem to have viewed them as demonic in nature, as a demon possessed a corpse for the purpose of working evil. The defense against the vampire usually involved the violation of the integrity or outright destruction of the corpse, with such practices as beheading, staking, burning, or cutting out and destroying the heart. This was also true in the Ottoman-ruled portion of the Balkans, where Muslim authorities recommended staking (in the belly rather than the heart), beheading, and burning of the corpse. However, authorities on both sides of the border saw this as a concession to popular superstition rather than an actually effective process.

The case of the Serbian vampire Arnaud Paole attracted attention through an account of a vampire hunt written by an Austrian military surgeon, Johannes Fluckinger, *Visum et Repertum* (1732). (An earlier report, by a physician named Glaser, remained in manuscript until the late nineteenth century and had little direct impact on the vampire legend.) Paole was a soldier who had recently settled in the area and claimed to have cured himself of a vampire's bite through eating soil from the vampire's grave. Several claimed to have been tormented by Paole after his death, and inspection of his grave showed that his body was undecayed and full of fresh blood—clear evidence of vampirism. The Fluckinger account spread what were to become several key aspects of the vampire legend,

An Early Account of Bloodsucking Vampires

"In a certain canton of Hungary, named in Latin Oppida Heidanum, beyond the Tibisk, vulgo Teiss, that is to say, between that river which waters the fortunate territory of Tokay and Transylvania, the people known by the name of Heyducqs believe that certain dead persons, whom they call vampires, suck all the blood from the living, so that these become visibly attenuated, whilst the corpses, like leeches, fill themselves with blood in such abundance that it is seen to come from them by the conduits, and even oozing through the pores. This opinion has just been confirmed by several facts which cannot be doubted, from the rank of the witnesses who have certified them."

Source: Augustin Calmet. 1850. *The Phantom World: The History and Philosophy of Spirits, Apparitions, Etc.* Translated by Henry Christmas. Philadelphia: A. Hart. Originally published 1746. Pp. 264–265.

including the blood dribbling from the vampire's lip, the spread of vampirism through the vampire draining a victim's blood, and the destruction of the vampire with a stake through the heart. Paole's head was cut off as well and the body burnt. The same treatment was given to four people believed to have been killed by Paole, to prevent them from rising from the grave as vampires. The ashes were thrown into the Morava river, an early example of the antipathy believed to exist between vampires and running water. A subsequent vampire outbreak a few years later involved people who had become vampires after their deaths by such actions in life as eating meat killed by Paole or smearing themselves with the blood of vampires in order to protect themselves against vampires. Vampirism in women was also believed to be inherited by offspring, including stillborn offspring. The work was published in France, Germany, and England leading to an increased interest in vampires across Europe and a lively debate about the reality of the vampire phenomenon, with dozens of studies of the subject published in the following decades. Enlightenment writers who wrote about vampires were more likely to look for "natural" rather than demonic explanations, although the spiritual world can be considered "natural" as long as neither God's power to work miracles nor demonic action is involved. One writer suggested that the power of Poglojowitz's imagination survived his death and killed those who he hated.

Pope Benedict XIV (1675–1758, Pope 1740–1758) discussed vampirism in his book on sainthood *On the Servants of God* (1743) to distinguish between the allegedly incorruptible body of the vampire, in reality an illusion caused by superstition, and the actually incorruptible bodies of the saints. The vampire raised numerous additional religious questions, such as the relation between the resurrection of the vampire's body and the resurrection of the body at the Last Judgment. The vampire's nourishment through blood can parallel the consumption of consecrated wine, the "Blood of Christ" during the Mass.

Like other forms of supernatural belief, belief in vampires was increasingly defined as "superstition" during the Enlightenment.

In 1755, the Hapsburg Empress Maria Theresa (r. 1740–1780) issued a decree declaring the belief in vampires to be superstitious and forbidding judges and ecclesiastics in her domains from involvement in vampire cases. This decree followed the case of an accused vampire named Rosina Polakin from the Moravian town of Hermersdorf. Polakin's exhumed body was undecayed and full of blood, as was expected of a vampire, and it was decapitated and burned. The curiosity aroused by the case caused a court physician, Gerard van Swieten, to elaborate a "natural" explanation of how a body could be thus preserved. Van Swieten, an opponent of popular superstition, was a moving spirit in both the anti-vampire decree and a suspension of witch trials in the Habsburg Empire the following year. Although this largely ended interest in vampires among the learned, it did not do so among the peasantry where belief in vampires has continued to the present day. Vampires were most associated with Hungary, Romania, Greece, and the Balkans, but vampire beliefs were found elsewhere in Eastern Europe, such as Poland, where people born with a caul were believed likely to become vampires.

Another tradition of "vampirism," although the term itself was not used by believers, can be found in eighteenth- and nineteenth-century New England, far from the vampire heartlands of Eastern Europe. New England vampirism was associated with tuberculosis, one of the leading causes of death in the period. Tuberculosis, a usually fatal "wasting away," was thought to be caused by a recently deceased relative who had not completely died and was somehow draining the vitality of the tuberculosis victim. Fresh blood in the heart of an exhumed corpse was considered evidence of guilt and the burning of the corpse a remedy. The last and by far the most widely publicized case was that of Mercy Brown of Rhode Island in 1892. A recently deceased nineteen-year-old woman, Brown, was blamed for the "wasting away" of her tubercular brother Edwin. Brown was exhumed, her heart and liver burned, and the ashes mixed into a drink given to Edwin. Edwin died two months later.

In the romantic era of the nineteenth century, there was a considerable literary interest in the vampire, which created much of the modern vampire mythology. Such widely known aspects of the mythos such as the connection of the historical Romanian Voivode Vlad Dracul (1428–1477), also known as Vlad Tepes or "Vlad the Impaler," with vampirism are the works of novelists such as Bram Stoker (1847–1912), author of *Dracula* (1897), rather than being rooted in actual folk-beliefs. Other influential creators of vampire mythology included John William Polidori (1795–1821), author of *The Vampyre* (1819), and Sheridan Le Fanu (1814–1872), author of *Carmilla* (1872), one of the earliest fictional treatments of a female vampire. The nineteenth century also saw the growing identification of vampires with bats.

See also: Blood; Caul; Death; Dogs; Garlic; Monsters; Silver

Further Reading

Barber, Paul. 1988. *Vampires, Burial and Death: Folklore and Reality*. New Haven, CT: Yale University Press; Bell, Michael E. 2008. "Vampires and Death in New England, 1784–1892." *Anthropology and Humanism* 31: 124–130; Bohn, Thomas M. 2019. *The Vampire: Origins of a European Myth*. Translated by Francis Ipgrave. New York: Berghahn; Braunlein, Peter J. 2012. "The Frightening Borderlands of Enlightenment: The Vampire Problem." *Studies in History and Philosophy of Biological and Biomedical Sciences* 43: 710–719; García

Marín, Álvaro. 2021. "Analysis of a 1725 Report of Vampirism in Kragujevac." *Journal of Vampire Studies* 1: 137–164; Vermeir, Koen. 2012. "Vampires as 'Creatures of the Imagination' in the Early Modern Period." In Yasmin Haskell, ed. *Diseases of the Imagination and Imaginary Disease in the Early Modern Period.* Turnhout: Brepols. Pp. 341–373.

Visions. See Apparitions

Volcanoes

Among the Earth's most dramatic and destructive phenomena, volcanoes have been the subject of numerous attempted explanations and superstitions. The suddenness and destructiveness of volcanoes led them to be associated with angry deities such as the Hawaiian goddess Pele. In many societies, people have believed that the way to protect themselves from volcanic eruptions was to propitiate the gods or spirits that ruled or dwelt in the volcano. The word "volcano" derives from Vulcan, the Roman god of fire, including volcanic fire. The best-known volcano cult today is the Japanese reverence for Mt. Fuji, which has not erupted since 1707. Not only is Mt. Fuji the home of a powerful goddess, but it is also a pilgrimage site, and to dream of it augurs good fortune.

Early non-supernatural explanations for volcanic eruptions include the Greek philosopher Aristotle's (384–322 BCE) theory that they were explosions of trapped air or wind that ignited due to friction. (Although Aristotle provides the earliest surviving example of this theory, it was probably not an original idea of his but was held by other ancient Greek philosophers.) Earthquakes were also caused by the movement of trapped air, and volcanoes and earthquakes were considered closely related. Another ancient Greek theory associated volcanoes with fire rather than air. The fifth-century BCE Greco-Sicilian philosopher Empedocles, who lived in close proximity to the volcano at Etna, believed that volcanic eruptions were outbreaks of elemental fire. The Athenian philosopher Plato (428–348 BCE) associated volcanoes with the mythological river of fire under the Earth, Pyriphlegethon.

In the Hellenistic and Roman periods, with the growth of alchemy, there was more interest in viewing volcanic fire as the product of a chemical reaction. The Roman philosopher Seneca (4 BCE–65 CE) claimed that the winds trapped in the Earth by its movement ignited vast quantities of underground sulfur and other combustible substances. Volcanoes attracted particular interest after the eruption of Vesuvius in 79 CE, which destroyed the towns of Pompeii and Herculaneum and killed Pliny the Elder.

The idea of volcanic eruptions as outbreaks of subterranean fire proved highly compatible with the Christian idea of hell as an underground lake of fire. Several volcanoes were identified as entrances, or gates, to hell or to purgatory. These included Mt. Etna and Iceland's Mount Hekla. (According to one legend, King Arthur lived at the bottom of Mt. Etna.) The idea that volcanoes and earthquakes did not exist on the newly created world but were consequences of the Fall of Adam and the entrance of sin into the world is found in several Christian writers, including the founder of Methodism John Wesley (1703–1791).

The Jesuit scientist and collector Athanasius Kircher (1602–1680) was lowered into the crater of Vesuvius in 1638 to conduct research. His *Mundus Subterraneus* (1665) posits an Earth riddled with channels and underground passages moving heat from a fiery central core to areas near the surface, where heat combines with water, also moving through underground passages, to create superheated steam that issues in the form of volcanic eruptions. Isaac Newton (1643–1727), a practicing alchemist, believed that volcanoes erupted from an internally heated Earth fueled by the reaction of iron and sulfur.

See also: Earthquakes; Elemental Systems

Further Reading

Sigurdsson, Haraldur, ed. 2015. *The Encyclopedia of Volcanoes.* Second Edition. London: Academic Press.

Voodoo Dolls. See Doll Magic

Weapon-Salve

In sixteenth- and seventeenth-century Europe, the weapon-salve was an ointment that could supposedly cure wounds caused by weapons. It worked not by application to the wound itself, but to the weapon that had caused the wound. The salve was magical, involving ingredients such as the moss from the skull of a hanged thief, powdered mummy, and warm human blood from the wounded person. The power of the weapon-salve raised the question of how it worked when the weapon and the salve were separated from the actual wound. Paracelsus (1493–1541), the German alchemist and medical maverick, described the weapon-salve and its effectiveness was supported by his followers and attacked by anti-Paracelsians. The Netherlands Paracelsian alchemist Johannes Baptista Van Helmont (1579–1644) became publicly involved in the controversy in 1621, claiming in *The Magnetic Cure of Wounds* that the success of the cure depended on the sympathy between the blood in the body of the wounded person and the blood remaining on the weapon, which Van Helmont claimed was necessary for the cure to work. Van Helmont tried to make the process more a matter of natural magic than witchcraft by claiming that the moss from any skull would work in the salve. Since magnetism worked at a distance, the weapon-salve could also work at a distance. Catholics such as Van Helmont who supported Paracelsianism or anything magical were viewed with great suspicion by Church authorities. Even more offensive was Van Helmont's suggestion that similar "magnetic" forces, rather than God's miraculous power, explained the healing power of relics. The weapon-salve tract contributed to Van Helmont's troubles with the Inquisition and local church courts that led to his imprisonment.

The English offshoot of the Continental weapon-salve controversy began with *Hoplocrisma-spongus, or a sponge to Wipe Away the Weapon-Salve* (1631) by William Foster, a Protestant minister. Foster's main purpose was to defend the good name of Protestantism, which he thought was injured by Protestants who endorsed a magical, quite possibly satanic, weapon-salve cure. Foster employed Aristotelian natural philosophy to deny the possibility of a magical cure, attacking a number of magicians, but principally the physician Robert Fludd (1574–1637). Fludd replied in *Doctor Fludd's answer to M. Foster, Or, the Squeezing of Parson Fosters sponge, ordained by him for the wiping away of the weapon-salve* (1631), defending the weapon-salve and the idea that entities separated in space could affect each other across distances through the power of "sympathy," employing the example of the magnet. Although Foster never replied to Fludd, other English anti-Paracelsians continued to use the weapon-salve to attack Fludd and Paracelsianism.

The English natural philosopher Sir Kenelm Digby (1603–1665) had a version of the weapon-salve he called the Powder of Sympathy, which could allegedly heal by being applied not only to the weapon that caused the wound but also to the bloody bandages. After his death, it figured in a plan for solving the most vexing problems of early modern science, finding time at sea. Wounded dogs would be taken on ships while bandages taken from the wounds would remain on shore, being dipped punctually at noon in London in a solution of the powder, which would cause the dogs on ship to yelp, marking the time. It is unknown if this was ever tried, and the suggestion was probably meant as a joke.

See also: Alchemy; Blood; Execution Magic; Natural Magic; Relics

Further Reading

Anonymous. 1668. *Curious Enquiries, Being Six Brief Discourses*. London: Randal Taylor; Debus, Allen G. 1977. *The Chemical Philosophy: Paracelsian Science and Medicine in the Sixteenth and Seventeenth Centuries*. New York: Science History Publications; Harline, Craig. 2003. *Miracles at the Jesus Oak: Histories of the Supernatural in Reformation Europe*. New York: Doubleday.

Weather. See Lightning; Rain; Rainbows; Sun

Weidan. See Alchemy

Wild Hunt

The myth of the wild hunt is found throughout northern, central, and western Europe and may have roots in pre-Christian Celtic and Germanic paganism. It takes many forms, but the common elements are a group of the souls of dead people usually with a leader. The crowd of dead was usually seen at night, although in some versions, the hunt was heard rather than seen. The hunters traveled sometimes by foot, sometimes on the back of spectral horses, and sometimes on the back of goats. Evidence for belief in nocturnal gatherings of the dead dates as far back as the eleventh century. The phenomenon went by many names, but the term "Wild Hunt" was invented much later, the work of the nineteenth-century German folklorist Jacob Grimm (1785–1863).

This period was a time of increased interest among Christian Europeans in the afterlife and the fate of the dead, as can be seen from the rising popularity of ghost stories. In twelfth-century northwestern Europe, a supposedly cursed ancient British king named Herla was described as leading the crowd of dead or, in another version, a horde of Herla's living followers kept alive by magic and doomed to die if they ever dismounted. In Germany, where the crowd of dead were referred to as the "Furious Army" and the legend seems to have persisted longest, the leader was sometimes called Wotan, the chief god of the pagan German pantheon, although he was viewed as an ancient king rather than a God. In Celtic Wales, a supernatural figure called Gwyn ap Nudd was often viewed as the leader of the dead. He was depicted as a conductor of the souls of fallen warriors to the otherworld or as the king of the fairies. The spread of Arthurian legendry in the high medieval period led to the leader being identified as King Arthur. Another legend was of a king or hunter who had somehow offended God, for example by hunting on Sunday, leading to him being cursed to hunt until the Judgment Day. Figures of evil in the Christian tradition, such as Herod, Cain, or even Satan himself, were also identified as leaders of the hunt. Unlike the related legend of the troupe of mostly benevolent witches led by Diana, female leaders of the hunt were rare; one example is Gudrun, the wife of the legendary hero Sigurd or Siegfried. The rank and file of the hunt could consist of sinful people generally, soldiers who had died in battle, executed criminals, and babies who had died before being baptized. Some scholars have argued that belief in the wild hunt contributed to the later development of the idea of the witches' sabbat, but direct references to the wild hunt are rare in witch cases.

The wild hunt was also viewed as an omen. As such, it could presage war or a change in the weather. Hearing the hunters or their dogs could also mark an individual for death. Various charms were employed to protect people from being carried away by the hunt; in Britain, wearing a sprig of rowan provided protection from the hunters.

See also: Diana; Witchcraft

Further Reading

Hutton, Ronald. 2014. "The Wild Hunt and the Witches' Sabbath." *Folklore* 125: 161–178.

Witch Bottle. See Excrement

Witch Cake. See Excrement

Witchcraft

Belief in witchcraft, the deployment of supernatural power by individuals to harm others in a community, is found in many cultures across the globe. The term "witch" can also apply to benevolent or neutral magical practitioners—"good witches"—but its most common use is to refer to evildoers. Witches can be of any class or gender, although women often predominate, and range in age from children to the very old. Although many witches were socially marginal, major political and social leaders were also accused of witchcraft. Sometimes witches were viewed as inheriting witchcraft from their parents and passing it down to their children and sometimes not. Witches were often associated with shape-changing, cannibalism, and deviant sexual practices such as incest. The harm allegedly done by witches includes physical harm like death and disease inflicted on people or their livestock, but it also includes more intangible harms such as the draining of the soul. Some anthropologists draw a distinction between witches proper, who harm others through an innate ability, and sorcerers, who use learned magical skills such as spells and potions. However, the distinction is not always easy to make in practice, and many witch figures operate in both realms. Responses to witchcraft ranged from judicial proceedings to community action including ritual degradation, violence, and defensive magic. Societies that believe in the danger of witchcraft frequently give birth to a class of specialist witch hunters or diviners who claim magical abilities to locate, identify, and neutralize witches.

In Western culture, the belief in witches has Greco-Roman and Jewish roots. Greco-Roman literature and mythology featured female witches with incredible powers, such as Circe in Homer's *Odyssey*, who turned men into pigs. Hecate the Moon goddess was also the goddess of witches. There are a few hostile references to witches in the Jewish Bible, the most famous being the Witch of Endor who summoned the spirit of the prophet Samuel for King Saul. In Christian debates about the reality of witchcraft, the Witch of Endor would be frequently invoked. Witches are also found in the Celtic, Germanic, Slavic, Basque, and other

A sensationalist engraving of a witch at work, from the seventeenth-century Dutch artist Jan van de Velde. (The Cleveland Museum of Art, Mr. and Mrs. Lewis B. Williams Collection)

Indigenous non-Mediterranean European traditions.

As opposed to the terrifying witches of classical literature, witches in the ancient Mediterranean world engaged in more mundane activities such as divination and the provision of aphrodisiacs. This tradition of witchcraft continued in Italy in particular into the modern era, with little interaction with the concept of satanic witchcraft that provoked European witch hunting.

The rise of Christianity at first had little impact on witches and witch hunting. Belief in witches continued into the European Middle Ages, but there was little persecution. The Church was more concerned with reining in exaggerated popular beliefs in witch's powers than crushing witchcraft. Documents such as the early tenth-century *Canon Episcopi*, or "canon of the bishops" warned Christians not to believe in the stories of women flying through the night. This began to change in the later Middle Ages, when persecutions of witches began. One classic statement of witch belief was the *Malleus Maleficarum* (1486), or *Hammer of Female Evil-doers.* This work was the work of the Dominican witch hunter Heinrich Institoris. Institoris would be influential in strongly associating witchcraft with women, asserting that women were more vulnerable to Satan's temptations due to their weakness and insatiable lust, a pattern he traced back to Eve in the Garden of Eden. Institoris's identification of witchcraft with the worship of Satan was also influential, as it meant that witchcraft was heresy, punishable by death.

The early modern period from the late fifteenth to the early eighteenth centuries

The Canon Episcopi

The "canon of the bishops" (so-called because it began with the word "bishops") was a piece of Church legislation about witches originating in the ninth century and surviving in its first copy from the early tenth century. In addition to recommending that bishops and priests zealously combat sorcery and maleficia, the canon also identified the belief that some women traverse great distances in the night led by the pagan goddess Diana as diabolical illusion and superstition. It also condemned belief in the reality of the power of witches and demons to work transformation. The canon emerged at a time when much of Europe had only recently been officially converted to Christianity, and the struggle against residual pagan beliefs was much on the minds of clerics. The clerical writers of the canon identified pagan deities with demons and thought women were particularly vulnerable to remaining pagan beliefs. Later writers during the witch hunt spent much time and effort attacking the canon's authority.

saw a series of efforts to entirely purge witches from European society involving leaders in Church and state, judicial experts, and demonologists, both Catholic and Protestant. The phenomenon was not entirely elite-driven, though, as many ordinary people believed in witches and blamed their troubles on the actions of a local witch. The witch was usually female (although there were a few regions where male witches predominated) and was defined as getting her powers from the figure of supernatural evil known as the devil or Satan rather than being powerful in her own right. The witch's main crime, according to many demonologists and legal scholars, was the agreement with Satan rather than the evil deeds wrought through her magic, although popular hostility to witches almost always focused on evil actions. Some witch hunters wrote that the worst witch of all was the "good witch," who concealed their true satanic nature with a false benevolence. Various kinds of non-Christian beliefs and practices, such as the Italian reverence for Diana or the "Lady of the East," originally a Moon goddess like Hecate, the Sicilian belief in the "Ladies from Outside," or the benandanti cult of Friuli, were swept up into the stereotype of the witch.

The mythology of witchcraft in the European witch hunt became increasingly elaborate, incorporating such concepts as the Satanic Pact, the sabbat or gathering of witches, the devil's mark, and familiars, usually conceived of as spirits but often taking the form of animals, most commonly toads and cats. Common forms of defensive magic included drawing blood from the witch, hanging a horseshoe over the threshold of a house, and the use of a "witch bottle," a bottle of urine buried under the threshold of a house or another place thought vulnerable to a witch's attacks. Defensive magic was frequently condemned by religious authorities, who thought it showed a lack of trust in God. Catholic authorities recommended religious defenses such as the use of holy water or consecrated oil, while Protestants had nothing to recommend for people to protect themselves against witches except prayer and repentance.

Europe's elites became more skeptical of witchcraft by the late seventeenth century, although this was first manifested in skepticism about the evidence in individual cases than about the concept of witchcraft itself. Some feared that the rejection of the belief in witches would lead to a turning against the supernatural that would eventually result in atheism. Western Europe preceded Eastern Europe chronologically in skepticism as it had preceded it in witch hunting. Eastern European regimes turned from witch hunting in the late eighteenth century, partly because the fascination with witches turned to the new figure of the vampire. Ordinary people in East and West often continued to believe in the danger of witches, and lynchings and other forms of anti-witch activity continued into the twentieth century.

Not all Christian cultures had the idea of the satanic witch, which was mainly characteristic of Catholic and Protestant societies. Russians believed in witches and their evil deeds, but had little interest in how witches derived their power and never extensively employed the concept of the Satanic Pact.

Like the Bible, the Quran endorses the existence of witches and their hostility to God and man. Sorcery was taught by demons to humanity. Belief in harmful magic is common in many Islamic societies and produces some of the same tensions as in Christianity. The wearing of talismans and amulets to protect from harmful magic is common in many Islamic societies, but religious authorities condemn these practices and recommend that people instead pray and trust in God. The degree of hostility to witches displayed by Islamic regimes varied and was not usually expressed in the form of witch hunts. Much of the Balkans was ruled by the Muslim Ottoman Turks during the early modern period, but saw little witch hunt activity, and Ottoman authorities discouraged witchcraft accusations between Christians. The Islamic regime today most associated with government-led witch hunting is the Kingdom of Saudi Arabia, which executed several people for sorcery in the early 2010s.

Outside the European Christian tradition, witches were usually viewed primarily as blameworthy for evil deeds rather than for how they acquired their powers. Navajo Native Americans believed in a kind of shape-shifting witch called a skinwalker that can possess a human being or animal to commit violent acts against people or livestock. In some versions of the skinwalker legend, the skinwalker was originally a medicine man who became corrupted. Similar beliefs about witches are found in other Southwestern Native communities.

Witch beliefs have been common in many African cultures. Africans believed that witches, who like European witches were usually but not invariably female, practiced cannibalism and incest and were responsible for the spreading of disease. Witches could also be responsible for large-scale misfortunes such as natural disasters. Some believed that it was possible for witches to perform evil deeds without being consciously aware of them, a belief shared by some accused witches themselves, making them more likely to confess. In many African societies, witches worked their evil by getting possession of something that came from a person's body, such as excrement, nail clippings, and sexual fluids. Africans developed many ways to defend themselves against witchcraft, from divination to determine the identity of a witch to defensive magic to violent action against witches. Witchfinders in many African societies singled out witches for punishment.

The arrival of Christian missionaries, mostly in the nineteenth and twentieth centuries, caused a coming together of African and European Christian witch stereotypes in many areas as Africans influenced by Christianity increasingly identified witches with the demonic. In some places in postcolonial Africa, this has led to legal proceedings against witches similar to those in early modern Europe as well as organized popular violence against suspected witches. Sub-Saharan Africa is where the majority of witch hunts take place today.

In the Americas, Native, European, and African traditions of witchcraft and magic met, creating rich witchcraft cultures. Native American traditional religious practices were often stigmatized as witchcraft. Since many European settlers thought of Natives as devil worshippers, Native American resistance such as the Pueblo revolt against Spain in 1680 was frequently blamed on witches acting at Satan's behest. The famous Salem witch trials not only basically involved European elements transplanted to the Americas but also included accusations that witches were allied with Native Americans. Societies with slavery saw African-based magical and religious practices such as obeah defined as witchcraft by whites. Some in the early modern Caribbean credited witchcraft with having driven the buzzards from Hispaniola, but African and Native witchcraft was usually viewed as threatening to whites.

Witchcraft beliefs are widespread in India. Witches are usually identified as female and are believed to get their powers from malignant supernatural forces. In nineteenth-century South India, there was a belief that a witch could be neutralized by pulling out his or her teeth, which would prevent them from repeating the mantra or spell by which they were working harm.

Belief in evil witches or sorcerers is also common in China, and on at least one occasion, it has led to a major witch hunt. In the late eighteenth century, many ordinary Chinese feared "soul stealers," sorcerers who would cut off the end of a queue, the long braided hair that Chinese men were required to wear, and repeat a few words to steal the soul of the person. Soulstealers, usually male but often socially marginal people such as wandering Buddhist monks or underemployed artisans, could also steal souls by possessing a person's written name. The person whose soul was stolen would then sicken and die, and the sorcerer would use the soul force for malign purposes. Persecution of soulstealers was taken up by the imperial government, leading to a witch hunt with many of the features of European witch hunts, despite its completely independent origins.

See also: Amulets and Talismans; Aphrodisiacs; Benandanti; Black Mass; Blood; Cats; Defensive Magic; Devil's Mark; Diana; Divination; Evil Eye; Execution Magic; Goats; Hand of Glory; Hares and Rabbits; Horseshoes; Hyenas; Lycanthropy; Mandinga Pouches; Necromancy; Obeah; Onions; Satanic Pact; Toads; Vampires

Further Reading

Burns, William E. 2003. *Witch Hunts in Europe and America.* Westport, CT: Greenwood; Evans-Pritchard, E. E. 1937. *Witchcraft, Oracles and Magic Among the Azande.* Oxford: Clarendon Press; Kivelson, Valerie. 2013, *Desperate Magic: The Moral Economy of Witchcraft in Seventeenth-Century Russia.* Ithaca, NY and London: Cornell University Press; Kuhn, Philip A. 1990. *Soulstealers: The Chinese Sorcery Scare of 1768.* Cambridge, MA: Harvard University Press; Levack, Brian P. 2016. *The Witchhunt in Early Modern Europe.* Fourth Edition. London and New York: Routledge; Stewart, Pamela J. and Andrew Strathern.

2004. *Witchcraft, Sorcery, Rumors and Gossip.* Cambridge, UK: Cambridge University Press.

Witch's Mark. See Devil's Mark

Wolves and Coyotes

Few wild animals have attracted more fear, or more legends and superstitions, than the wolf. The wolf's destructive power and enmity to humanity (and to its livestock) were legendary. In many cultures, wolves were symbols of danger and destruction. Norse mythology offers Skoll and Hatti, the wolves that chase the Sun and Moon, respectively, across the heavens, as well as Fenrir, who when the world ends at Ragnarok will slay Odin, king of the Gods. The wolf also features unfavorably in Rabbinic Judaism and Christianity. In the metaphor of Christians as sheep and Jesus or pastors (literally shepherds) as their guardians, the wolf plays the role of the devil, giving it in Christian eyes an indelible association with evil. Many beliefs concerning the wolf were related to its rivalry with humanity. From the ancient Greeks to the Middle Ages in the West, it was believed that if a human was seen first by a wolf, the human would lose the power to speak. Medieval Christians believed that such was the wolf's fear of humanity that it would never eat a human face. In Islamic dream interpretation, a wolf represents a thief or an enemy.

A kinder side to the wolf was seen in the common legends of she-wolves as foster mothers to humans. Although wolves were not the only animals credited with raising abandoned human babies, they were the most common. The most prominent example of canine fosterage was the twins who founded Rome, Romulus and Remus. This story appears as early as the fourth century BCE, although some ancient historians argued that the boys' foster mother was really a prostitute, referred to in Roman slang as a she-wolf, *lupa*. The violence and greed of the Romans was sometimes ascribed by their enemies to having "suckled at the she-wolf's teat." Subsequent legendary figures fostered by wolves include the fifth-century Irish saint Ailbe, who allowed his lupine stepmother to be fed with bread at his table, making him a rare case of a she-wolf's fosterling who maintained a relationship with his foster mother. Most children allegedly raised by wolves were not saints or kings, however. Stories of wolf-children, along with stories of children raised by other wild animals, appear in chronicles and journalism into the twentieth century. Rather than showing these children thriving in society as Romulus or Ailbe did, they tend to emphasize the essential and untamable wildness of "feral" children. A common trope is that such children could not walk erect, incapable of displaying the posture that distinguished humans from beasts.

Like other animals, parts of the wolf's body were regarded as having medical or magical uses. Roman natural historian Pliny the Elder (d. 79 CE) recounted the belief that a specific lock of hair from a wolf's tail was an aphrodisiac, but warned that it only had this power if taken from a living wolf. In France, a wolf's tooth was a protective amulet and could cure toothache when rubbed against the afflicted spot. In parts of Europe, wrapping a person in wolfskin could prevent epileptic fits, or hanging a wolfskin in a house would ward off flies. The connection between wolves and epilepsy was particularly strong in the western Balkans. Among Serbs in Kosovo, epileptics were cured by drinking water through the throat or larynx of a wolf, which they then ate. Wolves' hearts could either be eaten or dried and worn on the body. In Eastern Serbia, eating a wolf's testicles protects from epilepsy, and in Central Serbia, wearing a wolf's eyes performs the same function. Eighteenth-century Bosnians were told to drink water from a wolf's footprint to cure epilepsy, while a nineteenth-century Croatian almanac recommended eating either a powdered wolf heart or liver mixed with water or wolf meat. Torturing or tormenting a wolf, by contrast, would cause the torturer to contract epilepsy.

Coyotes, wild dogs closely related to the wolf and inhabiting North America, became the subject of a considerable body of legend. In many Native cultures, coyote figured as a powerful god and a morally ambiguous trickster. The Navajo viewed seeing a coyote as a bad omen and recommended turning back if a person saw a coyote while on a journey. Coyote and wolf skins were among the most commonly used by "skin-walkers," evil shape-changing witches.

See also: Dogs; Lycanthropy

Further Reading

Plas, Pieter. 2004. "Falling Sickness, Descending Wolf: Some Notes on Popular Etymology, Symptomatology, and 'Predicate Synonymy' in West Balkan Slavic Folk Tradition." *Zeitschrift fur Slawistic* 49: 253–272;
Steel, Karl. *How Not to Make a Human: Pets, Feral Children , Worms, Sky Burial, Oysters.* Minneapolis: University of Minnesota Press, 2019.

Wuxing. See Elemental Systems

Yellow Emperor

Huangdi, the "Yellow Emperor," was the mythical founder of the Chinese people, revered as a source of wisdom and authority. The traditional dates for his reign are 2698–2598 BCE. He is one of the legendary Five Emperors, the earliest human Chinese rulers. (the "Three Sovereigns" who ruled earlier were divine or semidivine.) The Five Emperors were treated as sages who had invented many aspects of Chinese life. The Yellow Emperor's reign was considered to be governed by Earth, one of the five elements or phases. He is also identified with the star Regulus.

The Yellow Emperor remains literally a name to conjure within modern China. This copy of the "Yellow Emperor's Sutra for Secret Charm Against Evil" dates from the twentieth century. (The Metropolitan Museum of Art, New York, Seymour and Rogers Funds, 1977)

The cult of the Yellow Emperor first appears in the Warring States period (475–221 BCE). Numerous legends gathered around his name. As a "culture hero," the Yellow Emperor was credited with the invention of such basics as huts and clothing as well as the calendar and *cuju*, an early form of soccer. Leizu, the principal wife among his four wives, was believed to have invented the cultivation and manufacture of silk. As a god, the Yellow Emperor is referred to as the Yellow Deity with Four Faces or the Great Deity of the Central Peak and represents the center of the cosmos.

The *Huangdi Neijing* or *Internal Canon of the Yellow Emperor*, also known as *The Yellow Emperor's Inner Classic* or *The Yellow Emperor's Classic of Internal Medicine*, is considered the foundation of Traditional Chinese Medicine. It was probably compiled from previous medical writings around 300 BCE. Influenced by Daoism, it suggests that the way to lead a healthy life is to follow the Dao and that diseases are caused by the imbalance of Yin and Yang forces in the body. It is structured in the form of dialogues between the emperor and his ministers. In addition to Yin and Yang, it includes many Chinese cosmological concepts such as the Five Elements and the correspondence of the microcosm and macrocosm. It also emphasizes the importance of environmental factors in maintaining health and includes one of the earliest discussions of acupuncture. The Yellow Emperor is also credited with *Huangdi Sijing* or *Four Classics of the Yellow Emperor*, a set of political treatises thought lost for millennia until discovered in the Mawangdui Archeological Site in 1973. He was a particularly important figure for Daoists, who ascribed many of their basic concepts to him. In the Han Dynasty (202 BCE–220 CE), the role of the Yellow Emperor as the authority figure for Daoists was gradually supplanted by Laozi, the author of the *Dao Dejing*, but *Huangdi Yinfujing, The Yellow Emperor's Hidden Talisman Classic*, a work of Daoist astrology and internal alchemy credited to the Yellow Emperor, was produced as late as the Tang Dynasty (618–907).

The Tomb of the Yellow Emperor in Shaanxi Province is a sacred site, at which Chinese governments over the millennia (including the current government) have made offerings.

See also: Acupuncture; Alchemy; Elemental Systems; Garlic; Macrocosm/Microcosm; Trepanation; Unicorn; Yin/Yang

Further Reading

Yang, Lihui and Deming An, with Jessica Anderson Turner. 2005. *Handbook of Chinese Mythology.* Santa Barbara, CA: ABC-CLIO.

Yijing

The *Yijing*, also known as the *I Ching*, or *Book of Changes*, is one of the Five Classics of China and the basis of a major divination system. In its earliest form, the *Yijing* dates to the Zhou Dynasty and was developed from tortoise-shell or oracle bone divination. The completed *Yijing* includes both the original divination text and a set of commentaries known as the "Ten Wings," placing it into cosmological context. It was designated one of the Classics by Emperor Wu of Han in 136 BCE. Along with other classics, it was carved into stone in the late second century CE to ensure that the text was not altered. The core of the *Yijing* is sixty-four hexagrams, each composed of six lines stacked vertically. Each line is either broken or unbroken. The total number of ways to arrange the broken and unbroken lines is sixty-four, and all sixty-four possibilities are represented in the *Yijing.* The broken lines are identified with yin, the unbroken with yang. Hexagrams can be divided into two trigrams, upper and lower, or into three pairs, the lowest two lines representing Earth, the middle two man, and the upper two heaven. (The eight trigrams, the sets of three lines, also play a central role in traditional Chinese cosmology.) Each line in each hexagram has a short phrase attached to it as an interpretation, and there is also an interpretation for each entire hexagram. Over the millennia, the *Yijing* has acquired a number of commentaries that treat it as a book of cosmological insight in addition to a diviner's tool. These commentaries, and commentaries upon commentaries, are often published alongside the original text of the *Yijing.* The *Yijing* was often associated with Confucius, who was sometimes credited with the authorship of the Ten Wings. There is a passage in the *Analects*, the standard collection of Confucius's sayings, that if he had an additional fifty years of life he would study the *Yijing*, but its authenticity is disputed. However, the *Yijing*, like the other Classics, was incorporated into Confucian scholarship culture. The Neo-Confucian Zhu Xi (1130–1200) was the author of one of the most influential commentaries on the *Yijing*, and it was one of the books on which examination candidates were tested.

Yijing divination involves generating six lines in order to create starting from the bottom a hexagram. Each line must be identified not only as broken or unbroken but also as "old" or changing, or "young" and unchanging. There are several ways to generate the lines of the hexagram. The traditional method, involving repeated sortings of a group of yarrow sticks (although sticks made of other kinds of wood can be substituted), produces a 3-1 ratio of unchanging to changing lines. Other methods include flipping coins and casting dice. For a full interpretation, following the reading of the hexagram, moving broken lines are changed to unbroken and vice versa, and the new hexagram so generated is also interpreted. The interpretation requires skill and intuition on the part of the diviner, as the application of the texts to the question being asked is not always obvious. Although the cosmological aspects of the text have become more prominent, it has never ceased being used as a divination system. Outside China, it has been widely adopted in East Asia and is the subject of editions and commentaries in Korea, Japan, and Vietnam. Both Korea and Vietnam have incorporated trigrams from the *Yijing* into national flags.

See also: Divination; Oracle Bones; Yin/Yang

Further Reading

Hon, Tze-Ki. 2019. "Chinese Philosophy of Change (Yijing)." In Edward N. Zalta, ed. *The Stanford Encyclopedia of Philosophy.* https://plato.stanford.edu/archives/sum2019/entries/chinese-change/; Smith, Richard J. 2012. *The I Ching: A Biography.* Princeton, NJ: Princeton University Press.

Yin/Yang

The categories of yin and yang shape traditional Chinese thought and have influenced East Asian culture generally. (Yin and yang are almost invariably named in that order.) Yin and yang represent the interdependence of opposite poles of a duality. Yin is considered the passive principle, yang the active one. Yin is identified with coldness, darkness, the female, and the Moon, among many other principles. Yang is identified with heat, light, the male, and the Sun. South, in the direction of the hotter lands, is the direction of yang, north the direction of yin. In the Five Elements system, metal and water are yin, wood and fire yang, while Earth represents their balance. Numerologically, even numbers are yin, odd ones yang. Yin and yang are not viewed statically, as polar opposites, but as existing in dynamic interaction. The cycle of twelve animals, or "Chinese Zodiac," also allocates its individual animals to either yin or yang.

Yin and yang are both forms taken by qi, energy. The interaction of yin and yang is often expressed as a temporal succession, as the year moves from yin-dominated winter reaching its peak on the winter solstice to yang-dominated summer reaching its peak on the summer solstice and back or as waves occur in a pool showing peaks and valleys. Another way the complementarity of yin and yang is expressed is sexual reproduction, which requires the participation of yin female and yang male. Yin and yang are not the most fundamental realities—the mainstream tradition of Chinese thought is monist and identifies the fundamental reality as the undifferentiated Dao.

The interaction of yin and yang creates the phenomenal world of the "ten thousand things." The cosmos is at its best when yin and yang are in harmony. The fact that heaven, Earth, and man were all subject to the forces of yin and yang was one of the things binding the entire cosmos together. The terms "yin" and "yang" occur in the oracle bone texts of the Shang period, but they refer simply to darkness and sunlight rather than cosmological forces. (Their earliest meaning was probably to refer to the shaded—yin—and sunny—yang—side of a hill on a bright day.) A "School of Yinyang" coexisted with the early Confucianists and Daoists and seems to have been mostly concerned with divination. None of its writings survive. The first surviving document to discuss the concepts at length is the *Zhuangzhi*, produced around the third century BCE. It is considered one of the two founding texts of Daoism alongside the *Daodejing*. The tradition of classical Confucian thought as embodied in the study of the Five Classics is also influenced by yinyang thinking, although the writings of Confucius himself and his most important early disciple Mencius do not mention yinyang. The broken lines of the *Yijing* are identified as yin, the unbroken ones as yang. Music and ritual are two central concepts for Confucian social thinking; the *Book of Rites*, one of the Five Classics, identifies music as yang, ritual as yin.

The categories of yin and yang have shaped Chinese beliefs about medicine, which focus on maintaining a proper balance of yin and yang in the body. This idea appears in *The Yellow Emperor's Inner Classic*, the foundational text of traditional Chinese medicine. Organs of the body are classified as yin or yang; yin organs store matter and qi, yang organs transport them. If a body has an excess of yang, the person should consume yin foods to bring the two forces back into balance and vice versa. Menstruating women, extremely yin, should avoid things that were yin and strive to bring their bodies back into balance by consuming yang foods and exposing herself to heat.

Weather and the environment are also shaped by yin and yang. Flooding indicates an excess of yin; drought an excess of yang. The Han Dynasty Confucian philosopher Dong Zhongshu (195–115 BCE) in *Luxuriant Gems of the Spring and Autumn* suggested that during times of drought the south gate should be closed and men should be sent into seclusion while women appear in public to foster yin; during floods, the opposite policy should be pursued to foster yang. Dong Zhongshu is credited with the integration of yinyang cosmological thought into Confucianism as well as the promotion of this type of Confucianism as the ideology of the Chinese state, a change that would shape Chinese thought for millenia. Feng shui, the art of domestic architecture and room arrangement, also draws upon these principles to ensure a proper balance of yin and yang in the interior space. Chinese martial arts have also drawn on yinyang cosmology in terms of a balance of yang action and yin stillness.

Chinese conceptions of gender are shaped by yin and yang, although yin and yang cannot be reduced to female and male. The association of yin with female and yang with male has produced some tension within yinyang thought. Traditional Chinese civilization is male-dominated, but yin and yang are equal and the fact that yin is almost always named first indicates that it might even have priority. Dong Zhongshu, whose approach to Confucianism is highly authoritarian, emphasized the importance of obedience by subordinates rather than Confucius's original emphasis on reciprocity between superiors and inferiors in a hierarchy. Dong associated yang with hierarchical superiors including husbands in relation to wives. In human terms, yang is superior and yin inferior in Dong's thought, which had a strong influence on subsequent Chinese thinking about gender. The association of yang with rulership and positive human qualities justified male domination, although traditional Chinese thought allows for exceptional cases of women with sufficient yang to exercise leadership.

The idea of maintaining a balance of yin and yang also has a political meaning. Maintaining the balance is part of the responsibility of the government. The disasters and prodigies that accompany a Chinese regime losing the mandate of heaven are caused by a lack of harmony between yin and yang.

Yinyang cosmology, like other aspects of Chinese culture, had an early influence on Korea, where yin is known as um. Such was the importance of yinyang to Korea that the yinyang symbol appears on the South Korean flag. Yinyang cosmology along with other aspects of Chinese culture entered Japan through Korea around the sixth century of the Common era. In Japan, yinyang became known as inyo or onmyodo. The "Bureau of yinyang" or Onmyoro was the department of the government charged with divination, the determination of auspicious and inauspicious days, and other services requiring a knowledge of cosmic forces. Founded around 701, it was only abolished in 1870. Over the centuries, knowledge of these techniques diffused outward from imperial officials into private practitioners known as onmyoji. Yinyang cosmology has also influenced Vietnamese culture.

See also: Alchemy; Astrology; Divination; Dragons; Elemental Systems; Foxes; Herbs; Macrocosm/Microcosm; Menstruation; Mirrors; Moon; Oracle Bones; Prodigies; Sex Change; Solstices and Equinoxes; Sun; Tea; Tobacco; Yellow Emperor; *Yijing*

Further Reading

Henderson, John B. 1994. *The Development and Decline of Chinese Cosmology.* New York: Columbia University Press; Wang, Robin. 2012. *Yinyang: The Way of Heaven and Earth in Chinese Thought and Culture.* Cambridge, UK: Cambridge University Press.

Yoga

From early times in India, yoga referred to a group of techniques for gaining enlightenment and supernatural power, including but not limited to the physical practices that the term usually denotes in the West. Although yoga is usually associated with Hinduism, yoga techniques and documents are found in the Jain, Buddhist, and even Islamic traditions as well.

The word "yoga" in the modern sense first appears in the *Katha Upanishad*, probably dating to between the fifth and third century BCE. The foundational document of the tradition is the *Yoga Sutras* of the sage Patanjali, which were written sometime between 500 BCE and 400 CE, although the roots of yoga can be found earlier in Hindu, Buddhist, and Jain traditions. The *Yoga Sutras*, mostly a compendium of earlier sources, treat yoga principally as a form of spiritual liberation as the true person is liberated from nature and thereby from suffering. However, it also states that the liberated practitioner can attain various powers, or *siddhis.* Physical practices include adopting postures and motions and breath control. (According to legend, a yoga master could breathe only once every twelve years.) *Siddhis* include health, virility, and peace of mind. Patanjali later became identified as the incarnation of a serpent god, in some versions half-human and half-serpent. The *Visuddhimagga*, a Sri Lankan Theravada Buddhist text from the fifth century CE, speaks of siddhis more supernatural in form, such as flying and walking on water or through walls. Siddhis also include the power to turn invisible and to know one's future and past lives. These powers were to be attained along the way of a directed process of meditation and purification although they were not the goal. In addition to the physical disciplines nowadays associated with the word "yoga," original yoga writings also refer to guided meditation and devotional practices.

The *Bhagavad-Gita*, a component of the epic *Mahabharata* and a central Hindu religious text, also influenced yoga, particularly among worshippers of Vishnu (Vaishnavites). The *Bhagavata Purana*, a text produced by Vaishnavites sometime between the eighth and tenth centuries CE, speaks of the powers granted by yoga and meditation as including the knowledge of past, present, and future; protection from poison, heat, and cold; and the ability to move the body across great distances with the power of the mind. The yoga master can also choose the moment of his or her death. In the later Middle Ages, the emphasis of yoga writers moved from Patanjali's freeing of the soul of bondage to nature and toward a union with God.

In Indian popular tradition, however, the image of the yogi was often not that of the enlightened sage, but that of a monster or evil wizard possessing terrifying magic powers. More benevolent yogis still appeared as master magicians more than saints.

Further Reading

White, David Gordon. 2014. *The Yoga Sutra of Patanjali: A Biography*. Princeton, NJ: Princeton University Press.

Yokai

Yokai is a Japanese term used to refer to supernatural, monstrous, or demonic entities. They come in a myriad of forms including the demonic horned oni, the kitsune fox-spirit, the water-dwelling kappa, and the mountain-inhabiting tengu. In medieval Japan, a belief even developed that abandoned tools could become yokai.

Yokai appear in texts as old as the eighth-century CE Japanese chronicles. They frequently appeared in Japanese art. This tradition culminated in the production of elaborate visual catalogs of yokai in the late eighteenth century, including the popular work of the scholar and woodblock artist Toriyama Sekien (1712–1788), *The Illustrated Demon Horde's Night Parade* (1776), *The Illustrated Demon Horde from Past and Present, Continued* (1779), *More of the Demon Horde from Past and Present* (1780), and *A Horde of Haunted Housewares* (1784). In addition to yokai from folklore, Toriyama added creatures from his own imagination.

One of the earliest forms of yokai was the oni, traditionally pictured as humanoid, with reddish skin, two horns, and wearing a tiger-skin loincloth. Oni were powerful and terrifying, dwelling apart from humanity and visiting the human realm on the night of seisbun, the festival commemorating the end of winter and the beginning of the new year. They appear in early eighth-century chronicles, some of the earliest surviving writing in Japanese, as monsters capable of devastating entire provinces in their rampages. Some of the earliest oni writings treat them as invisible. Oni were worthy opponents of gods, and the thunder god, Raijin, was even pictured as an oni. Oni were identified as the cause of destructive phenomena such as lightning and epidemics.

One common form of yokai was the kappa, a water-dwelling monster usually in the form of a small child with a shell and a

A modern representation of the tengu, a long-nosed yokai. (Bidouze Stephane | Dreamstime.com)

depression on the top of its head filled with water. The term "kappa" seems to have originated in the area around Tokyo and then applied to water-dwelling monsters throughout Japan in the early modern period. Some believed kappa were seasonal, moving from the oceanside to the mountains in the summer and returning in the fall. They were frequently believed to be aged otters or more commonly soft-shelled water-dwelling turtles who had metamorphosed into monsters. An alternative theory was that the kappa were originally dolls. Kappa were malicious and pulled humans and animals, most commonly horses, into water to drown. They also challenged humans to sumo wrestling bouts. The way to defeat a kappa was to spill the water out of the depression on its head. Kappa started appearing in written texts in the fifteenth century and become common in the seventeenth century.

The tengu was a birdlike yokai inhabiting mountain regions. Strange and unexplained noises on the mountains are frequently viewed as the work of tengu. Unlike other yokai, who can appear in male or female form, tengu are virtually always male. References to tengu are found as early as the eighth century CE. Tengu were described as having a birdlike face and possessing the ability to fly. In medieval Buddhist literature, tengu appeared as evil monsters whose goal was to prevent the spread of Buddhism. They were also frequently charged with kidnapping people and spreading disease. A new form of tengu appeared in the early modern period, more humanlike with a face distinguished from a human face by having a long nose rather than resembling a bird's. These humanlike tengu dominate the tengu beliefs of the time, and the original avian tengu were forced into a supporting role.

The term "yokai" was popularized by a Japanese scholar, Inoue Enryo (1858–1919). Enryo viewed belief in yokai as superstition that should be eradicated as part of Japanese modernization. However, yokai flourish in Japanese culture to this day and have spread due to their prominent role in anime and manga.

See also: Foxes; Ghosts; Lightning; Monsters

Further Reading

Foster, Michael Dylan. 2015. *The Book of Yokai: Mysterious Creatures of Japanese Folklore.* Berkeley: University of California Press; Komatsu, Kazuhiko. 2017. *An Introduction to Yokai Culture: Monsters, Ghosts and Outsiders in Japanese History.* Translated by Hiroko Yoda and Matt Ali. Tokyo: Japan Publishing Industry Foundation for Culture.

Z

Zhi. See Unicorn

Selected Bibliography

Primary Sources

Al-Akili, Muhammad M. 1992. *Ibn Seerin's Dictionary of Dreams according to Islamic Inner Traditions.* Philadelphia: Pearl.

Albertus Magnus. 1973. *The Book of Secrets of Albertus Magnus of the Virtues of Herbs, Stones and Certaine Beasts. Also a Book of the Marvels of the World.* Edited by Michael R. Best and Frank H. Brightman. Oxford: Clarendon Press.

Al-Jawbari, Jamāl al-Dīn ʿAbd al-Raḥīm. 2020. *The Book of Charlatans.* Edited by Manuela Dengler. Translated by Humphrey Davies. New York: New York University Press.

Bacon, Francis. 1999. *The Essays or Counsels Civil and Moral.* Edited with an introduction and notes by Brian Vickers. Oxford: Oxford University Press.

Browne, Thomas. 1981. *Sir Thomas Browne's Pseudodoxia Epidemica.* Edited by Robin Robbins. Oxford: Clarendon Press.

Ciruelo, Pedro. 1977. *Pedro Ciruelo's A Treatise Reproving All Superstitions and Forms of Witchcraft, Very Necessary and Useful for all Good Christians Zealous for Their Salvation.* Translated by Eugene A. Maio and D'Orsay W. Pearson. Rutherford, NJ: Fairleigh Dickinson University Press.

Clare, John. 1825. "Popularity in Authorship." *The European Magazine New Series* 1: 300–303.

Fontane, Felix. n.d. *The Golden Wheel Dream-book and Fortune-teller.* New York: Dick & Fitzgerald.

Kieckhefer, Richard. 1998. *Forbidden Rites: A Necromancer's Manual of the Fifteenth Century.* University Park: Pennsylvania State University Press.

Kirk, Robert. 2008. *The Secret Commonwealth of Elves, Fauns and Fairies.* Garden City, NY: Dover.

Paracelsus. 1991. *Four Treatises of Theophrastus von Hohenheim, Called Paracelsus.* Baltimore: Johns Hopkins University Press.

Pliny the Elder. 1855–1857. *The Natural History of Pliny.* Translated, with copious notes and illustrations, by the late John Bostock and H. T. Riley. Six Volumes. London: H. G. Bohn.

Seneca, Lucius Annaeus. 1910. *Physical Science in the Time of Nero; Being a Translation of the Quaestiones Naturales of Seneca.* Translated by J. Clarke. London: MacMillan.

Skinner, Stephen and Daniel Clark, eds. 2019. *Ars Notoria: The Grimoire of Rapid Learning by Magic with the Golden Flowers of Apollonius of Tyana.* Translated by Robert Turner. Singapore: Golden Hoard Press.

Reference Works

Burns, William E. 2003. *Witch Hunts in Europe and America.* Westport, CT: Greenwood.

Burns, William E., ed. 2018. *Astrology through History: Interpreting the Stars from Mesopotamia to the Present Day.* Santa Barbara, CA: ABC-CLIO.

Hanegraaff, Wouter J., ed. 2006. *Dictionary of Gnosis and Western Esotericism.* Leiden: Brill.

Hazlitt, W. Carew. 1965. *Faiths and Folklore of the British Isles: A Descriptive and Historical Dictionary.* New York: Benjamin Blom. Originally published 1905, as a revision of John Brand, *The Popular Antiquities of Great Britain* published in 1813.

Opie, Iona and Moira Tatern, eds. 1989. *A Dictionary of Superstitions.* Oxford: Oxford University Press.

Schwartz, Donald Ray. 2000. *Noah's Ark: An Annotated Encyclopedia of Every Animal Species in the Hebrew Bible.* Northvale, NJ: Jason Aronson.

Yang, Lihui and Deming An, with Jessica Anderson Turner. 2005. *Handbook of Chinese Mythology.* Santa Barbara, CA: ABC-CLIO.

Focused Studies

Arnold, Martin. 2018. *The Dragon: Fear and Power.* London: Reaktion Books.

Aveni, Anthony. 2002. *Behind the Crystal Ball: Magic, Science and the Occult from Antiquity through the New Age.* Boulder: University Press of Colorado.

Aveni, Anthony. 2017. *In the Shadow of the Moon: The Science, Magic and Mystery of Solar Eclipses.* New Haven, CT and London: Yale University Press

Barber, Paul. 1988. *Vampires, Burial and Death: Folklore and Reality.* New Haven, CT: Yale University Press.

Bascom, William. 1991. *Ifa Divination: Communication between Gods and Men in West Africa.* Bloomington and Indianapolis: Indiana University Press.

Bloch, Marc. 1973. *The Royal Touch.* Translated by J. E. Anderson. London: Routledge.

Bohn, Thomas M. 2019. *The Vampire: Origins of a European Myth.* Translated by Francis Ipgrave. New York: Berghahn.

Briggs, Katharine. 1967. *The Fairies in Tradition and Literature.* London: Routledge and Kegan Paul.

Brogan, Stephen. 2015. *The Royal Touch in Early Modern England: Politics, Medicine and Sin.* Woodbridge, Suffolk; Rochester, NY: Boydell & Brewer.

Brown, Michael F. 1986. *Tsewa's Gift: Magic and Meaning in an Amazonian Society.* Washington, DC: Smithsonian Institution.

Bulkeley, Kelly, Kate Adams and Patricia M. Davis, eds. 2009. *Dreaming in Christianity and Islam: Culture, Conflict and Creativity.* New Brunswick, NJ: Rutgers University Press.

Burns, William E. 2002. *An Age of Wonders: Prodigies, Politics and Providence in England 1657–1737.* Manchester: Manchester University Press.

Camporesi, Piero. 1995. *Juice of Life: The Symbolic and Magic Significance of Blood.* Translated by Robert R. Barr. New York: Continuum.

Dan, Joseph. 2007. *Kabbalah: A Very Short Introduction.* Oxford: Oxford University Press.

Darnton, Robert. *Mesmerism and the End of the Enlightenment in France.* 1968. Cambridge, MA and London: Harvard University Press.

Daston, Lorraine and Katherine Park. 1998. *Wonders and the Order of Nature 1150–1750.* New York: Zone.

Davidson, Jane P. 2012. *Early Modern Supernatural: The Dark Side of European Culture, 1400–1700.* Santa Barbara, CA: Praeger.

Davies, Owen and Francesca Matteoni. 2017. *Executing Magic in the Modern Era: Criminal Bodies and the Gallows in Popular Medicine.* Cham: Palgrave Macmillan.

Debus, Allen G. 1977. *The Chemical Philosophy: Paracelsian Science and Medicine in the Sixteenth and Seventeenth Centuries.* New York: Science History Publications.

Driediger-Murphy, Lindsay G. and Esther Eidinow, eds. 2019. *Ancient Divination and Experience.* Oxford: Oxford University Press.

Dummett, Michael, Thierry DePaulis and Ronald Decker. 1996. *A Wicked Pack of Cards: The Origins of the Occult Tarot.* New York: Saint Martin's Press.

Dundes, Alan ed. 1981. *The Evil Eye: A Folklore Casebook.* New York: Garland.

Dym, Warren. 2010. *Divining Magic: Treasure Hunting and Earth Science in Early Modern Germany.* Leiden: Brill.

Eamon, William. 1994. *Science and the Secrets of Nature: Books of Secrets in Medieval and Early Modern Culture.* Princeton, NJ: Princeton University Press.

Ebeling, Florian. 2007. *The Secret History of Hermes Trismegistus: Hermeticism from Ancient to Modern Times.* Translated from the German by David Lorton. Ithaca, NY and London: Cornell University Press.

Eco, Umberto. 1997. *The Search for the Perfect Language.* Translated by James Fentress. Malden, MA: Blackwell Publishing.

El-Zein, Amira. 2009. *Islam, Arabs and the Intelligent World of the Jinn.* Syracuse, NY: Syracuse University Press.

Evans-Pritchard, E. E. 1937. *Witchcraft, Oracles and Magic among the Azande.* Oxford: Clarendon Press.

Faraone, Christopher A. 2001. *Ancient Greek Love Magic.* Cambridge. MA: Harvard University Press.

Farley, Helen. 2009. *A Cultural History of Tarot: From Entertainment to Esotericism.* London: I. B. Tauris.

Fogel, Edwin Miller. 1915. *Beliefs and Superstitions of the Pennsylvania Germans.* Philadelphia: American Germanica Press.

Foster, Michael Dylan. 2015. *The Book of Yokai: Mysterious Creatures of Japanese Folklore.* Berkeley: University of California Press.

Gardiner, Noah. 2020. *Ibn Khaldun versus the Occultists at Barquq's Court: The Critique of Lettrism in al-Muqaddimah.* Ulrich Haarmann Memorial Lecture no. 18. Berlin: EB Verlag.

Genuth, Sara Schechner. 1997. *Comets, Popular Culture, and the Birth of Modern Cosmology.* Princeton, NJ: Princeton University Press.

Gerbi, Antonello. 1973. *The Dispute of the New World: The History of a Polemic, 1750–1900.* Translated by Jeremy Moyle. Revised Edition. Pittsburgh: University of Pittsburgh Press.

Ginzburg, Carlo. 1985. *Night Battles: Witchcraft and Agrarian Cults in the Sixteenth and Seventeenth Centuries.* Translated by John and Anne Tedeschi. New York: Penguin.

Ginzburg, Carlo. 1991. *Ecstacies: Deciphering the Witches's Sabbath.* Translated by Raymond Rosenthal. New York: Pantheon Books.

Ginzburg, Carlo and Bruce Lincoln. 2020. *Old Theiss a Livonian Werewolf: A Classic Case in Comparative Perspective.* Chicago: University of Chicago Press.

Goodare, Julian and Martha McGill, eds. 2020. *The Supernatural in Early Modern Scotland.* Manchester: Manchester University Press.

Graf, Fritz. 1997. *Magic in the Ancient World.* Translated by Franklin Philip. Cambridge, MA and London: Harvard University Press.

Grant, Edward. 1994. *Planets, Stars and Orbs: The Medieval Cosmos, 1200–1687.* Cambridge, UK and New York: Cambridge University Press.

Green, Richard Firth. 2016. *Elf Queens and Holy Friars: Fairy Beliefs and the Medieval Church.* Philadelphia: University of Pennsylvania Press.

Hall, Alaric. 2007. *Elves in Anglo-Saxon England: Matters of Belief, Health, Gender and Identity.* Woodbridge, Suffolk; Rochester, NY: Boydell & Brewer.

Hall, David. 1989. *Worlds of Wonder, Days of Judgment: Popular Religious Belief in Early New England.* New York: Knopf.

Hanafi, Zakiya. 2000. *The Monster in the Machine: Magic, Medicine and the Marvelous in the Time of the Scientific Revolution.* Durham, NC: Duke UniversityPress.

Harkness, Deborah E. 1999. *John Dee's Conversations with Angels: Cabala, Alchemy and the End of Nature.* Cambridge, UK and New York: Cambridge University Press.

Harline, Craig. 2003. *Miracles at the Jesus Oak: Histories of the Supernatural in Reformation Europe.* New York: Doubleday.

Harris, Victor. 1966. *All Coherence Gone: A Study of the Seventeenth-Century Controversy over Disorder and Decay in the Universe.* London: Cass.

Henderson, John B. 1994. *The Development and Decline of Chinese Cosmology.* New York: Columbia University Press.

Hutton, Ronald, ed. 2016. *Physical Evidence for Ritual Acts, Sorcery and Witchcraft in Christian Britain.* New York: Palgrave Macmillan.

Idel, Moshe. 1988. *Kabbalah: New Perspectives.* New Haven, CT: Yale University Press.

Iversen, Erik. 1993. *The Myth of Egypt and its Hieroglyphs.* Reprint. Princeton, NJ: Princeton University Press.

Iwasaka, Michiko and Toelken Barre. 1994. *Ghosts and the Japanese: Cultural Experience in Japanese Death Legends.* Logan: Utah State University Press.

Janacek, Bruce. 2011. *Alchemical Belief: Occultism in the Religious Culture of Early Modern England.* University Park: Pennsylvania State University Press.

Johnson, J. Cale and Alessandro Stavru, eds. 2019. *Visualizing the Invisible with the Human Body: Physiognomy and Ekphrasis in the Ancient World.* Berlin/Boston: De Gruyter.

Jortner, Adam. 2017. *Blood from the Sky: Miracles and Politics in the Early American Republic.* Charlottesville: University of Virginia Press.

Kieckhefer, Richard. 1989. *Magic in the Middle Ages.* Cambridge, UK: Cambridge University Press.

Knight, Michael Muhammad. 2016. *Magic in Islam.* New York: Tarcher Perigee.

Komatsu Kazuhiko. 2017. *An Introduction to Yokai Culture: Monsters, Ghosts and Outsiders in Japanese History.* Translated by Hiroko Yoda and Matt Ali. Tokyo: Japan Publishing Industry Foundation for Culture.

Kuhn, Philip A. 1990. *Soulstealers: The Chinese Sorcery Scare of 1768.* Cambridge, MA: Harvard University Press.

Long, Kathleen P. 2006. *Hermaphrodites in Renaissance Europe.* Aldershot: Ashgate.

Lovejoy, Arthur O. 1936. *The Great Chain of Being: A Study of the History of an*

Idea. Cambridge, MA: Harvard University Press.

Lynn, Michael R., ed. 2022. *Magic, Witchcraft and Ghosts in the Enlightenment.* New York: Routledge.

Maarouf, Mohammed. 2007. *Jinn Eviction as a Discourse of Power: A Multidisciplinary Approach to Moroccan Magical Beliefs and Practices.* Leiden: Brill.

Malinowski, Bronislaw. 1948. *Magic, Science and Religion; and other Essays.* Boston: Beacon Press.

Maxwell-Stuart, P. G. 2010. *Astrology from Ancient Babylon to the Present.* Chalford, Gloucestershire: Amberley.

McGill, Martha. 2019. *Ghosts in Enlightenment Scotland.* Suffolk: Boydell and Brewer.

Midelfort, H. C. Erik. 2005. *Exorcism and Enlightenment: Johann Joseph Gassner and the Demons of Eighteenth-Century Germany.* New Haven, CT: Yale University Press.

Moellering, H. Arnim. 1963. *Plutarch on Superstition: Plutarch's De Superstitione, Its Place in the Changing Meaning of Deisidaimonia and In the Context of his Theological Writings.* Revised Edition. Boston: Christopher Publishing House.

Monod, Paul Kleber. 2013. *Solomon's Secret Arts: The Occult in an Age of Enlightenment.* New Haven, CT and London: Yale University Press.

Moran, Bruce T. 2005. *Distilling Knowledge: Alchemy, Chemistry and the Scientific Revolution.* Cambridge, MA: Harvard University Press.

Nummedal, Tara. 2019. *Anna Zieglerin and the Lion's Blood: Alchemy and End Times in Reformation Germany.* Philadelphia: University of Pennsylvania Press.

Ogden, Daniel. 2021. *The Werewolf in the Ancient World.* Oxford and New York: Oxford University Press.

Otten, Charlotte F. ed. 1986. *A Lycanthropy Reader: Werewolves in Western Culture.* Syracuse, NY: Syracuse University Press.

Parish, Helen and William G. Naphy, eds. 2002. *Religion and Superstition in Reformation Europe.* Manchester and New York: Manchester University Press.

Peek, Philip M., ed. 1991. *African Divination Systems: Ways of Knowing.* Bloomington: Indiana University Press.

Pinto-Correia, Clara. 1997. *The Ovary of Eve: Egg and Sperm and Preformation.* Chicago: University of Chicago Press.

Platt, Peter G., ed. 1999. *Wonders, Marvels and Monsters in Early Modern Culture.* Newark: University of Delaware Press.

Pregadio, Fabrizio. 2019. *The Way of the Golden Elixir: An Introduction to Taoist Alchemy.* Third Edition. Mountain View, CA: Golden Elixir Press.

Rankin, Alisha. 2021. *The Poison Trials: Wonder Drugs, Experiment and the Battle for Authority in Renaissance Science.* Chicago: University of Chicago Press.

Rochat de la Vallee, Elisabeth. 2009. *Wuxing: The Five Elements in Classical Chinese Texts.* London: Monkey Press.

Ruggiero, Guido. 1993. *Binding Passions: Tales of Magic, Marriage and Power at the End of the Renaissance.* New York: Oxford University Press.

Ryrie, Alec. 2008. *The Sorcerer's Tale: Faith and Fraud in Tudor England.* Oxford: Oxford University Press.

Saif, Liana. 2015. *The Arabic Influences on Early Modern Occult Philosophy.* New York: Palgrave Macmillan.

Schimmel, Annemarie. 1993. *The Mystery of Numbers.* New York: Oxford University Press.

Schmitt, Jean-Claude. 1999. *Ghosts in the Middle Ages: The Living and the Dead in Medieval Society.* Chicago: University of Chicago Press.

Smith, Richard J. 1991. *Fortune-Tellers and Philosophers: Divination in Traditional Chinese Society.* Boulder, CO: Westview Press.

Smith, Richard J. 2012. *The I Ching: A Biography.* Princeton, NJ: Princeton University Press.

Souza, Laura de Mella e. 2003. *The Devil and the Land of the Holy Cross: Witchcraft, Slavery and Popular Religion in Colonial Brazil.* Translated by Diane Grosklaus Whitty. Austin: University of Texas Press.

Steel, Karl. 2019. *How Not to Make a Human: Pets, Feral Children, Worms, Sky Burial, Oysters.* Minneapolis: University of Minnesota Press.

Stewart, Pamela J. and Andrew Strathern. 2004. *Witchcraft, Sorcery, Rumors and Gossip.* Cambridge, UK: Cambridge University Press.

Strong, John S. 2004. *Relics of the Buddha.* Princeton, NJ: Princeton University Press.

Swain, Simon, ed. 2007. *Seeing the Face, Seeing the Soul: Polemon's Physiognomy from Classical Antiquity to Medieval Islam.* Oxford and New York: Oxford University Press.

Thomas, Keith. 1971. *Religion and the Decline of Magic.* New York: Charles Scribner's Sons.

Thurston, Edgar. 1912. *Omens and Superstitions of Southern India.* London: T. Fisher Unwin.

Van Bladel, Kevin. 2009. *The Arabic Hermes: From Pagan Sage to Prophet of Science.* Oxford: Oxford University Press.

Verheyden, Joseph. 2013. *The Figure of Solomon in Jewish, Christian and Islamic Tradition: King, Sage and Architect.* Leiden: Brill.

Vyse, Stuart M. 1997. *Believing in Magic: The Psychology of Superstition.* New York: Oxford University Press.

Walker, D. P. 1981. *Unclean Spirits: Possession and Exorcism in France and England in the late Sixteenth and early Seventeenth Centuries* Philadelphia: University of Pennsylvania Press.

Wang, Robin. 2012. *Yinyang: The Way of Heaven and Earth in Chinese Thought and Culture.* Cambridge, UK: Cambridge University Press.

Wang, Xing. 2020. *Physiognomy in Ming China; Fortune and the Body.* Leiden and Boston: Brill.

White, David Gordon. 1991. *Myths of the Dog-Man.* Chicago: University of Chicago Press.

White, David Gordon. 2014. *The* Yoga Sutra of Patanjali: *A Biography.* Princeton, NJ: Princeton University Press.

Wilde, Jane Francesca Agnes. 1887. *Ancient Legends, Mystic Charms, and Superstitions of Ireland.* London: Ward and Downey.

Yates, Frances Amelia. 1964. *Giordano Bruno and the Hermetic Tradition.* Chicago: University of Chicago Press.

Index

Page numbers in **bold** indicate the location of main entries.

About the Author

William E. Burns is a historian who lives in the Washington, DC, area. His many books include *Witch Hunts in Europe and America* (Greenwood, 2003). He also edited *Astrology through History: Interpreting the Stars from Ancient Mesopotamia to the Present* (ABC-CLIO, 2018), a CHOICE Outstanding Academic Title.